AF615047

WIND POWER

WIND POWER

Recent Developments

Edited by D.J. De Renzo

NOYES DATA CORPORATION
Park Ridge, New Jersey, U.S.A.
1979

Library of Congress Catalog Card Number: 79-14069
ISBN: 0-8155-0759-3
Printed in the United States

Published in the United States of America by
Noyes Data Corporation
Noyes Building, Park Ridge, New Jersey 07656

Library of Congress Cataloging in Publication Data

Main entry under title:

Wind power.

(Energy technology review ; no. 46)
Bibliography: p.
Includes index.
1. Wind power--United States. I. De Renzo, D.J.
II. Series.
TK1541.W56 621.4'5 79-14069
ISBN 0-8155-0759-3

FOREWORD

This book presents in concise form most of the important developments which have occurred in wind power technology in the United States in recent years and contains substantially new material since our previous book on this subject was issued in 1975.

Again here are condensed new vital data on wind power that are scattered and difficult to pull together. New useful devices and constructions are interpreted and explained by actual case histories. This compact information should provide a sound background for action toward combating the energy shortage through the use of wind power, which now can be utilized even by public utilities.

Because the information in this book is taken from multiple sources, it is possible that certain portions of this book may disagree or conflict with other parts of the book. This is especially true of monetary values and opinions of future potential. We chose to include these different points of view, however, in order to make the book more valuable to the reader.

Cost figures provided are those given in the report cited, the date of which is always given. When the dates of the cost figures themselves are given, we have included them.

Advanced composition and production methods developed by Noyes Data are employed to bring these durably bound books to you in a minimum of time. Special techniques are used to close the gap between "manuscript" and "completed book." Industrial technology is progressing so rapidly that time-honored, conventional typesetting, binding and shipping methods are no longer suitable. We have by-passed the delays in the conventional book publishing cycle and provide the user with an effective and convenient means of reviewing up-to-date information in depth.

The expanded table of contents is organized in such a way as to serve as a subject index and provides easy access to the information contained in this book which is based on various studies produced by and for diverse governmental agencies under grants and contracts. These primary sources are listed at the end of the volume under the heading *Sources Utilized.* The titles of additional publications pertaining to topics in this book are found in the text. An interesting and useful feature is an appendix containing a list of manufacturers, researchers and distributors currently involved in the development of wind energy conversion systems. The discerning reader will realize that such a list can never be quite complete nor up-to-date.

Some of the illustrations in this book may be less clear than could be desired; however, they are reproduced from the best material available to us.

CONTENTS AND SUBJECT INDEX

INTRODUCTION

Wind energy conversion has long been recognized as a potentially abundant source of clean and inexhaustible power. In 1973, the impending worldwide shortage of nonrenewable energy sources and the nation's increasing dependence upon imported fuels led to a renewed effort to investigate the feasibility of converting the wind into useful energy. Before significant benefits from wind power could be realized, however, many technological and environmental problems had to be solved. It was also recognized that high cost was the greatest barrier to the use of wind systems. These costs had to be reduced to make wind power cost-competitive with other power sources. While much has been accomplished, the development of economical wind energy conversion systems capable of providing many years of reliable, automatic, relatively maintenance-free service remains a primary challenge.

This book presents some of the more important developments which have occurred in the United States since 1975 when Noyes Data Corporation published its first book on wind power. The first chapter gives an overview of the status of wind power conversion in 1975 while subsequent chapters provide more up-to-date information on the assessment of the geographical distribution of available wind power in the United States, the technological developments in wind power conversion systems and the applications of wind power, on a large scale, to electric utilities and, on a small scale, to farm and rural use.

The status of legal, social and environmental issues is discussed in the last chapter. Finally, an appendix is included to provide the reader with the names and addresses of many of the manufacturers, researchers and distributors currently involved in the development of wind energy conversion systems.

OVERVIEW

Material for this chapter has been based upon a report prepared by the Mitre Corporation (PB 249 936).

VIABILITY

Wind Resources

The winds of earth are caused primarily by unequal heating of the earth's surface by the sun. During the day, the air over the oceans and lakes remains relatively cool, since much of the sun's energy is consumed in evaporating water, or is absorbed by the water itself. Over land, the air is heated more during the day since the land absorbs less sunlight than the water, and evaporation is less. The heated air over land expands, becomes lighter, and rises. The cooler, heavier air from over the water moves in to replace it. In this way, local breezes on a shoreline are created.

During the night, these local seashore breezes reverse themselves, since the land cools more rapidly than the water, and so does the air above it. The cool air blows seaward to replace the warm air that rises from the surface of the water. Similar local breezes occur on mountainsides during the day as heated air rises along the warm slopes heated by the sun. During the night, the relatively cool heavy air on the slopes flows down into the valleys.

Likewise, circulating planetary winds are caused by the greater heating of the earth's surface near the equator than near the poles. This causes cold surface winds to blow from the poles to the equator to replace the hot air that rises in the tropics and moves in the upper atmosphere toward the poles.

The rotation of the earth also affects these planetary winds. The inertia in the cold air moving near the surface toward the equator tends to twist it to the west, while the warm air moving in the upper atmosphere toward the poles tends to be turned to the east. This causes large counterclockwise circulation of the air around low pressure areas in the northern hemisphere and clockwise circulation in the southern hemisphere. Since the earth's axis of rotation is inclined at an angle of 23.5° to the plane in which it moves around the sun, seasonal variations in the heat received from the sun result in seasonal changes in the strength and direction of the winds at any given location on the earth's surface.

Sufficient energy is continually being transferred from the sun to the winds of the earth to maintain an estimated total power capacity of over 10^{11} gigawatts (GW) in these winds. Of this, one panel of experts, i.e., the 1972 Solar Energy Panel of the National Science Foundation and the National Aeronautics and Space Administration, has estimated that the power that is potentially available from surface winds over the continental U.S., the Aleutian Arc, and the eastern seaboard is equivalent to about 10^5 GW of electricity. This is more than 30 times the estimated total power consumption of the U.S. by 1980 and more than 100 times the estimated U.S. electrical generating capacity by that time.

U.S. Energy Status

Present U.S. energy consumption is the equivalent of about 2,500 billion kWh/yr. Of this, about 19% is used for heating and cooling of buildings, 25% for industrial process heat, 24% for transportation, 25% for producing electricity, 5% for producing petrochemicals, and 2% for exported energy products.

Present sources of U.S. energy include oil (40%), natural gas (33%), coal (20%), hydropower (4%), and nuclear (2%). The remaining 1% is supplied by synthetic fuel, oil shale, geothermal energy, solar energy, and wind energy. About 14% of the present U.S. energy consumption is in the form of imported oil.

Figure 1.1: Comparison of Wind Energy and Total U.S. Energy Consumption

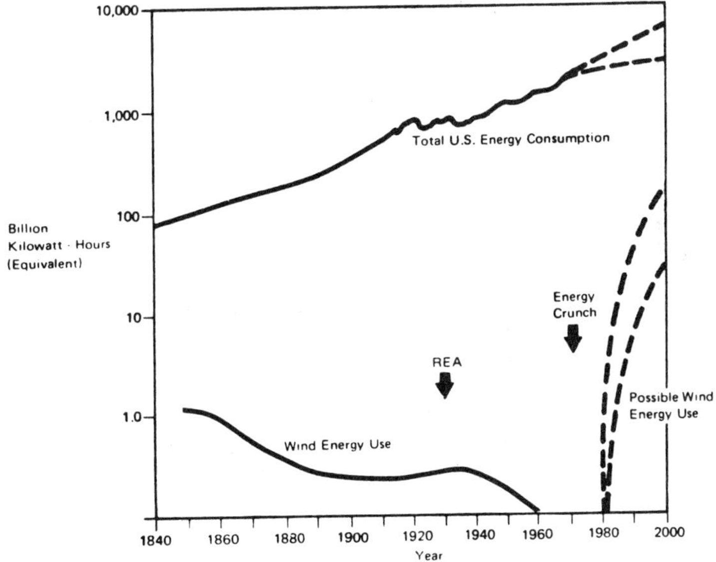

Source: PB 249 936

Although U.S. oil and natural gas production is expected to decrease rapidly in the next few decades, U.S. energy consumption is expected to increase to the equivalent of about 6,000 billion kWh/yr by the year 2000, unless stringent conservation measures are applied (see Figure 1.1).

Federal Program

Wind energy supplied a significant amount of the energy consumed in rural areas of the U.S. until the Rural Electrification Administration (REA) introduced electrical cooperatives in the mid-1930s. However, with the emergence of the energy crunch, there has been a renewed interest in the use of wind energy for these and other applications.

As a result of this interest the federal government has undertaken an accelerated Wind Energy Conversion System (WECS) program with the objective of stimulating the development of WECS capable of producing a significant amount of the U.S. energy needs by the year 2000.

To accomplish this, a series of experimental WECS units of from 100 kW to several megawatts rated capacity will be developed in the next few years. It is anticipated that these developments will be supported by an extensive research and technology effort, leading to incrementally installed 10 to 100 MW multi-unit demonstration systems, jointly funded with public utilities and other users, by the early 1980s. While the emphasis in this federal program will be placed on the use of WEC systems to produce electricity, a number of other applications such as fuel generation and the production of direct heat, crop-drying, and fertilizer manufacturing will also be demonstrated.

There are a number of possible constraints to the rate at which WEC systems can be expected to be developed and applied. These possible constraints are being intensively addressed in the federal research and technology program. They include uncertainties in capital, operations, and maintenance costs; the economic viability of various types of WECS applications; the availability of capital to build WEC systems; the environmental and public acceptability of such systems; and various legal and institutional problems that may arise from the use of WEC systems.

Economics

An important measure of the economic viability of WECS is the cost of electricity produced by such systems. This cost is, currently, still relatively large compared to that of fossil fueled systems. However, the unit price of wind turbines is expected to decrease as the size of the units is increased and as mass production techniques are introduced.

Present estimates indicate that WECS, if located at sites with reasonably high average wind velocities, would be expected to produce electricity at busbar prices of 20 to 30 mills/kWh. This would be competitive with conventional diesel electrical generation plants using oil priced from \$10 to \$11 per barrel or higher.

Wind systems without energy storage may be tied into a conventional system, powered, for instance, by a diesel generating plant and used to save fuel when the wind is blowing. On the other hand, wind systems with suitable energy storage capabilities can be used as a complete substitute for a conventional electric generating plant. In either case, the break-even points might be expected to occur at fuel prices of about $10 to $11 per barrel of oil for sites with average wind velocities of about 15 mph. To achieve these break-even prices at more moderate wind velocities will require development of systems with improved performance or lower costs.

Public Acceptance

Public acceptance of wind energy conversion systems is an important consideration in planning for the widespread application of wind energy. Preliminary studies have shown that the environmental impact of such systems is relatively small compared to conventional electric power systems. Wind-powered systems do not require the flooding of large land areas or the alteration of the natural ecology, as for hydroelectric systems. Furthermore, they produce no waste products or thermal or chemical effluents, as fossil fueled and nuclear fueled systems do.

Conventional wind turbine systems that generate several megawatts of power require large exposed rotors several hundred feet in diameter, located on high towers. The rotors of such systems, being passive, are practically noiseless. However, special precautions will be necessary to prevent them from causing interference with nearby TV or radio receivers, and some safety measures may be required to prevent damage or injury from possible mishaps in cases where there is danger that the rotors might break or shed ice.

The only other concerns with conventional wind machines are those of esthetics. Large numbers of units and interconnecting transmission lines will be required in the future if such systems are to have any significant impact on U.S. energy demands. Particular attention is being given, therefore, to the development of attractive designs for the towers, rotors, and nacelles of these conventional systems to avoid visual pollution.

HISTORY

Past History

The first wind turbines were probably simple vertical-axis panemones, such as those used in Persia as early as about 200 B.C. for grinding grain. The use of these vertical-axis mills subsequently spread throughout the Islamic world. Later, horizontal-axis windmills, consisting of up to ten wooden booms, rigged with jib sails, were developed. Such primitive types of windmills are still found in use today in many Mediterranean regions.

By the 11th century A.D., windmills were in extensive use in the Middle East and were introduced to Europe in the 13th century by returning Crusaders. During the subsequent Middle Ages in Europe, most manorial rights included the right to refuse permission to build windmills, thus compelling tenants to have their grain ground at the mill of the lord of the manor. Planting of trees near windmills was banned to ensure "free wind."

By the 14th century, the Dutch had taken the lead in improving the design of windmills and used them extensively thereafter for draining the marshes and lakes of the Rhine River delta. Between 1608 and 1612, Beemster Polder, which was 10 feet below sea level, was drained by 26 windmills of up to 50 hp each, operating in two stages.

Later, Leeghwater, the renowned hydraulic engineer, drained Schermer Polder in four years. Fourteen windmills pumped water into a storage basin at a rate of 1,000 cubic meters per minute. Thirty-six mills then pumped the storage basin water into a ring canal that emptied into the North Sea.

The first oil mill was built in Holland in 1582. The first paper mill was constructed in 1586 to meet the enormous demands for paper that resulted from the invention of the printing press. By the end of the 16th century, sawmills had been introduced to process timber imported from the Baltic regions. By the middle of the 19th century, some 9,000 windmills were being used in the Netherlands for a wide variety of purposes.

With the introduction of the steam engine, during the Industrial Revolution, the use of windpower in Hollard started to decline. By the turn of the 20th century, only about 2,500 windmills were still in operation in the Netherlands and, by 1960, fewer than 1,000 were still in working condition.

The Dutch introduced many improvements in the design of windmills and, in particular, the rotors. By the 16th century, the primitive jib sails on wooden booms had given way to sails supported by wooden bars on both sides of the stock. Later, the bars were moved to the trailing edge of the rotor to improve the aerodynamic design. More modern designs substituted sheet metal for the cloth sails, used steel stocks, and introduced various types of shutters and flaps to control the speed of the rotor in high winds. Larger industrial mills could deliver up to 90 hp in high winds (see Figure 1.2).

Figure 1.2: Types of Sail

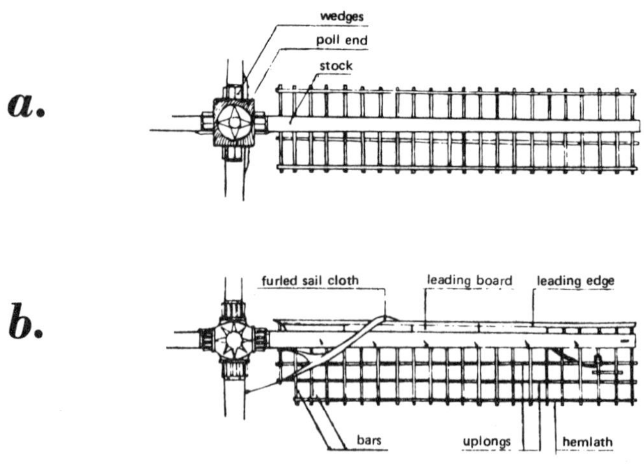

(continued)

Figure 1.2: (continued)

c.

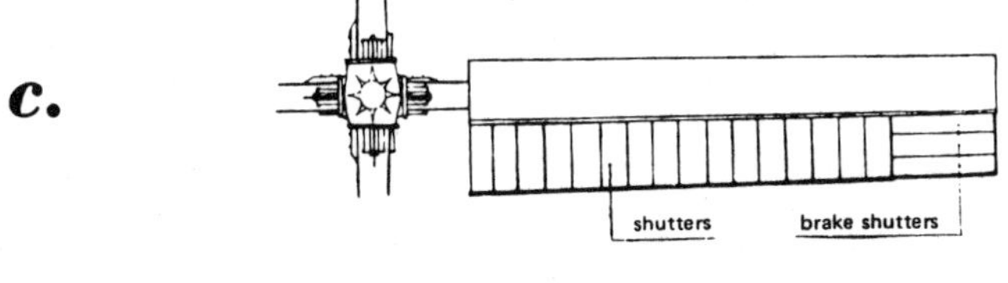

d.

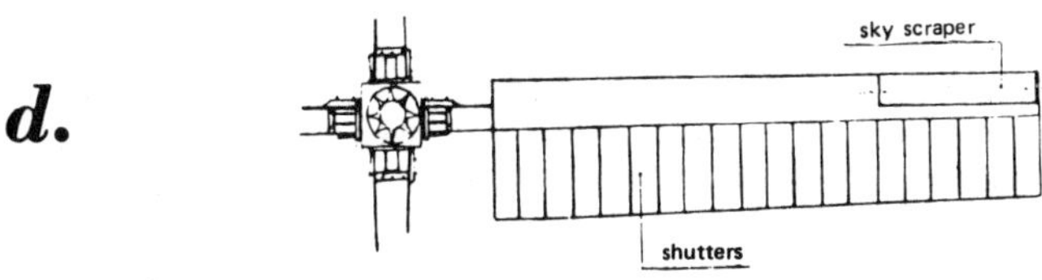

(a) Oldest type, double sided (about 1600)
(b) Normal old fashioned Dutch type (one leading board taken away)
(c) Shuttered type with air brake
(d) Shuttered type with sky scraper

Source: PB 249 936

Recent Experience

U.S. Experience: Since the mid-19th century, more than 6 million small windmills, of less than 1 hp each have been built and used in the U.S. to pump water, generate electricity, and perform similar functions. Over 150,000 are still in operation.

Types commonly used to pump water have metal fan blades, 12 to 16 feet in diameter, that are mounted on a horizontal shaft and have a tail vane to keep the rotor facing into the wind. The shaft is connected to a set of gears and a cam that move a connecting rod up and down. This, in turn, operates a pump at the bottom of the tower. A 12 foot diameter rotor of this type develops about ⅙ hp in a 15 mph wind and can pump about 35 gallons of water a minute to a height of about 25 feet.

Small windmills used to generate electricity usually have two or three propeller-type blades that are connected by a shaft and gear train to a dc generator. They usually incorporate some type of energy storage system, often consisting of a bank of batteries.

Many water-pumping windmills are still in use in the western U.S. for watering stock on ranges in remote areas. However, most of the wind powered electrical generators were displaced by centralized electric power after the Rural Electrification Administration was instrumental in providing cooperative utilities for most U.S. farms in the 1930s.

The largest operational wind powered electric system that has been built to date was the Smith-Putnam machine. After a lengthy study in the 1930s, Palmer Putnam concluded that a large machine was required to minimize the cost of the electricity generated. With the assistance of the eminent Cal Tech aerodynamicist Theodore Von Karman, and various members of the MIT staff, he designed a large wind turbine to feed power into the existing electrical network of the Central Vermont Public Service Company. The S. Morgan Smith Company of York, Pennsylvania, constructed and operated the plant in the early 1940s. The 2-bladed 175 foot diameter propeller-type rotor weighed 16 tons and operated at a rated speed of 28 rpm to produce 1.25 MW of ac power.

In March 1945, after intermittent operation over a period of several years, one of the blades broke off near the hub where a known weakness had previously been identified but had not been corrected because of wartime material shortages. A comprehensive economic study indicated that the plant, even if repaired, could not compete effectively, at that time, with conventional electrical generation plants, so the project was abandoned.

Danish Experience: In Denmark, by the end of the 19th century, there were about 3,000 industrial windmills and about 30,000 other types used for homes and farms. These supplied a total amount of power equivalent to about 200 MW. In 1890, the Danish government initiated a large program to develop improved large scale wind powered electric generators. By 1910, several hundred systems designed by Professor LaCour were built. These consisted of 80 foot towers that supported a 75 foot diameter 4-bladed rotor that operated a mechanical gear-and-shaft drive train connected to an electrical generator on the ground. These generators were rated at from 5 to 25 kW.

During World Wars I and II, improved wind powered electrical generation plants of this type supplied a small percentage of the Danish demand for electricity. During this time, it was found that the cost of the electricity delivered by these wind powered systems was about the same as the equivalent fuel cost for diesel-electric plants.

After World War II, the Danes developed and operated three experimental systems, rated at 12, 45, and 200 kW, which were tied into existing public utility networks. These operated successfully until the early 1960s. Their work on this project was stopped when they found that, by that time, the cost of electricity supplied by the wind powered systems was about twice the equivalent fuel cost of steam powered plants.

Russian Experience: In 1931, the Russians built an advanced 100 kW wind turbine, near Yalta on the Black Sea, which was tied by a 6,300 V line, to a 20 MW steam powered station 20 miles away. The annual output was found to be about 280,000 kWh/yr, which yielded a plant utilization factor of 0.32.

The generator and controls were located on top of a 100 foot tower. The speed of the rotor was regulated by controlling the pitch of the blades. The heel of the inclined strut was mounted on a carriage that ran on a circular track to keep the rotor facing into the wind. At one time in the 1930s, the Russians considered building a 5 MW system, but apparently this project was never implemented.

British Experience: In England, in the late 1940s and during the 1950s, considerable work was done on wind powered electrical generation plants under the leadership of E. Golding and A. Stoddard. Wind measurements made at about 100 sites in the British Isles during this period are among the best documented results available on wind characteristics in a given geographical region.

In 1950, the North Scotland Hydroelectric Board commissioned the John Brown Company to construct an experimental wind turbine on Cape Costa in the Orkney Islands. This unit was designed to generate 100 kW in winds of 35 mph. It operated for short periods in 1955, coupled to a diesel powered utility network, but was shut down because of operational problems.

In the 1950s, the Enfield Cable Company built a unique 100 kW wind powered generator designed by a Frenchman named Andreau and operated it at St. Albans, England, and in Algeria. This unit consisted of a hollow tower 85 feet high and had a hollow rotor 80 feet in diameter with openings at the blade tips. This caused a pressure differential that drove the air from openings near the base of the tower, up the tower, through an air turbine in the tower, and out of the rotor blade tips. The efficiency of the unit was found to be low compared to more conventional horizontal-axis wind powered rotors.

French Experience: The French built and operated several large wind powered electric generators in the period from 1958 to 1966. These included three horizontal-axis units, each with three propeller-type blades. A unit of this type was operated intermittently near Paris from 1958 to 1963. This unit was designed to generate 800 kW in winds of 37 mph. The rotor (100 foot diameter), generator, and gear assembly weighed 160 tons and was mounted on a 100 foot tower. The rotor was operated at a constant speed of 47 rpm and was coupled to a 50 Hz, 60 kV utility grid, using a synchronous alternator, operating at 1,000 rpm and generating 3,000 V. This voltage was increased to 60,000 V by voltage transformers that were connected to the main grid by a 15 km transmission line.

Two other units were constructed at St. Remy-des-Landes in Southern France. The smaller unit had a 70 foot diameter rotor operated at 56 rpm. An asynchronous generator with a nominal speed of 1,530 rpm generated 132 kW in winds of 28 mph or more. The larger of these two units was rated at 1,000 kW in winds of 37 mph and weighed 96 tons, excluding the tower. The capital cost of the Paris unit was about $1,155/kW, while that of the St. Remy units was about $1,000/kW (in 1960 dollars). The French also built and tested several experimental vertical-axis panemones during this period.

German Experience: The Germans, under the direction of Professor Ulrich Hutter, introduced a number of improvements in the design of wind powered generators, including light weight constant speed rotors that were controlled by variable pitch propeller blades. These machines used light weight fiber glass and plastic blades, with the generator mounted on a tower consisting of a small diameter hollow pipe, supported by guy wires. The largest unit generated 100 kW in 18 mph winds. These units operated successfully for more than 4,000 hours during the period from 1957 to 1968. This work resulted in some of the most advanced wind turbines that have yet been built. The light weight fiber glass blades resulted in less bearing and blade failures than were experienced with the heavier type machines built in other countries.

TYPES OF WIND ENERGY COLLECTORS

Many types of wind energy collectors have been devised. It is said that more patents for wind systems have been applied for than for nearly any other type device. Basically, almost any physical configuration which produces an asymmetric force in the wind can be made to rotate, translate, or oscillate, and power can be extracted. Machines using rotors as wind energy collectors may be classified in terms of the orientation of their axis of rotation, relative to the windstream, i.e.:

> horizontal-axis rotors (head-on)—for which the axis of rotation is parallel to the direction of the windstream, typical of conventional windmills;
>
> crosswind horizontal-axis rotors—for which the axis of rotation is both horizontal to the surface of the earth and perpendicular to the direction of the windstream, somewhat like a water wheel;
>
> and vertical-axis rotors—for which the axis of rotation is perpendicular to both the surface of the earth and the windstream.

In addition, a number of types of translational wind machines have been devised, including the sailing ship itself; sailing ships that carry water-driven turbines mechanically connected to an electric generator as well as land vehicles driven by sails or solid airfoils on a closed track or roadway, with their wheel mechanically linked to an electric generator. Other types of translational devices have been designed to produce power by oscillating in the windstream (see Figure 1.3). Still other types of wind energy conversion devices are being developed that use no moving parts, such as devices that use differential cooling in a windstream to generate electricity by means of the Thomson thermoelectric effect.

Horizontal-Axis Rotors

Head-On: Horizontal-axis rotors can be either lift or drag devices. Lift devices are generally preferred, since, for a given area, many times more force can be developed by lift than by direct drag. In addition, a drag device generally cannot move faster than the wind velocity. Thus, a lifting surface can obtain high tip-to-wind speeds and a higher power-output-to-weight ratio and a lower cost-to-power output ratio.

Systems can be designed with different numbers of blades, ranging from one-bladed devices with a counterweight, to devices with large numbers of blades (i.e., up to 50 or more). Such systems often use canted blades to reduce the bending loads on the roots of the blades. Some horizontal-axis rotors are designed to be yaw-fixed, i.e., they cannot be rotated around a vertical axis perpendicular to the windstream. Generally, this type would only be used where there are prevailing winds from one direction. Most types are yaw-active and will track the changing direction of the wind. Small systems are usually designed to yaw using a tail vane, whereas larger systems are normally servo operated.

A number of means are used to prevent a propeller from overspeeding in high winds, including feathering of the blades, flaps rotating with the blades, or flaps on the blades themselves, as well as devices that turn the propeller sideways to the wind using pilot vanes set parallel to the blades.

Figure 1.3: Types of Wind Energy Collectors

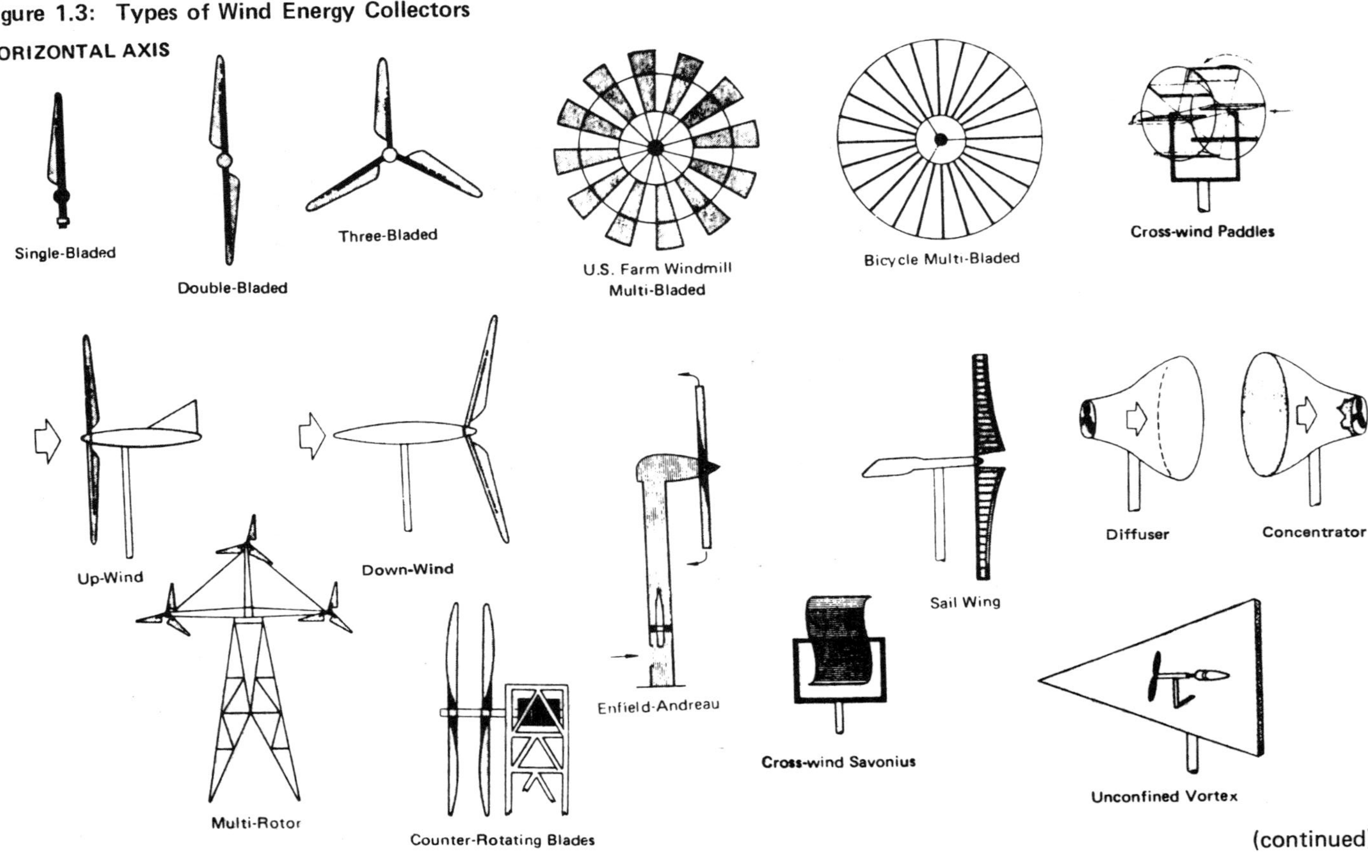

(continued)

Figure 1.3: (continued)

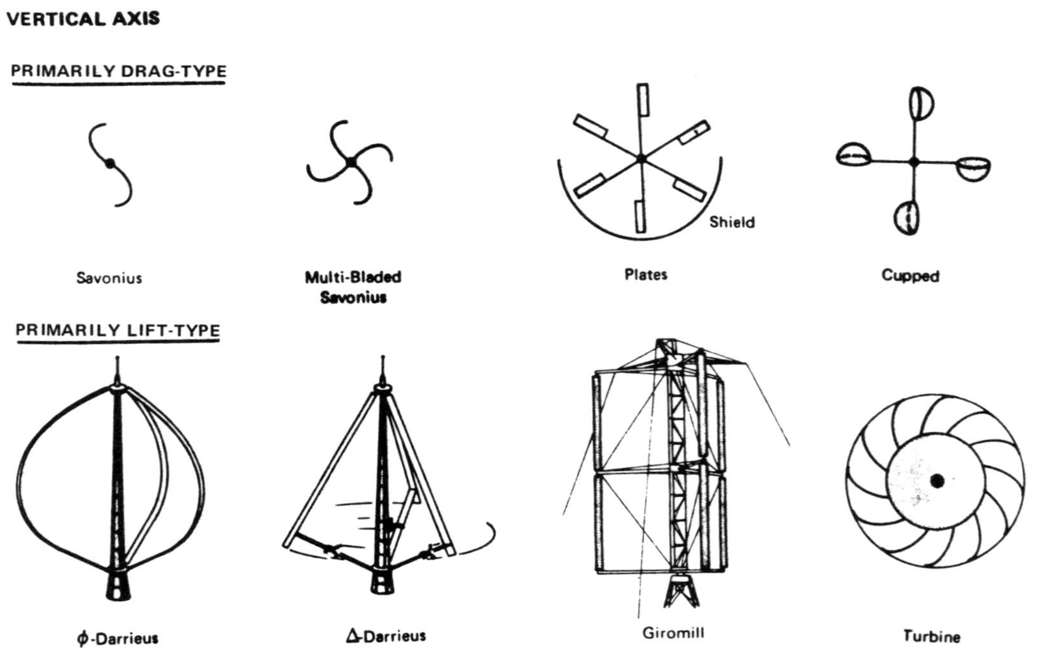

VERTICAL AXIS

PRIMARILY DRAG-TYPE

Savonius

Multi-Bladed Savonius

Shield

Plates

Cupped

PRIMARILY LIFT-TYPE

ϕ-Darrieus

Δ-Darrieus

Giromill

Turbine

(continued)

Figure 1.3: (continued)

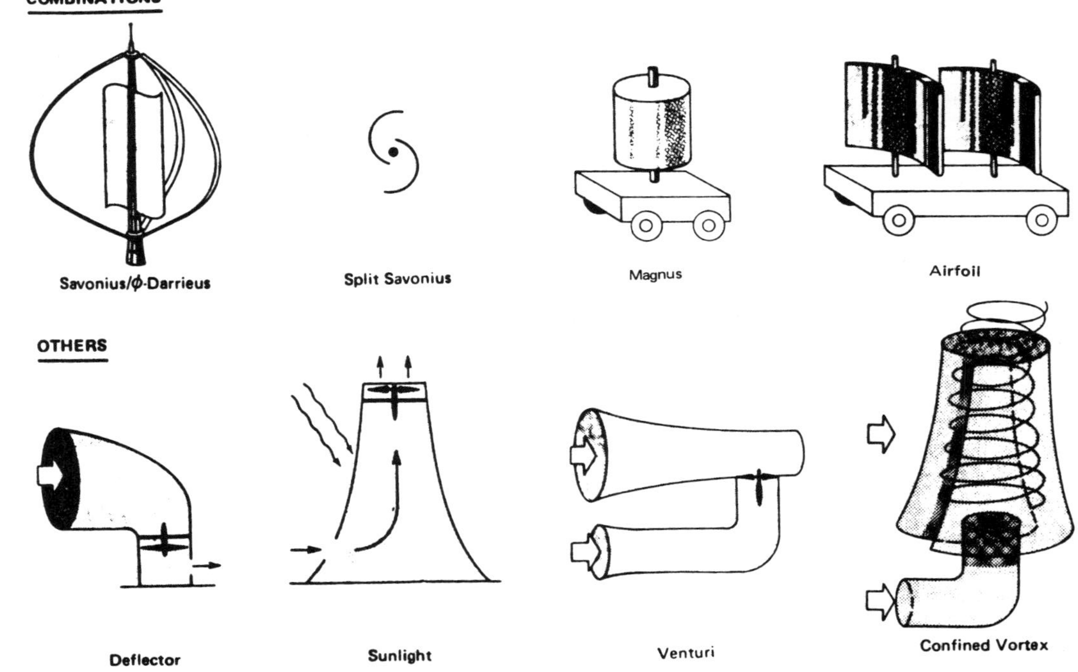

Source: PB 249 936

Some blades are directly coupled to the output of the system through a shaft on which the rotor is mounted. Others use a circular rim attached to the blade-tips to drive a secondary shaft that is mechanically connected to an electric generator or other form of power output.

Some horizontal-axis rotors are designed with blades that rotate in front of the towers, with respect to the direction of the windstream. These are called up-wind rotors. On the other hand, rotors that are designed with the blades rotating in back of the tower are called downwind rotors.

Modern descendants of the ancient jib sail rotors, designed by Princeton University, are manufactured from metal tubing and cloth. These have the general shape of solid propellers and are known as sailwings. Sailwings consist of a rigid tubular leading edge to which short bars are attached to form a rigid tip and root. A cable is stretched between the free ends of the tip and root bars to serve as a trailing edge for the blade.

The blade surface usually consists of a sleeve of dacron sailcloth that is slipped over the tubular leading edge and the cable and stretched by tightening the cable. Sailwings such as these have been designed that have a lift-to-drag ratio comparable to conventional rigid propeller blades, but which are 50% lighter in weight than their rigid counterparts. The sizes of sailwing rotors are limited to about 30 feet in diameter by the strengths of the material used.

Various types of horizontal-axis systems have also been developed with counter-rotating blades. Others have been proposed with multiple rotors on a single tower to reduce tower costs for a given power output of the system. Others use tapered shrouds to concentrate and/or diffuse the windstream as it passes a horizontal-axis turbine, thereby increasing the windstream velocity and reducing turbulence. Still others are designed to produce vortices around the turbine to concentrate the windstream and increase its angular velocity. Figure 1.4 shows the output power of various sizes of horizontal-axis wind turbines.

Figure 1.4: Typical Family of Horizontal-Axis Wind Turbines for Average Wind Speeds of 17 mph

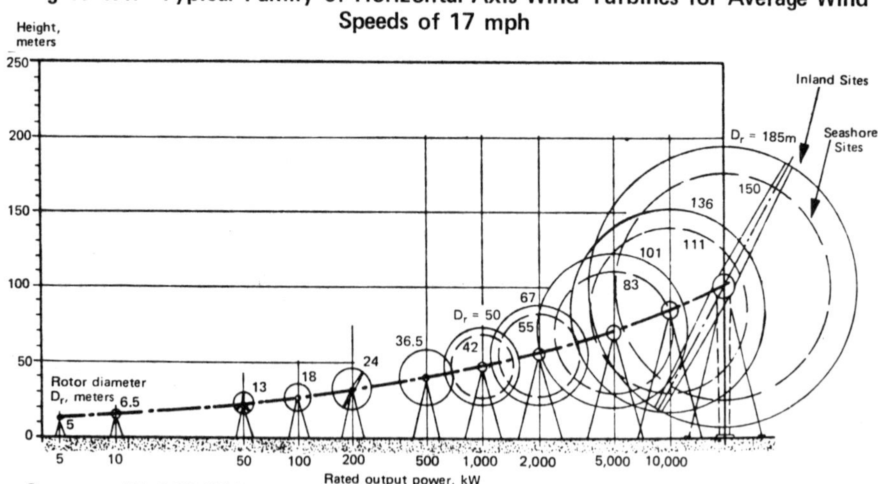

Source: PB 249 936

Crosswind: Various horizontal crosswind devices have been developed but have not been found to be very effective, since they must be turned into the wind as the wind direction changes, the same as conventional head-on horizontal-axis rotors. Relatively complicated means must be used to collect the output power from such devices, which generally result in a loss of efficiency of the system. In general, there appear to be no significant advantages of crosswind horizontal-axis rotors over either head-on horizontal-axis rotors or vertical-axis rotors.

Vertical-Axis Rotors

In general, vertical-axis rotors have a major advantage over horizontal-axis rotors. They do not have to be turned into the wind as the direction of the windstream varies. This reduces the design complexity of the system and decreases gyro forces on the rotors, when yawing, that stress the blades, bearings, and other components used in horizontal-axis rotor systems.

Various types of vertical-axis panemones have been developed in the past that use drag forces to turn rotors of different shapes. These include those panemones that use plates, cups, or turbines as the drag device, as well as the Savonius S-shaped cross-section rotors which actually provide some lift force but are still predominantly drag devices. Such devices have relatively high starting torques, compared to lift devices, but relatively low tip-to-wind speeds and low power outputs per given rotor size, weight and cost.

The Darrieus-type rotor was invented by G.J.M. Darrieus of France in the 1920s and has been under extensive development by the National Research Council of Canada since the early 1970s. It is now considered to be a potential major competitor to the propeller-type systems.

Figure 1.5: Typical Family of Vertical-Axis, Darrieus-Type Wind Turbines for Average Wind Speeds of 17 mph

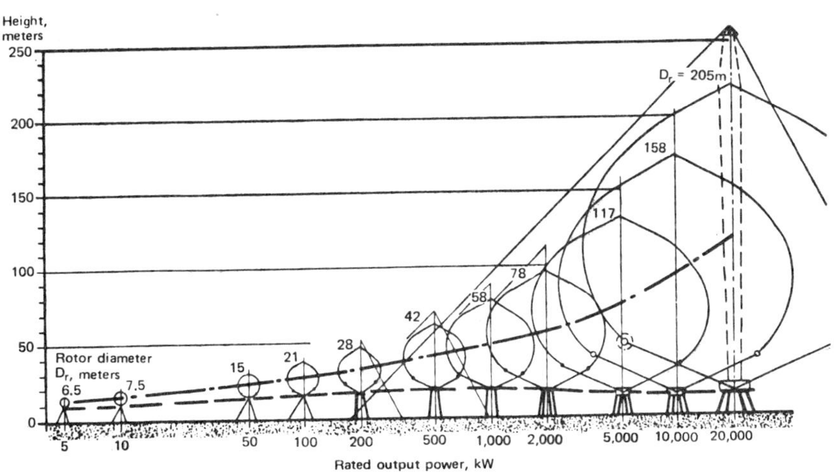

Source: PB 249 936

Darrieus-type rotors are lift devices, characterized by curved blades with airfoil cross sections. They have relatively low starting torques, but high tip-to-wind speeds and, therefore, have relatively high power outputs per given rotor weight and cost. Various types of Darrieus rotor configurations have been conceived, including the ϕ-Darrieus, the Δ-Darrieus, the Y-Darrieus, and the $\diamond$-Darrieus. Such Darrieus rotors can be designed to operate with one, two, three, or more blades.

Darrieus rotors can also be combined with various types of auxiliary rotors to increase their starting torques. However, such additions increase the weight and cost of the system, so tradeoffs in these characteristics must be considered in developing an optimum design for a given application. Figure 1.5 illustrates the relationship between Darrieus rotor size and power output.

Other types of vertical-axis rotors include Magnus Effect rotors typified by the Madaras or Flettner designs that consist of spinning cylinders. When operated in a windstream, translational forces are produced perpendicular to the windstream by the Magnus Effect. Such a device can be used as a sail to propel ships or land vehicles. (See Figures 1.6 and 1.7.)

Figure 1.6: The Magnus Effect

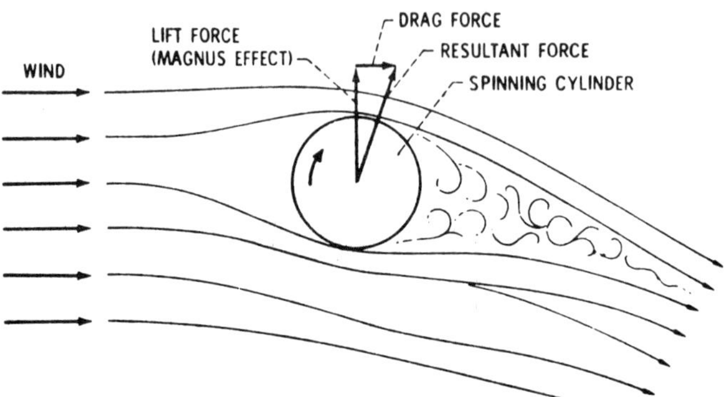

Figure 1.7: The Madaras Concept for Generating Electricity

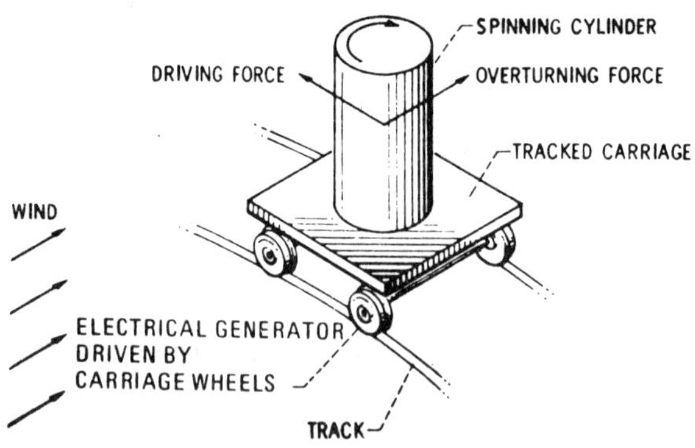

Source: PB 249 936

Still other types of vertical-axis rotors have been conceived that use ducts and/or vortex generator towers, augmented by shrouds or diffusers to deflect the horizontal windstream to a vertical direction and increase its velocity, sometimes with the addition of heat, either by direct solar insolation or by burning a fuel of some type, in the final extension. The latter becomes, in effect, a gas turbine.

In conclusion, while many types of configurations can be conceived, the primary consideration is the amount of energy, per unit system cost, that is derived at a given wind speed.

FUTURE POTENTIAL

There is an urgent need to develop alternative clean, competitive sources of energy as soon as possible. Sharp increases have occurred in the price of energy derived from conventional energy sources and systems. The rapidly growing dependence of the U.S. on imports of foreign oil and the extended public awareness of the potential impact of conventional energy systems on the environment have created an intensified demand for viable alternative energy sources. Wind energy conversion systems are one such alternative.

Federal Support

Federal support is being provided for the WECS program to accelerate the design, construction and operation of the large types of WECS units (i.e., 1 MW, or more, rated capacity) that will be needed to tie in economically with public utility networks. In addition, the application of WECS to many types of farm functions appears increasingly competitive as the price of conventional fuel and electricity rises. Federal incentives may be needed to spur these applications.

The key hurdles to be cleared in public utility applications are the demonstration of reliable MW-scale wind energy collection units with expected lifetimes of 20 to 30 years and a reduction in the capital costs of these units to a level of $250 to $300 per kW, through improved engineering designs and mass production.

Other important problems that are foreseen involve transferring the Federally developed WECS technology to industry, public utilities, farm operators, and other users, and obtaining the necessary capital required to build WECS manufacturing facilities and operating plants rapidly enough to meet the program goals.

The planned Federal WECS program involves a significant amount of R & D on such subjects as WECS mission and regional analysis; wind characteristics; advanced system concepts, components and subsystems; and possible environmental, societal, legal, or institutional problems that might constrain rapid implementation.

In addition, the Federal government plans to construct and demonstrate a number of WEC systems for farm applications, including the generation of electricity, the pumping of water for irrigation purposes, and the heating of water or air for space heating and cooling.

Most of the initial large scale WECS units will be of the two-bladed horizontal-axis type, although one-bladed rotors and various vertical-axis alternatives are also being considered for inclusion in the program. Performance and testing standards for various types of WECS units that might be used in the future are also being studied.

ERDA-NASA

The first large scale WECS unit built for the ERDA under the Federal program is located at the NASA-Lewis Research Center facility at Plus Brook, Ohio, and was put into operation in September 1975. It has a horizontal-axis, downwind rotor, 125 feet in diameter, designed for a rated capacity of 100 kW in 18 mph winds. The expected power coefficient for the rotor blades is 0.375 and the power train efficiency is 0.75. Cut-in wind speed is 8 mph. The rotor is designed to operate at a constant speed of 40 rpm, in winds over 6 mph, by changing the system load and the pitch of the blades. The blades are fully feathered in wind speeds greater than 60 mph and the system is designed to withstand velocities as high as 150 mph.

The rotor, transmission train, generator and controls are mounted on a bedplate on top of a 100 foot tower of pinned truss design. The controls consist of a pitch change mechanism, as well as a yaw control mechanism for rotating the bedplate at the top of the tower at a speed of ⅙ rpm. Torque is transmitted from the rotor hub to the electric alternator through a gear box with a speed ratio of 45:1. The alternator is rated at 125 kVA at 1,800 rpm and weighs about 1,425 lb.

This initial 100 kW system is designed to serve as a test bed for improved WECS components and subsystems and will be used to collect performance data that can be applied in the design of other wind generators of all sizes that will be built during the course of the WECS demonstration program. Performance data that will be collected include power output at various wind speeds; loads, stresses, and vibrations in components such as the blades, hub, and tower; and the stability and effectiveness of the control systems incorporated in the WECS units.

The objectives of the initial studies of this 100 kW system were to identify WECS components and subsystems for which life cycle costs and maintenance can be reduced and to acquire a basis for estimating mass production costs of future WECS systems of various sizes and types.

Future Systems

Because of the resulting economies of scale, future wind power systems, using blades or other types of rotors operating in ambient windstreams, will be as large as possible, i.e., initially 1 to 3 MW and eventually, perhaps up to 20 MW output. A possible limitation on rotor size will be the strength that can be achieved in the design of bearings, blades, and other stressed components.

These large future systems will be designed, therefore, with rotors that are as lightweight as possible. The rotors will probably use lift forces, rather than drag forces, in their normal mode of operation, because of the higher tip-to-wind speeds that can be achieved, resulting in higher power coefficients. Offshore systems may be used to take advantage of the higher and more persistent wind velocities at such sites. Research is also being done on systems that incorporate towers designed to produce tornado-type vortices using buildings or other types of man-made or natural baffles designed to increase the wind velocity and the pressure gradient across the wind turbine that extracts power from the system. (See Figure 1.8.)

Figure 1.8: Vortex Tower for Omnidirectional Winds

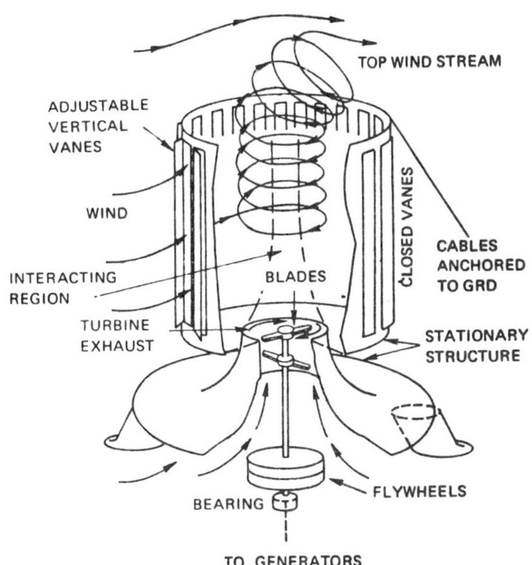

Source: PB 249 936

APPLICATIONS

Energy extracted from the wind is initially energy in the form of rotary, translational, or oscillatory mechanical motion. This mechanical motion can be used to pump fluids or can be converted to electricity, heat, or fuel. Some of the most effective applications will be able to use the energy that is derived without further processing. However, if required, the energy can be stored by the use of compressed fluids, pumped-hydro systems, batteries, hydrogen, flywheels, hot water, etc. Figure 1.9 illustrates the uses of wind energy.

Pumping Applications

A typical pumping application is one that uses horizontal-axis windmills. An example is the ancient jib sail design, that is still used to pump irrigation water in the valley of Lasithi on the island of Crete. There are so many windmills in this area that Lasithi is often called the "valley of 10,000 windmills." Large numbers of water-pumping windmills were used on American farms in the late 1800s.

Figure 1.9: Applications of Wind Power

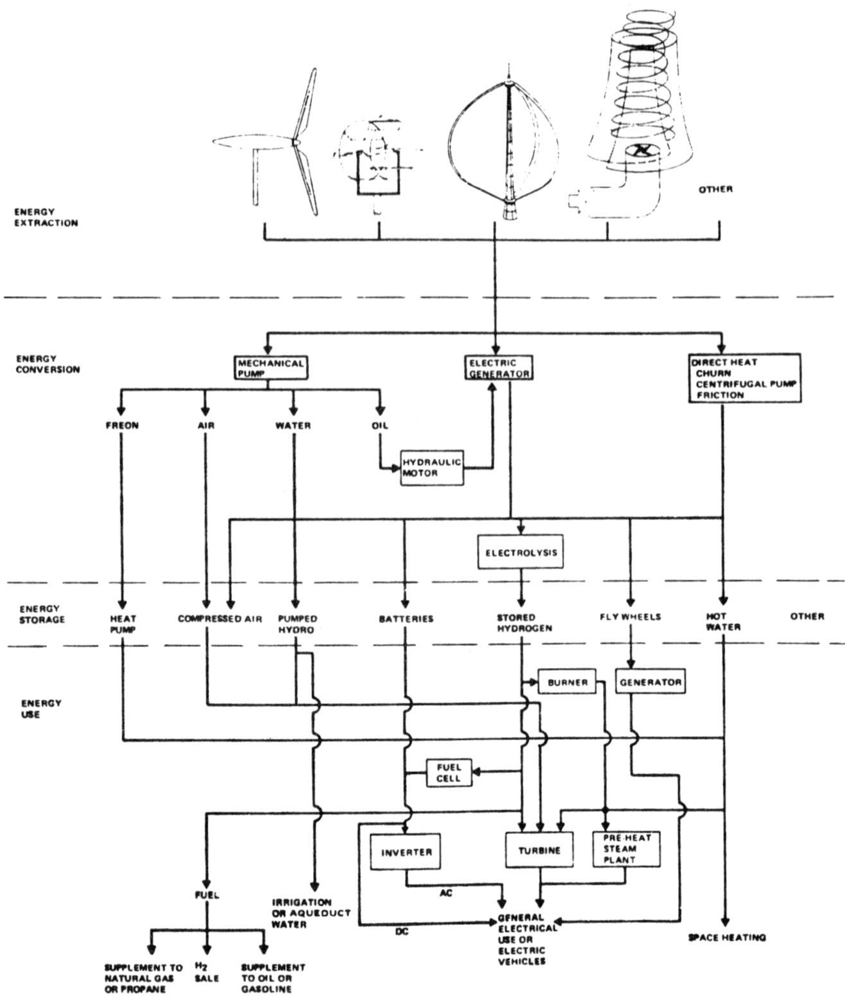

Source: PB 249 936

A more modern version of an irrigation application is the windmill pumping system designed by the Brace Research Institute of Canada to pump water under high pressure for irrigation sprinklers. Other applications include the pumping of water for aqueducts or for pumped hydro storage of energy. In aqueduct systems, large scale wind-driven units can provide power for the pumping of water from the main reservoir to auxiliary reservoirs in other parts of the aqueduct system. This can be done either by the direct mechanical pumping of the water or through the generation of electricity by the wind units, and the subsequent use of this energy to operate electrical water pumps incorporated in the aqueduct system.

In pumped hydro applications, the wind units can be used to supply power to pump water from an auxiliary reservoir, below a hydro-electric dam, back into the main reservoir above the dam. This enables the water that is stored in the main reservoir to be replenished when the wind is blowing, thereby adding to the capacity of the hydro-electric system to generate base load electrical power.

Wind power can also be used to compress air for use in various applications, including the operation of a gas turbine for generating electricity during the peak demand periods of a public utility system. For this type of application, conventional gas turbines can be modified to separate the compressor, generator, and power stages by clutches. In one mode of operation, the motor-generator, operating as a motor and powered by a wind machine, drives the air compressor. The compressed air is fed into a storage tank or into a large cavern or depleted natural gas well. Under this mode, the power turbine is inoperative, and no fuel is consumed.

In a second mode of operation, when the demand for power exceeds the supply of the base load utility system, the compressor is disengaged, and the power turbine is connected to the generator. The burner that drives the power turbine is fed fuel and compressed air from storage to generate power for the utility system. (See Figure 1.10.)

The temperature of air is raised when it is compressed without loss of heat, i.e., adiabatic compression. If the air is stored in a well insulated high temperature container, it will retain most of its heat, i.e., adiabatic storage. In this case, less heat will need to be added to the air, when it is eventually used to drive a turbine at a given efficiency, than if its heat had been allowed to escape from the storage container and the temperature of the air had been allowed to drop to the ambient temperature, i.e., isothermal storage.

Adiabatic storage is obviously better, from the standpoint of energy conservation, than isothermal storage. Wind powered pumps can also be used to save fuel and electricity by compressing the working fluids used in heat pumps for space-heating applications.

Figure 1.10: Wind Assisted Gas-Turbine Generating Unit

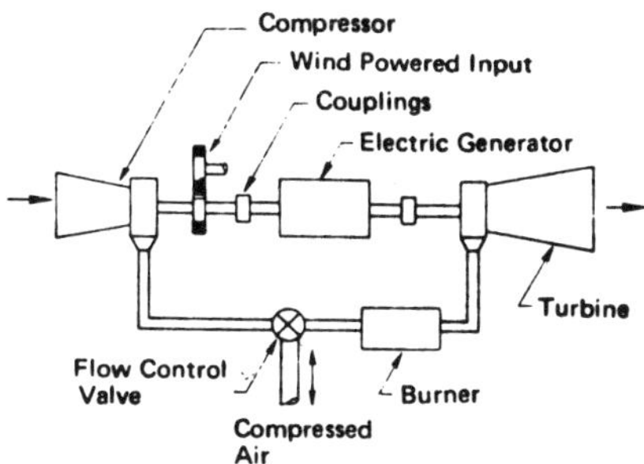

Source: PB 249 936

Direct Heat Applications

Mechanical motion derived from wind power can be used to produce heat from the friction of solid materials, or by the churning of water or other fluids, or by the use of centrifugal pumps. This heat may then be stored in materials having a high heat capacity, such as stones, eutectic salts, etc, or the heat may be used directly for such applications as space-heating, industrial processes, crop drying, etc.

Figure 1.11: Alternatives for Converting and Storing Wind Energy

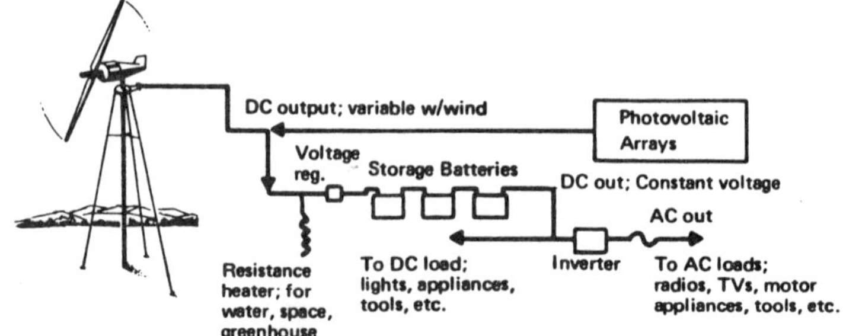

SYSTEM WITH BATTERY STORAGE

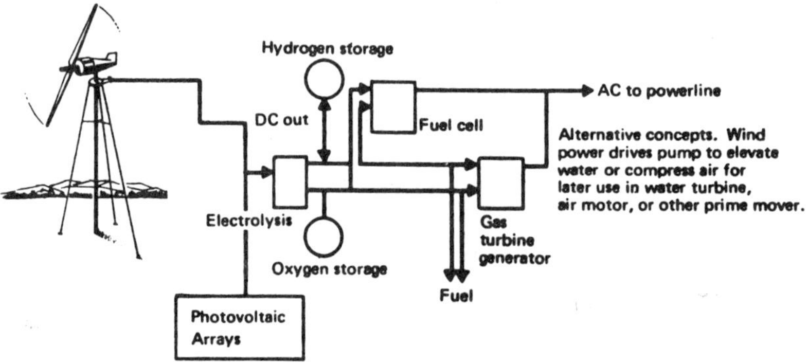

SYSTEM WITH HYDROGEN STORAGE

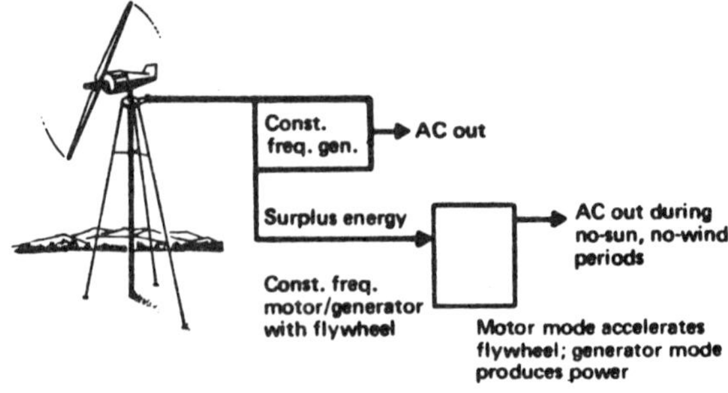

SYSTEM WITH FLYWHEEL STORAGE

Source: PB 249 936

Electric Generation Applications

Wind power can be used to drive synchronous ac electrical generators, and the energy can be fed directly into power networks through voltage step-up transformers. In other applications, wind power can be used to generate dc electrical power that, in turn, can be used for dc appliances or space heaters, such as resistance heaters, or can be stored in batteries and then inverted for use by ac loads.

In these types of applications, the energy can also be stored in the form of the mechanical motion of a flywheel or as hydrogen and oxygen gases derived from the electrolytic dissociation of water. The hydrogen and oxygen can be stored in liquid form in tanks, or in gaseous form in tanks, caverns, depleted natural gas wells, etc. Figure 1.11 illustrates these alternatives for converting and storing wind energy.

The stored hydrogen can be used either as a fuel for direct space-heating or industrial process heat, or it can be reconverted to electricity through the use of fuel cells, gas turbine generators that burn hydrogen, or by other means.

Interconnected Systems

There appear to be important advantages to using wind derived energy in combination with energy derived from other sources, such as conventional fuels, sunlight, ocean thermal differences, bioconversion fuels, etc.

Since the wind blows intermittently in most locations, there may be a need to store wind energy over long periods of time, perhaps up to 10 days or more, if the energy is being used for isolated applications requiring continuous power. The cost of providing sufficient storage capacity for such applications can be reduced if the wind derived power is interconnected with other sources of power.

For instance, since in most locations the wind often blows when the sun is not shining, and vice versa, a system using wind energy collectors and sun energy collectors, e.g., solar photovoltaic arrays or solar thermal collectors, in combination, can be expected to require less energy storage capacity than systems that use these types of collectors singly.

Large numbers of dispersed WECS units tied into the same grid network can also be used to reduce storage requirements for base load applications served by the grid, since wind speeds can vary considerably over large areas at any given moment in time.

SITING

Wind Distribution

Currently available maps showing patterns of average wind power over the U.S. provide only very rough estimates of this power. Many are based mainly on measurements obtained near ground level at airports. However, airport locations are, generally, purposely chosen to avoid sites where local topography might result in high wind speeds. Other pattern maps are produced by extrapolating high altitude wind measurements down to a standard height above ground level. An example of the latter type is shown below in Figure 1.12. Its shaded areas indicate where average wind speeds are estimated to equal or exceed 18 mph at 150 feet altitude over the U.S. Many of these areas are near large population centers such as New York, New England, Western Texas, Denver, Colorado Springs, Los Angeles and San Francisco. Others are areas served by large utility networks such as those of the Bonneville Power Administration and the Tennessee Valley Authority.

A rough integration of the average power available within the 18 mph contour surrounding the high Great Plains region indicates that with conservative assumptions regarding the operating efficiencies and proper spacing of wind machines, it is expected that the power that could be extracted from the winds in that region, alone, is several times the present U.S. electrical power demand.

Figure 1.12: Areas in the U.S. Where Annual Average Wind Speeds Exceed 18 mph

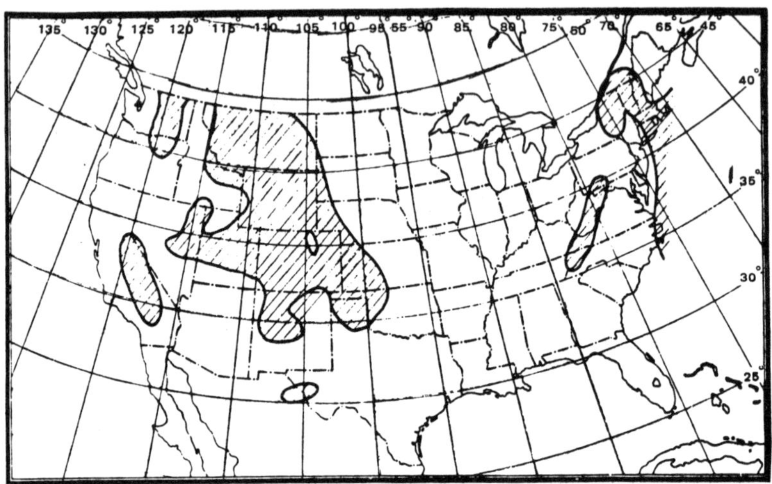

Source: PB 249 936

Wind Characteristics

The amount of power available in a freely flowing windstream of cross-sectional area A is equal to this area times the velocity of the windstream V times the kinetic energy of a unit volume of the windstream (or, one-half the air density, times the subtended area of the windstream, times the cube of the wind speed).

The power per unit cross-sectional area of the windstream (i.e., the wind power density) is therefore equal to one-half the air density, times the cube of the wind speed.

Power = volumetric flow rate x kinetic energy per unit volume

$$P = (AV) \times \left(\frac{\rho V^2}{2}\right)$$

$$\text{or } P = \frac{\rho A V^3}{2}$$

As a result, the power densities of winds at sea level increase from about 5 watts per square foot for wind speeds of 10 mph, to over 140 watts per square foot for wind speeds of 30 mph, and to about 650 watts per square foot for winds of 50 mph. At higher altitude sites the air density is less, so the wind power density is lower. (See Figure 1.13.)

The total available wind power in a freely flowing windstream will increase with the subtended area of the windstream. For instance, at sea level, a 20 mph windstream with a cross-sectional area of 100 square feet will contain about 4 kW of power, while over 40 MW of power will be available in a windstream with a cross-sectional area of 1 million square feet. (See Figure 1.14.)

Figure 1.13: Change of Wind Power Density with Wind Speed and Altitude

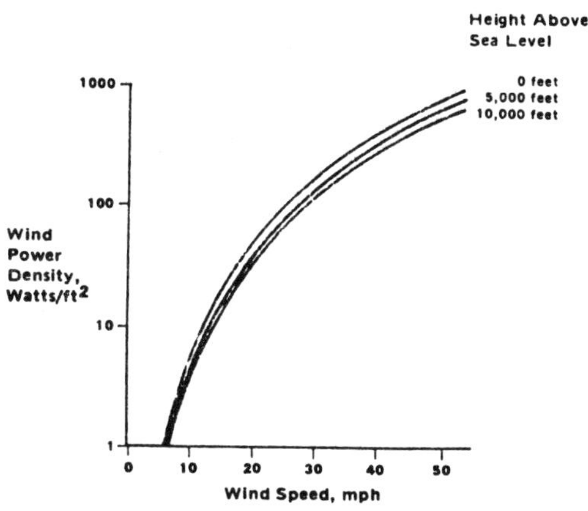

Figure 1.14: Power in Windstream Versus Cross-Sectional Area and Wind Speed

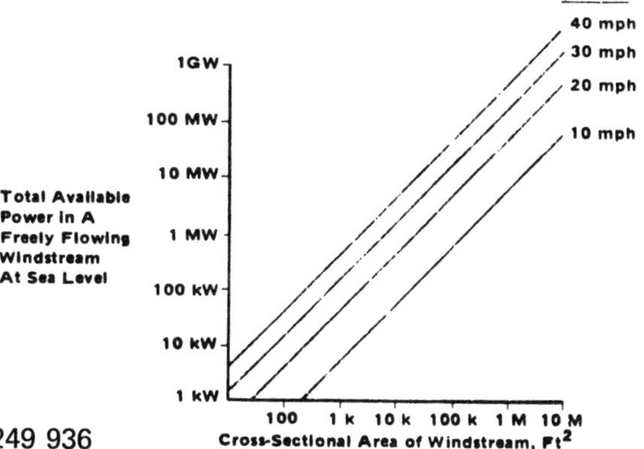

Source: PB 249 936

The wind at a given site usually varies frequently in direction and its speed may change rapidly under gusting conditions. Its average velocity also usually changes significantly with the season of the year. In most locations, its amplitude will often be two or three times higher in the winter than in the summer. (See Figures 1.15 and 1.16.)

Graphs showing the number of hours per year that the windstream, at a given site, attains a specified hourly mean wind speed are known as "annual average velocity duration curves." (Figure 1.17.) Curves showing the distribution of annual average wind power per unit subtended area, as a function of windspeed, are called "annual average power density distribution curves."

It is estimated by Sandia Laboratories that, because of the cubic relationship between wind power and wind velocity, coupled with the fact that the wind gusts and is seldom steady, the actual wind power available at a given site can be two or three times that calculated on the basis of average annual wind speeds at that site. Therefore, depending on the responsiveness of a wind machine to these changes in wind speeds, the estimated performance of the machine, if based on annual average wind speeds, may be conservative. (Figures 1.18 and 1.19).

Figure 1.15: Typical Wind Speed Record

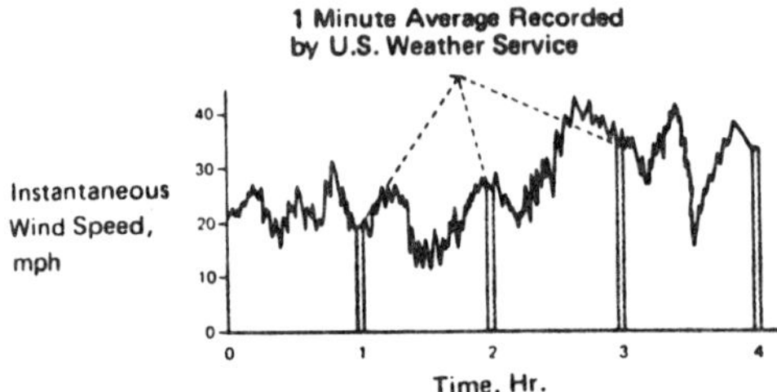

Source: PB 249 936

Figure 1.16: A Wind Rose Shows the Hours per Year That the Wind Blows From Each Direction

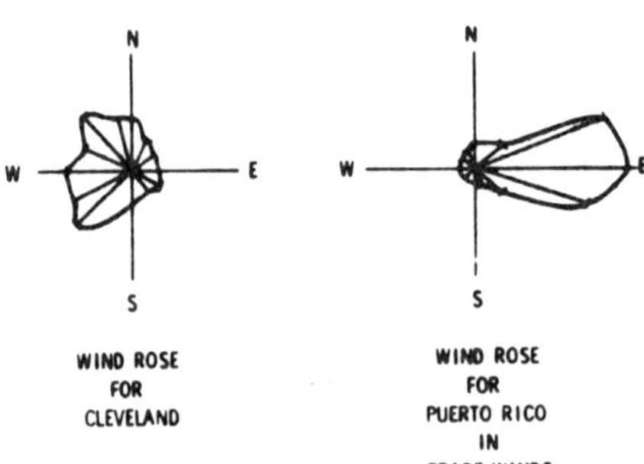

Source: PB 249 936

The annual average wind energy density distribution is equal to the annual average power density distribution, times the number of hours per year that the corresponding wind speeds occur. (Figure 1.20). Plots of the distribution of the annual average energy density of winds of various speeds at a given site show that most of the energy content of the wind is at speeds above the average wind speed, and that the contribution to the total annual average energy content of winds of all speeds is usually small for winds of speeds greater than about three times the average wind speed. (Figure 1.21).

Figure 1.17: Annual Average Velocity Duration Curves for Three Sites

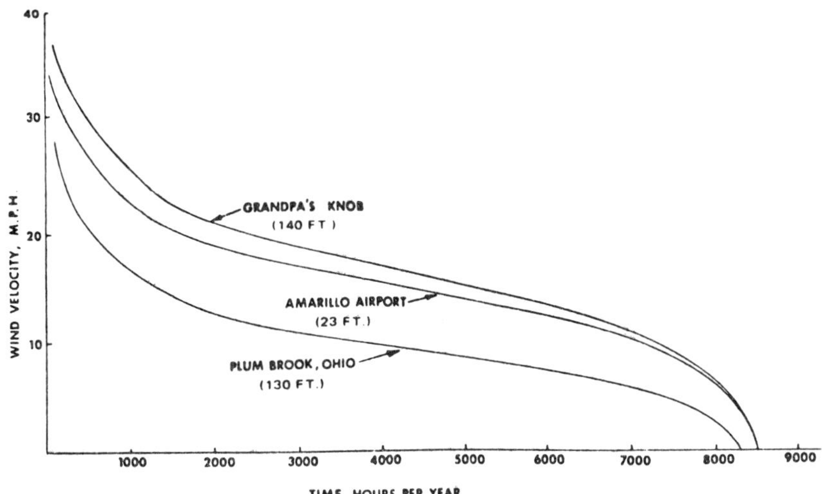

Figure 1.18: An Example of Gusting

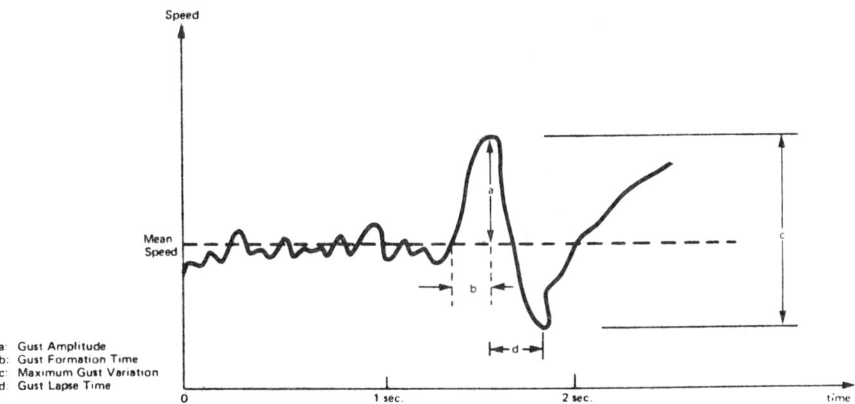

Source: PB 249 936

Figure 1.19: Comparison of Actual and Calculated Wind Power at Different Sites

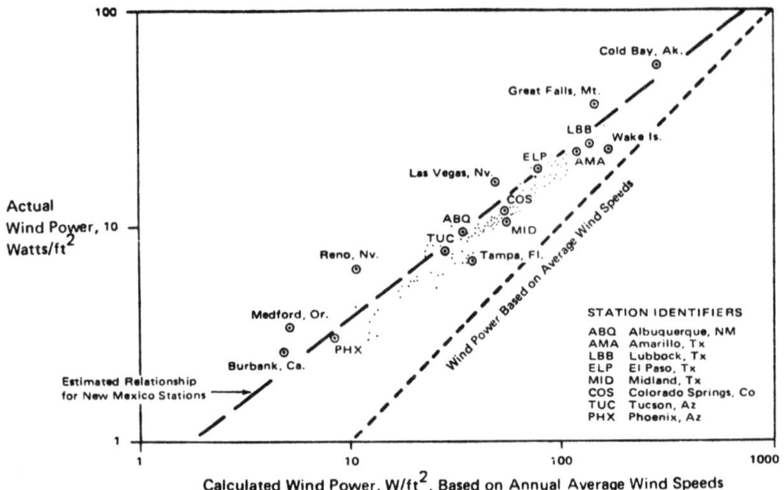

Figure 1.20: Typical Distribution of Annual Average Energy Density of Winds of Various Speeds

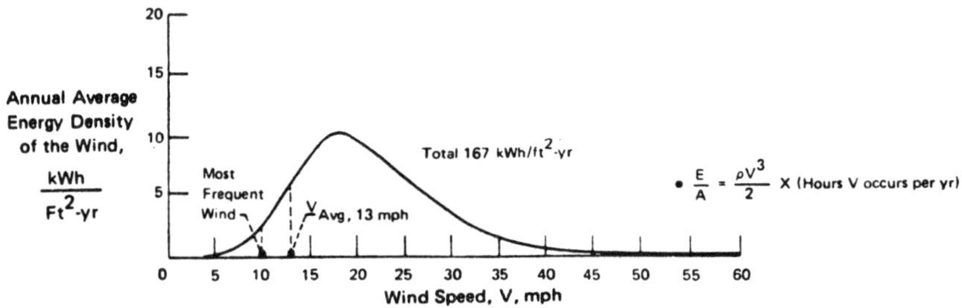

Figure 1.21: Typical Distribution of Annual Average Energy Density of Winds for Sites with Low and High Annual Average Wind Speeds

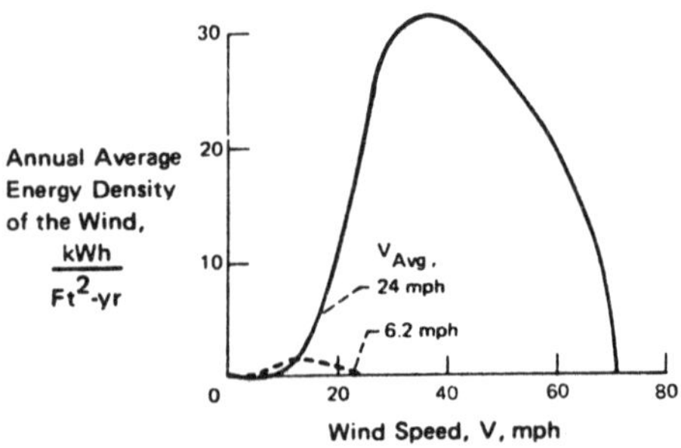

Source: PB 249 936

Choice of Site

Care should be exercised in choosing a suitable site for a wind machine since there is a significant amount of shear and compression in a normally horizontal windstream as it passes over the surface of the earth. This shear results in lower wind speeds near the surface than at heights that are great enough for free flow to occur. The free flow velocity, at heights that are large enough to be unaffected by the surface shear, is usually significantly larger than that of winds either at the surface or at standard anemometer heights, where the wind is normally measured. A common rule of thumb is that wind speeds near the earth's surface will increase as the one-seventh power of the height above the earth's surface.

The wind shear, and consequently the available wind power, is also affected by the roughness of the earth's surface in a given location, i.e., whether it contains buildings, trees, wind machines, or other obstacles. Some wind tunnel tests have been made in Sweden to determine the minimum permissible spacing of wind rotors in a given geographical area to prevent interaction of the rotors. From this work it appears that the spacing should be at least 6 rotor diameters. Other estimates conclude that 8 to 12 diameter spacing may be desirable.

Another significant effect is that the streamlines of a windstream are compressed and its flow is accelerated as it passes over a hill or through a narrow valley. It is often possible to increase the average power output of a wind machine by siting it to take advantage of the increased average wind speed that results from anomalies such as these.

Wind Features	Effects
Unsteadiness	Power and energy are unsteady and are sometimes zero; energy storage or backup system may be required
Omnidirectionality	To maximize the energy output, propeller rotors must always be vectored to face the wind
Spatial nonuniformity due to ground roughness and terrain variations	WTG systems must be ruggedly built to resist gusts and high winds
Horizontally from place to place	Suitable sites may be hard to find
Vertically with altitude	On level terrain, rough ground results in low winds under 200 ft; taller towers are required on rough ground than on smooth ground

Summary of Features of Suitable Site

High annual average wind speed (consult local National Weather Service Station)
No tall obstructions upwind for a distance depending on the height
Top of smooth well rounded hill (with gentle slopes) on flat plain or island in a lake or sea
Open plain, open shoreline
Mountain gap that produces a funneling

Wind Surveys

The National Climatic Center, NOAA, collects wind characteristics data from about 600 weather stations throughout the U.S. from which average wind speeds and directions and monthly and annual average wind energy and wind power distributions can be obtained. However, because of local anomalies, it is usually worthwhile to make detailed wind surveys before choosing a site for a wind machine. Typical wind measurements at potential sites for a large scale unit usually consist of the following:

- Instrumentation:
  - 3 cup anemometer and wind direction sensor
  - Height of instruments:
    - 30 ft for preliminary data
    - 100 ft for longtime data
  - Data recording systems:
    - Strip chart
    - Magnetic tape
  - Type of data:
    - Wind speed and direction—hourly averages
  - Data reporting:
    - Wind frequency curves
    - Daily, weekly, monthly

PERFORMANCE

Power Extraction

The power, P, that can be extracted from a windstream by an unshrouded wind turbine is equal to the efficiency, η, of the system used to extract the power, times the wind's volumetric flow rate at the turbine, $V_T A_T$, times the sum of the change in pressure energy, Δp, plus the change in kinetic energy, ΔQ, of a unit volume of air that passes through the turbine:

$$P = \eta V_T A_T [\Delta p + \Delta Q]$$

For a conventional wind turbine, operating under optimum steady state conditions in an initially free-flowing windstream of constant initial velocity, V_o, the velocity of a unit volume of air will decrease monotonically as it approaches the turbine, passes through it and recedes from it. As the air receives kinetic energy from the surrounding winds, its velocity will increase until it again reaches its initial velocity, V_o.

The sum of the change in pressure energy and the change in kinetic energy of the unit volume of air will be constant as it approaches the turbine upstream. The pressure energy will increase, and the kinetic energy will decrease, until the pressure reaches a maximum at the interface with the turbine. As the air passes through the turbine, it will impart kinetic energy to the turbine, and the pressure energy will drop to a level below atmospheric pressure. As it recedes from the turbine, its pressure will rise until it again reaches atmospheric pressure. After passing the turbine, its kinetic energy will continue to decrease until its pressure again reaches atmospheric pressure.

The cross-sectional area of the windstream that passes through the system will be inversely proportional to its velocity. Its streamlines will, therefore, expand as it approaches the turbine, passes through it and recedes from it. As the disturbed flow receives kinetic energy from the surrounding winds, the effects of the turbine will diminish through the dispersal of the disturbance in the windstream.

The power coefficient, C_p, of such a system is defined as the power delivered by the system divided by the total power available in the cross-sectional area of the windstream subtended by the wind turbine. There are significant tradeoffs to be considered in the design of wind turbines. Turbines with a small total blade area will maintain a high volumetric flow rate of the airstream, but the pressure drop will be relatively small, and the resulting power output and power coefficient will be small.

For turbines with a large total blade area, the change in pressure and the reduction in windstream velocity will be large, but the volumetric flow rate of the windstream will be low. This will also result in a low power output and a low power coefficient. The optimum total blade design, therefore, is one that maximizes the product of the volumetric flow rate and the pressure drop across the turbine.

It can be shown from momentum theory that the maximum amount of energy that can be extracted by a wind turbine from a windstream in the process described above, is eight-ninths of the kinetic energy of the windstream passing through it. Under these conditions, the maximum loss in wind speed will be two-thirds of the initial velocity of the windstream, V_o.

For a system of 100% efficiency, the maximum power density that can be extracted by the turbine from the windstream in the above process will, therefore be:

$$\frac{P_{max}}{A} = \frac{2}{3} V_o \left[\frac{8}{9} \frac{(\rho V_o^2)}{2} \right] = 0.593 \frac{\rho V_o^3}{2}$$

where, $\dfrac{\rho V_o^3}{2}$ = the ambient power density of a unit volume of the windstream

The factor 0.593 is known as the Betz coefficient (from the name of the man who first derived it). It is the maximum fraction of the power in a windstream that can be extracted by a turbine in the windstream.

Typical Power Coefficients

The power coefficient of an ideal wind machine rotor varies with the ratio of blade tip speed to free flow windstream speed and approaches the maximum of 0.59 when this ratio reaches a value of 5 or 6. Experimental evidence indicates that two-bladed rotors of good aerodynamic design, running at high rotational speeds, i.e., where the ratio of the blade tip speed to free flow speed of the windstream is 5 or 6, will have power coefficients as high as 0.47.

Likewise, a Darrieus rotor can be expected to reach a maximum power coefficient of about 0.35 at a ratio of peripheral speed to windstream speed of about 6. Other designs have been found to have lower maximum power coefficients. These maxima occur at lower speed ratios than the high speed two-bladed horizontal-axis rotors and the Darrieus vertical-axis rotors. Figure 1.22 illustrates the performance of various types of wind machines.

Figure 1.22: Typical Performances of Wind Machines

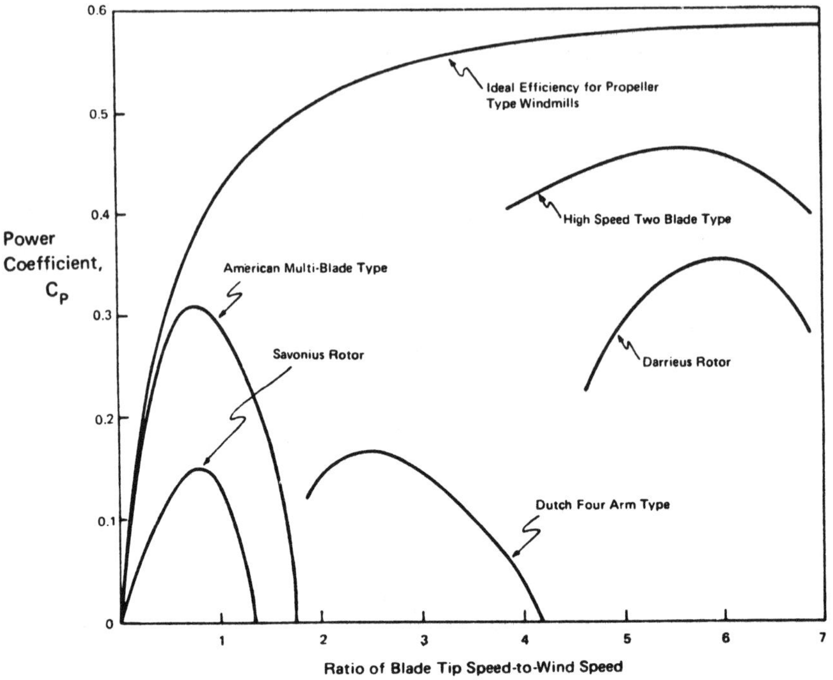

Source: PB 249 936

Output Performance

The torque developed by a horizontal-axis rotor blade of fixed pitch angle varies with both the shaft speed and the wind velocity. If the rotational speed of a blade is too slow, in a windstream of a given velocity, the blade will stall, and the output torque of the wind machine will decrease.

Therefore, to develop maximum power output from a windstream, as the speed of the windstream varies, either the pitch angle of the blade or the rotational speed of the blade must be varied. Many of the more modern wind machines are designed with variable pitch blades which are adjusted to operate them at as near a constant rotational speed as possible, for wind speeds above a minimum or cut-in value and for varying output loads.

At wind speeds lower than the rated value, the rotor speed must vary with the wind speed in order to extract maximum possible power. This does not match the optimum operating conditions for either synchronous or induction ac generators that can be driven by the wind machine rotor. This mismatch problem has led to several interesting concepts for interfacing the rotor with its electrical output.

One approach is to allow the rotor speed to vary optimally with the wind speed and to employ a variable speed constant-frequency generating system to obtain the constant frequency power needed for an electric utility grid. Both differential (i.e., mechanical or electrical frequency changing devices) or nondifferential (i.e., static frequency changers or rotary devices) may be used to accomplish this.

In other cases, such as the initial ERDA-NASA 100 kW wind machine, power output at wind speeds greater than its rated value of 18 mph has been sacrificed to simplify system design. For a power coefficient, C_p, of 0.35, this machine has a theoretical maximum power output as shown by the dotted line in the diagram (Figure 1.23) and would reach an output of 4,935 kW at wind speeds of 60 mph.

However, a synchronous ac generator with a fixed ratio gear train has been chosen for this design which must be operated at constant input speed, i.e., fixed wind rotor speed. Aside from friction losses, the theoretical maximum power output is maintained between the designed cut-in wind speed of 8 mph and the rated wind speed of 18 mph by varying the output load on the system to maintain a constant rotor speed of 40 rpm. For wind speeds above 18 mph the rotor speed is maintained at 40 rpm by varying the pitch of the rotor blades. In this manner a constant power output of 100 kW is maintained between the rated wind speed of 18 mph and the cut-off wind speed of 60 mph, at which point the rotor blades are feathered to protect the machine in high winds.

Figure 1.23: Power Output of 125 ft Diameter Rotor, Rated at 100 kW in Winds of Over 18 mph

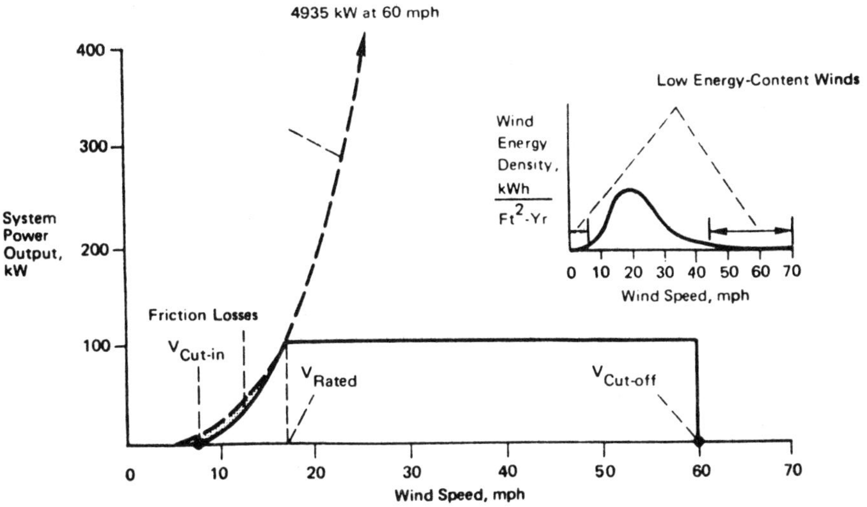

Source: PB 249 936

This type of design does not result in large losses of available wind power since, for example, when the average wind speeds are about 12 mph, the annual energy content of the wind is low for wind speeds below 8 mph and above 30 or 40 mph.

A power density duration curve shows the number of hours per year that the available winds at a given site will provide a power density of a specified amount for a wind machine, of a specified power coefficient, located at that site. The area under such a power density curve is the total amount of available energy per year, per unit swept area of a wind machine located at that site. For example, theoretically, a WECS with a power coefficient of 0.35, located at a site with the power density duration curve shown in the diagram (Figure 1.24), would produce more than 100 W of power output for each square foot of rotor area during 876 hours of the year.

Figure 1.24: Theoretical Power Density Duration Curve for a WECS with $C_p = 0.35$

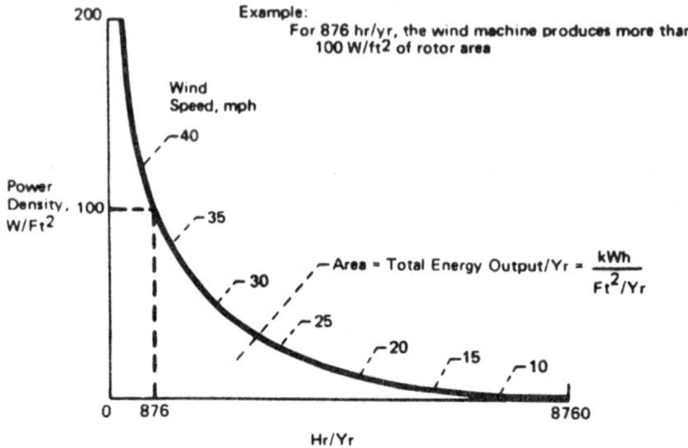

Source: PB 249 936

The actual power density of such a WECS with a synchronous ac generator driven by a gear train with a fixed ratio would be less than for the theoretical case shown. For instance, if such a WECS was designed for a cut-in wind speed of 15 mph, a rated wind speed of 30 mph and a cut-off wind speed of 60 mph, its actual output power density would vary according to the dotted line in Figure 1.25, and its actual energy density output per year would be represented by the shaded area under this dotted line.

Since the power density of a windstream varies with both the velocity of the windstream and the swept area of the rotor, the power output of a wind machine can be increased both by locating it in areas of high wind velocity and by increasing the diameter of its rotor.

However, the specific size and the siting of such a wind machine should be chosen to minimize the cost, to the consumer, of the energy produced and must take into account such factors as the transmission and distribution costs of the system and the physical limitations on rotor blade size.

Figure 1.25: Actual Annual Power Density Output of a WECS

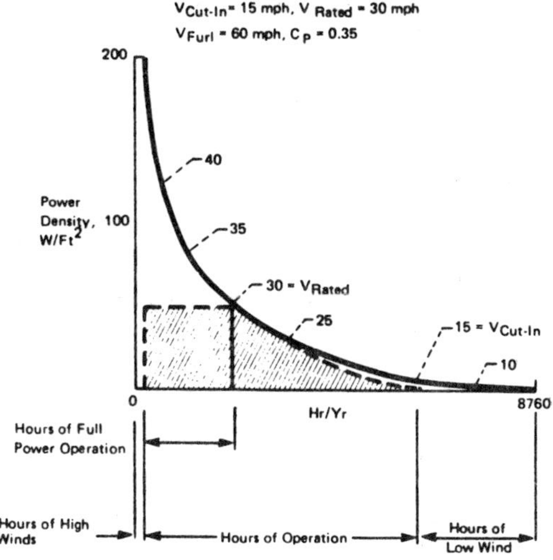

Source: PB 249 936

Extraneous Loading

Various types of extraneous loadings that occur in the operation of any wind machine can have a significant effect on blade motions. These include gyro forces, blade balance, wind shear, wind shifts, wind gusts, gravity forces, and tower shadow. These loadings can cause cyclic motions and vibration in the blade, the tower, the bearings, and other system components and can have a serious impact on system reliability, lifetime, and performance. They need to be carefully analyzed, therefore, when designing wind machines.

Ducted Systems

Various types of ducted systems using wind concentrators, diffusers, and other types of man made or natural structures can be designed to increase the amount of wind energy received by a system, thereby increasing both the wind speed and the pressure drop across the wind turbine. This, in turn, increases the amount of kinetic energy that it can extract from the wind by a turbine of a given diameter.

Such measures can, therefore, be used to decrease the size of a wind turbine required for a given power output and increase the rotational speed of the turbine. This decreases the amount of gearing required to interconnect the turbine and an electrical generator or other output device.

Ducted systems using wind concentrators and diffusers have been built that increase the wind speed through the turbine by a factor of 1.5 compared to the free flow wind speed and produce up to 3.5 times the power generated by an ideal wind turbine having a diameter equal to that of the ducted turbine and operating at the same ambient free flow wind speed. In practice, specific designs have to be analyzed to determine if the additional power output outweighs the cost of the diffuser. In general, inlet ducts or entry cones have not been found to be effective, whereas exit diffusers are still the subject of considerable research.

Vortex Generators

Both unconfined and confined vortex generators are being studied that spin the wind to increase the power output of a turbine located in or near the vortex. Unconfined vortex systems are being examined at both the Polytechnical Institute of New York and West Virginia University. Such systems use wing-like structures to deflect the wind and create an unconfined vortex around the turbine. It is estimated that such systems can be designed to provide up to six times the power output of a conventional system, with the same rotor diameter, that is located in a freely flowing windstream.

A confined vortex system is being developed by Grumman Aerospace Corporation in which the pressure drop across a ducted turbine and the wind speed through it are further enhanced by using additional ambient wind to generate a confined tornado-like vortex in a tower located at the exit of the duct. (See Figure 1.8.)

Such confined vortices can be created by a variety of structures designed to collect the wind over a wide area and spin it as it enters an enclosure, such as a circular tower, that is interconnected to the exit of the duct in which the turbine is located. For a typical system, the diameter of the tower might be three times the diameter of the turbine, and the height of the tower might be three times the diameter of the tower, or nine times the diameter of the turbine.

For such a tower the speed, V_o, and pressure, p_o, of the wind entering the tower may be larger than that of the ambient wind, depending on how effectively the ambient wind is ducted into the tower. Inside of the tower, the pressure and speed of the spinning wind near the tower wall will be approximately those of the entering wind.

Dr. James Yen of the Grumman Aerospace Corporation predicts, on the basis of aerodynamics theory, that the angular velocity of the wind in a large scale vortex can reach a maximum value of about ten times that of the angular velocity of the wind at the tower wall, at a radius that is about one-seventh the inside radius of the tower wall. He indicates that the vertical velocity of the wind inside the tower will be close to zero at the tower wall and will reach a maximum in the core of the vortex. He further predicts that the vertical velocity in the core will increase as the diameter of the vortex is increased and, for large scale systems, could reach values that are at least 7 or 8 times the angular velocity of the wind at the tower wall, which is as indicated above, about the speed of the wind entering the tower.

Such scaling increases in vertical velocity are indicated in his initial experimental results, but need to be verified by tests on larger scale models and prototypes. On the basis of the above types of analyses, Dr. Yen estimates that large scale wind energy systems, using ducted turbines interconnected to vortex generators, may be designed with power outputs that are 100 to 1,000 times those of conventional wind turbines, of the same diameter, operating at the same ambient free flow wind speed.

SYSTEM DESIGN

System Characteristics

Wind Energy Conversion Systems are primarily characterized by the intermittent nature of the wind, which is their source of energy, and by the fact that their power output varies with both the swept area of their rotors and the cube of the wind velocity. Because of this cubic relationship between output power and wind velocity and the fact that the wind is not steady, the actual power output of wind machines, over a long period of time, may be significantly higher than the power output estimated from the average wind speed over that time.

Exact siting is critical since the wind velocity increases with height above the ground level. Local anomalies in wind speed occur both because of increased rates of flow over or around smooth hills and other objects and because of decreased rates of flow caused by nearby trees, buildings, and other obstructions.

Furthermore, there are limitations on the size and power output of a WECS because of limitations on strengths of materials, which in turn, limit the size of blades, bearings, tower heights, and other critical system components. As a result, if a large capacity WECS for generating electricity (e.g., 100 MW power capacity) is required, either the system must be composed of a number of interconnected, dispersed WECS units of relatively small size, e.g., up to several megawatts each, or large scale wind powered vortex generator systems might be developed for this type of application.

System Requirements

In general, WECS applications can be categorized in terms of the type and amount of output power required and the requirements of the WECS energy consumer for an uninterruptable supply of power. In many applications, the power supply can be interrupted for short periods of time without disrupting the process utilizing the power, e.g., grinding grain or other materials, pumping water for irrigation or livestock, pumping water into a reservoir, drying crops or other materials, generating electrolytic hydrogen for fuel, etc.

If the consumer does require an uninterrupted source of power, the WECS must be interconnected with a standby source of power, e.g., a motor generator unit, a conventional public utility network, etc., or the WECS must be provided with a means for storing its output energy, e.g., batteries, pumped hydro storage, compressed air storage, flywheels, stored electrolytic hydrogen for use in regenerating electricity, etc.

Evaluation Criteria

In most WECS applications, the objective will be to design the system to minimize the life cycle cost of the system and, in turn, the price of the energy produced by the system. This will require that the capital cost of the WECS, including any energy storage facilities required, be minimized for a specified energy output capacity and that the system be designed to minimize operations and maintenance costs during the life cycle of the WECS by adequate design of the system components, e.g., blades, generator, bearing, gears, tower, storage facilities, etc. Other criteria that should be used to determine the viability of a WECS include:

the energy payback time (the time required for the WECS to generate sufficient energy from the wind to equal the amount of energy expended in manufacturing the WECS, as well as operating and maintaining it during this energy payback period);

the energy gain of the system (the amount of energy generated by the WECS during its lifetime, divided by the amount of energy required to manufacture, operate, and maintain it during its lifetime); and

various possible environmental, esthetic, legal, financial, institutional, or other types of constraints that might have an impact on the public acceptability of the WECS or its acceptability to public utilities, farm owners, home owners, or other possible users.

Size of Units

Conventional wind machines using unshrouded rotors are generally categorized, in terms of size, as follows:

small scale units—wind machines with rotors up to 35 ft in diameter and power output capabilities up to 10 kW_e;

medium scale units—wind machines with rotors up to about 100 ft in diameter and power outputs between 10 and 100 kW_e; and

large scale units—wind machines with rotors larger than 100 ft in diameter and power outputs up to several megawatts.

The shaft horsepower vs rotor diameter and wind speed for wind machines of 75% efficiency is shown in the table below.

Rotor Diameter (ft)	Wind Velocity, mph			
	10	20	30	40
	horsepower			
50	6	48	162	384
100	24	190	638	1,520
150	53	424	1,432	3,400
200	96	768	2,550	6,040
300	212	1,700	5,728	13,600

Cost of Energy Produced

The minimization of the cost of energy produced by a conventional WECS of a specified energy output capacity has been studied by E.W. Golding (see *The Generation of Electricity by Wind Power*, Philosophical Library, New York, 1956).

Even though the capital cost of conventional wind machines increases with the size and rated output of the machine, the capital cost per kilowatt decreases as the size and output power increases. The interconnection and transmission costs may also decrease as size is increased since fewer numbers of units are required. Economically, therefore, it is usually best to use one large unit in an application if possible rather than a number of small units. However, as indicated previously, there are constraints on the sizes that can be built and operated reliably.

Golding also indicates that the capital cost of a conventional WECS unit should decrease as the rated wind speed is increased since for a given output capacity of such a wind machine the size and weight of the blades, rotor, gear train, and other system components will decrease when the machine is designed for higher rated wind speeds. According to his estimates, the cost of large scale (MW size) mass produced conventional WECS will be about one-half of that of the mass produced medium scale size (i.e., 100 kW size) units.

On the other hand, the load factor of a WECS unit will increase as the ratio of the average-to-rated wind speed is increased (i.e., for a given location, with a specified average wind speed, as the designed rated wind speed of the WECS unit is decreased). There is, therefore, a tradeoff between equipment cost and load factor that results in a minimum energy cost of a WEC system.

From the data provided by Golding and others, it is found that with an assumed fixed charge rate of 15% and an O & M cost of 2 mills per kilowatt hour, the minimum cost of energy produced by large scale conventional WECS, at sites with average wind speeds of about 15 mph, will be about 20 mills per kilowatt hour and will occur at designed rated wind speeds that are roughly 1.8 times the average wind speeds; and that the capital cost per average kilowatt of capacity, (i.e., the capital cost per rated kilowatt, divided by the load factor) will be about $1,200 per average kilowatt of capacity.

Economic Viability

It was estimated that the first experimental 100 kW conventional WECS unit built by NASA would have a capital cost of about $500,000, or about $5,000 per rated kilowatt. Ultimately, production units of this type are expected to have a capital cost of about $150,000 per unit, or about $1,500 per rated kilowatt. These cost estimates apply to units that have a rated wind speed of 18 mph and are designed to operate optimally, i.e., in terms of the busbar cost of electricity produced by the WECS, at sites with average wind speeds of about 10 to 12 mph. (All costs are given in terms of 1975 dollars.)

These NASA cost estimates for 100 kW machines and Golding's estimates of the relative costs of large scale and medium scale machines can be used to estimate breakeven costs for large scale conventional wind machines and fossil fueled systems when the wind machine is interconnected to a fossil fueled system and used to save fuel oil when the wind is blowing.

For instance, for a fossil fueled electric generating plant, the incremental cost of the fuel oil alone adds about 2 mills per kilowatt hour to the price of the electricity generated by the plant, for each dollar per barrel increase in the cost of the fuel oil. The break-even cost occurs at the point where the price of fuel oil adds an increment to the price of electricity, as generated by the wind machine system. For the optimally designed large scale wind machine, designed say, for a 28 mph rated wind speed and operating in average wind speeds of 15 mph, this break-even point occurs when the price of oil rises to about $10 or $11 per barrel, as indicated in Figure 1.26.

Figure 1.26: Typical Break-Even Costs for MW Scale Wind Machines Used as Fuel Savers

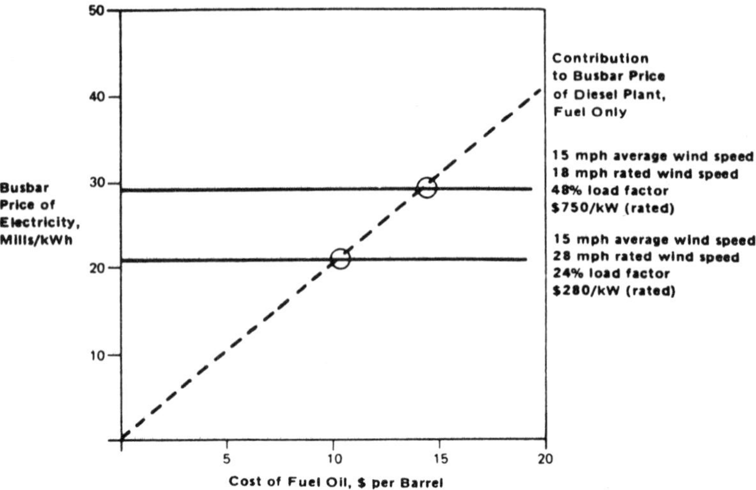

Source: PB 249 936

Other break-even points for other conditions can be determined in a similar manner. For instance, the break-even point for a typical wind machine designed for an 18 mph rated wind speed and operated in winds with average speeds of 15 mph would occur when the price of oil rises to about $14 or $15 per barrel.

The above types of analyses can also be applied to large scale wind powered vortex generator systems. For such systems, the capital cost per kilowatt and the resulting price of the energy generated may be somewhat lower than for the conventional WECS because of the smaller sized turbines that can be used for a given output capacity, coupled with the possibility of incorporating the vortex generator tower into a multipurpose building, thereby reducing construction costs for the electrical power generation system.

Storage Options

If adequate energy storage capacity is added to a wind energy conversion system, it can be operated independently to supply base loads without interconnection to a conventional system. In this case, it can be used as a complete substitute for a conventional electric generating plant. Systems capable of storing energy generated from the wind can be categorized in terms of electrochemical energy storage systems (e.g., batteries or systems that store hydrogen generated by electrolysis), thermal energy storage systems (e.g., systems that store heat produced by mechanical motion), kinetic energy systems (e.g., flywheels or superconducting electromagnetic systems), and potential energy systems (e.g., pumped hydro systems or compressed air systems).

A survey of energy storage systems has evaluated some of these available storage technologies in terms of the minimum economic sizes for utility applications, estimated total capital costs, i.e., for the energy conversion unit, as well as the energy-storage container unit, expected storage system lifetime, the capabilities for dispersed storage and the estimated turnaround efficiencies of the units. See the table below.

	Batteries	Hydrogen Storage	Thermal	Fly-wheels	Electro-magnetic	Pumped Hydro	Compressed Air Storage
Minimum economic sizes–utility application, MWh	10	10	600	10	10,000	10,000	100
Estimated capital costs, $/kW (rated)	180	300	100-400	400	500-600	200-300	230
Expected life, yr	10-20	30	20	30	30	50	20
Dispersed storage capability	yes	yes	yes	yes	no	no	no
Turnaround efficiency, %	70-80	40-60	high	80	90-95	70	45

The total capital costs, as estimated by the survey group, are based on an energy storage cycle time of about ten hours. This may not be sufficient for some wind energy base load applications where storage times of several days may be required to compensate for short time variations in wind velocities, or where storage times of six months or more may be required to allow for seasonal variations in wind strength. Under these conditions, there may be distinct cost advantages for storage systems (e.g., hydrogen or compressed air) that use depleted natural gas wells or other large natural underground structures for the storage of energy generated from the wind.

If we assume a fixed charge rate of 0.15, a turnaround efficiency of 0.60, a load factor of 0.9, and an O & M cost of 2 mills/kWh, the incremental increase in the consumer price of electricity that would result from the inclusion of energy storage in a WECS would vary from about 9 mills/kWh for a storage system with a capital cost of $200/kW (rated), to about 17 mills/kWh for a storage system with a capital cost of $400/kW (rated).

In the previous example, where it was estimated that the optimal MW scale conventional wind system operating in winds with average speeds of 15 mph could produce electricity priced at 20 mills/kWh, the addition of a wind energy storage system costing $200/kW (rated) would increase the busbar cost of the wind generated electricity to about 30 mills/kWh.

If the storage system costs $400/kW (rated), the busbar price would be increased to about 38 mills/kWh. Under these conditions, the break-even point with a diesel plant, operating at a 75% load factor and costing between $270 and $390 per rated kilowatt, would occur when the price of fuel oil reached a level of about $11 per barrel in the case of the wind system with energy storage costing $200/kW (rated) and about $15 per barrel for the case with a $400/kW (rated) storage system.

Since both the capital cost per rated kilowatt and the load factor of a WECS that incorporates a given amount of energy storage increase with the system capacity, the price of energy produced by such systems would not be expected to increase significantly when the capacity is increased to provide load factors equivalent to those of conventional fossil fueled or nuclear fueled systems.

SYNTHESIS OF NATIONAL WIND ENERGY ASSESSMENTS

Material for this chapter has been based upon a report prepared by Pacific Northwest Laboratories (BNWL-WIND-5).

The geographical distribution of the total available wind power over the United States has previously been described by Reed (1). Reed's report identified large-scale regions of high wind power potential over the United States and described seasonal variations in wind power. However, a need for an expansion and refinement of the initial national wind power assessment by Reed was evident. Reed's analysis did not adequately consider some important factors, e.g., atmospheric density variations with elevation, and height of the wind measurements above ground.

As a component of the National Wind Energy Mission Analysis for the Energy Research and Development Administration (ERDA), two more recent national wind power assessments were made: an assessment by the Lockheed-California Company (2) and an assessment by General Electric (3). A study has also been made [Justus et al (4)] showing the geographical distribution over the continental United States (based on wind summaries for 135 surface stations) of annual capacity factor values (ratio of average power output to rated power) for two types of wind-powered generators.

The main purpose of a study done by Batelle, Pacific Northwest Laboratories is to synthesize the national wind energy assessments by Sandia (Reed), Lockheed, and General Electric so that a consistent and representative analysis of the geographical distribution of the total wind power on a national scale can be demonstrated. Inconsistencies and discrepancies in the geographical distributions and estimates of wind power are apparent among the analyses, as shown in Table 2.1. The techniques, assumptions, data sets, and analysis methods used account for most of these differences.

It was realized that more than just an extraction of the best features of existing assessments was needed. For example, where different data sets or different techniques give considerably different wind power estimates, an examination of the criteria used was essential to determine which estimates were more realistic. Much attention is given in this study to the representativeness of the wind speed distribution summaries for typical wind turbine sites and how to best estimate the wind power for exposed areas in regions of different terrain types. Stations for which wind data are available are often located in sheltered areas and are not representative of exposed windier sites more suitable for wind turbines.

Table 2.1: Annual Average Wind Power at 50 m for Various Geographical Areas (W/m^2)*

	Sandia**	General Electric	Lockheed***
Texas Panhandle	450-650	250-450	400-990
Eastern Kansas	300-400	200-300	180-990
Iowa	250-350	150-200	290-673
Eastern Dakotas	350-450	200-250	300-812
Southern Wyoming	400-850	250-500	523-1,030
Lake Michigan	200-250	150-200	210-840
Indiana	200-250	100-200	200-565
Western New York	200-350	150-300	140-1,000
Atlantic Coast (NY to NC)	250-300	400-600	190-904
Central Gulf Coast (LA to AL)	150-200	300-450	170-220
Texas Coast	300-350	250-300	490-570
Oregon Coast	200-300	400-600	800-1,610
West New Mexico Mountains	NA	400-700	190-340
Arizona Mountains	NA	100-200	100-430

*These values were estimated from the national analyses made by each of the assessments. At the time of this report, the final reports were not available for the General Electric and Lockheed assessments. Thus, these values are subject to change.

**Sandia's analysis was assumed to be for a 10 m height, and values were extrapolated to 50 m using a $1/7$ power law.

***Isodyns were not drawn. Hence, the lowest and highest values in each area were selected.

Source: BNWL-WIND-5

In a national wind energy assessment, perhaps the most important items are the finalized maps showing the geographical distribution of wind power. The maps should be easy to interpret and realistically identify areas of relatively high or low wind power. The wind power estimates should be given for a typical wind generator hub height. Hub heights range from about 15 m for small WECS to over 50 m for large WECS. Over regions of gently rolling, hilly, and mountainous terrain, the estimates should be given for the exposed terrain features, representative of potential wind turbine sites.

TECHNIQUE FOR ESTIMATING WIND POWER

As discussed in the first chapter, the total power available from the wind passing through a unit area normal to the wind is

$$P = \tfrac{1}{2}\rho V^3$$

where ρ is the atmospheric density in kg/m^3, V is the wind speed in m/sec, and

P is the power in W/m^2. Since wind power is proportional to the cube of the wind speed, small increases in wind speed result in significant increases in wind power. An increase in the mean wind speed by a factor of 2 yields roughly 8 times the wind power. The wind speed at a site is affected by the surrounding environment, such as hills, trees, and buildings, and typically increases with height in the boundary layer. Therefore, selection of a site and height of the wind generator above the ground are two very important factors to consider for obtaining optimum wind power.

Because wind power varies as the cube of wind speed, the distribution of speed, rather than the mean speed, is needed to calculate the power resource. Determination of the wind speed frequency distributions typically requires the use of hourly or three hourly wind speed data. Frequency distributions have been compiled by the National Climatic Center (NCC), for selected time periods for many of the surface stations in the U.S. which make wind measurements. However, anemometer heights and locations varied considerably at many of the stations for the period of record over which these summaries were made. These variations can cause considerable errors in the wind power estimates, since a change in the anemometer height or location may give a significant difference in the wind power estimate.

Only mean wind speed summaries are readily available for many stations and for desired periods of record for which the anemometer heights are constant over the period. An empirical method developed by Widger (5) and tested on six New England sites indicated that the actual wind power can be approximated within ±20% by using only the mean wind speed. On the other hand, Baker and Hennessey (6) show that accurately estimating the shape of the wind speed distribution is critical and that the method suggested by Widger is not usable if the wind speed distribution is highly skewed or deformed. They found that the actual wind power was underestimated by over 50% at some sites in the Pacific Northwest, using Widger's technique.

Wind energy per unit area is just the power times the time period; that is

$$E = P \cdot \Delta t$$

where P is the power per unit area, W/m^2; Δt is the time period in hours; and E is the energy per unit area, Wh/m^2. One $MWh/m^2/yr$ is equivalent to approximately 114 W/m^2 in wind power.

TYPES OF WIND DATA SETS AND SUMMARIES AVAILABLE

All meteorological data recorded by the National Weather Service, the Federal Aviation Administration, and the military services are retained at the National Climatic Center (NCC), Asheville, NC. A comprehensive report on the availability of data and summarizations of land and marine surface wind measurements, upper level wind measurements, gusts, turbulence, and tower data has been done by Changery (7). The NCC is compiling an index of surface stations for which summaries of wind speed distributions are available.

Several types of surface wind speed distribution summaries available from the NCC are: (1) the wind speed frequency distributions for 758 stations originally compiled for Sandia Laboratories for the purpose of national wind power estimation. (Hereafter, this data set shall be referred to as the Sandia tape.) (2) STAR program wind speed frequency distributions for over 300 stations, and (3) wind speed distribution summaries for Marsden squares and Navy Marine Areas based on ship measurements. Also, hourly wind data (three hourly since 1965) are digitized for over 1,000 stations in the United States. (STAR programs are created from these digitized data.) The utilization of the above types of data and upper air data to assess the national wind power potential is briefly described below.

The Sandia tape of wind speed distributions is used in the national wind energy assessments by Sandia, General Electric, and Lockheed. As indicated in some of these assessments, these summaries have several major limitations: (1) periods of record vary considerably from station-to-station, ranging from 2 to 30 years or more, (2) anemometer heights often vary during the period of record for which the summaries are compiled, and (3) the number of wind speed classes vary from station-to-station, ranging from 3 up to as many as 16. All of these factors—period of record, anemometer height, and number of wind speed classes—can lead to significant errors in wind power estimates for individual stations, so these summaries must be used with caution.

The STAR (Stability Array) program gives the frequency distribution of wind speed versus wind direction for individual stability classes and usually for all classes combined. Thus, summaries without stability are available. These summaries are somewhat better in that the wind speed classes are identical for all stations and many of the summaries are for recent time periods (1960s and 1970s) when anemometer heights are generally constant over the time period. Although the STAR program summaries have not been used in any of the national wind energy assessments, they would be a valuable supplement. These summaries have been generated from digitized hourly (or 3 hourly) data.

Wind speed distribution summaries for Marsden squares (areas of 1° latitude by 1° longitude) and Navy Marine Areas are useful for identifying relative patterns of wind power over offshore regions, but there are some uncertainties on the typical heights and overall accuracy of shipboard measurements and the application of this data to properly assess wind power in near coastal areas, i.e., exposed coastal sites and sites 5 to 10 km seaward.

Hourly (or 3 hourly) digitized winds appear to be the best set of data for use in obtaining representative wind power estimates. The period of record can be selected so that the anemometer height is constant, and errors associated with estimating the wind power from wind speed classes are eliminated since the wind power can be computed directly from the actual observations. However, the utilization of digitized data to compute wind power is more expensive than that of summarized data but certainly not unreasonable. Justus (8) used this 3-hourly digitized data for studies of wind power over the Great Plains and New England.

Low-level rawinsonde data for 150 m, 300 m, and higher levels are summarized into wind speed classes in intervals of 5 m/sec for 65 rawinsonde stations in the U.S. The summaries are given separately for 00Z and 12Z observation times and, mostly, for a 5-year record period. [These data were originally summarized by the NCC for a study by Holzworth (9)]. Flow patterns are usually more coherent at these levels, and topography and roughness effects are much smaller. However, the chief limitations on the use of this rawinsonde data are that (1) the twice-daily observations may not be sufficient (e.g., they cannot account for the influence of diurnal variations), (2) the stations are too widely scattered, (3) the wind speed intervals are very broad, and (4) there are uncertainties regarding the accuracy of the rawinsonde wind speeds for 150 m and their extrapolation down to 50 m.

For estimating upper air wind power, useful data sources are the climatological maps by Crutcher (10) and statistical parameters of upper winds at 850 mb (near 1500 m) and 700 mb (near 3000 m). Another data source which can be used to estimate upper air wind power over the U.S. is gridded 12 hourly upper air data for standard levels (e.g., 850, 700, 500 mb), stored on tape at the National Meteorological Center (NMC), Suitland, MD. Upper air wind power is applied to estimate the wind power potential over mountainous regions.

GENERIC FACTORS TO CONSIDER

There are many factors which need to be considered in obtaining representative estimates of wind power and its geographical distribution. One important factor that should be considered is the reliability of the wind speed data. How good is the anemometer exposure, the accuracy of the measurements, and the manner in which the data is registered? For aviation purposes, errors of just 1 m/sec may not matter that much. However, the difference in wind power between a 4.4 m/sec (10 mph) and 5.6 m/sec (12.5 mph) wind speed is 100%. Thus, minor variations in wind speed due to instrument exposure or accuracy can result in substantial errors in the wind power estimates.

Other primary considerations in the accuracy of wind power estimates include atmospheric density variations with elevation and temperature, height of wind measurements above ground, terrain influences and local height variations in topography, representativeness of the available wind data to exposed wind turbine locations, vertical extrapolation of wind power, influence of the number of wind speed classes, techniques for estimating wind power over mountainous and offshore areas and areas of sparse data, techniques for using rawinsonde data, and methods of interpolation and analysis used for producing the finalized maps.

This last item, if not properly done, can itself result in major misrepresentations in the geographical distributions of wind power. Some of these factors, such as intrument exposure, are uncertainties which cannot be accounted for in the wind power estimates but should be recognized as sources of errors. Some of the more important generic factors are discussed below.

Atmospheric Density

Wind power is directly proportional to atmospheric density, which decreases with height and temperature. At sea level the standard density, in the U.S. Standard Atmosphere, is 1.225 kg/m^3. Figure 2.1 shows the relationship between height, density, and percent reduction in wind power. Wind power would be overestimated for much of the Great Plains and western mountains (e.g., 17% at Cheyenne, WY) by using sea level density in the calculation.

Figure 2.1: The Percent Reduction of Wind Power with Elevation Due to Atmospheric Density*

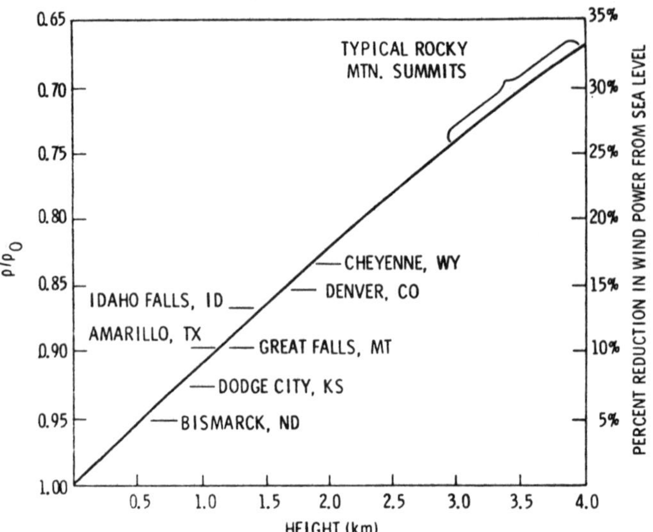

*Density ratios are given with respect to standard sea level density, ρ_0 (1.225 kg/m^3).

Source: BNWL-WIND-5

Figure 2.2 shows the effect of temperature variations on wind power. The national assessment by Lockheed (2) indicated that mean monthly atmospheric densities differ from the standard density at that elevation by generally less than 5% and rarely greater than 10%, based on computations for 7 stations well distributed in elevation and location throughout the conterminous United States.

Figure 2.2: The Variation of Wind Power and Atmospheric Density with Temperature*

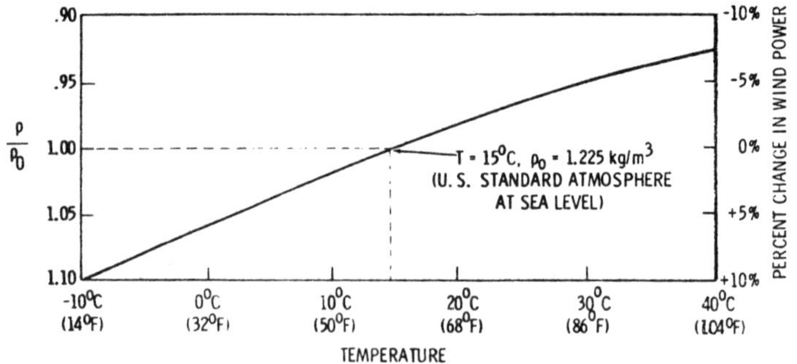

*Density ratios are given with respect to sea level density for the U.S. Standard Atmosphere at 15°C.

Source: BNWL-WIND-5

Neglecting seasonal variations in temperature results in the summer wind power being slightly overestimated and the winter wind power being underestimated. Thus, the annual wind power would be underestimated for regions where the colder months are the windier months.

Anemometer Heights

Anemometer heights at National Weather Service stations have varied considerably over the years. Typical heights of the wind sensors above the ground among the stations have ranged from 4 to 60 m. Nevertheless, 10 m is often assumed to be the standard anemometer height. Many anemometers have been located on airport terminals or buildings. Within towns and cities, anemometers may have been placed on buildings to obtain better exposure to the winds. However, the building itself may cause an obstruction to the air flow at the anemometer location, depending on the height of the anemometer above the roof of the building, the shape of the building, and the location of the anemometer on the building top. Thus, the exposure of an anemometer should not be assumed to be unobstructed to air flow until certain site characteristics are properly evaluated.

At the majority of stations, anemometer heights and locations were often changed during the period of time for which summarized wind speed data and/or frequency distributions are available. Wind power is very sensitive to changes in anemometer height and is typically 30 to 60% greater at 20 m than at 10 m. Thus, aside from all other factors, the temporal and spatial variations in anemometer heights make it very difficult to obtain representative estimates of wind power utilizing available wind speed distribution summaries.

Beginning in the late 1950s and throughout the 1960s, the National Weather Service (NWS) relocated anemometers to approximately 6 m (20 ft) above the ground. The relocations were completed by 1965 for most of the stations.

A history of the anemometer heights for most NWS stations in the U.S. is included with Local Climatological Data (LCD), a publication available from the NCC. A report by Coty et al (11) lists anemometer heights for 478 of the stations on the Sandia tape and presents the wind power tabulations for these stations. However, the anemometer heights varied over the period of record or were not known for many of the stations, and therefore, were "assumed" to be 10 m at those stations. If the actual heights were much greater than 10 m (e.g., 15 to 20 m), this assumption could result in the wind power being considerably overestimated for the 10 m level and consequently for all other levels.

Influences of Minor Variations in Topography

Less than 20% of the United States can be considered flat terrain, that is, with local relief varying no more than 30 m (100 ft). Figure 2.3 shows the geographic distributions of these areas. [This figure was constructed from a map in the National Atlas of the United States (12)]. Areas for which local relief is defined are square units 9.6 km (6 mi) across. Thus, most of the nonmountainous area of the U.S. is classified as gently rolling to hilly terrain, with local relief variations greater than 30 m.

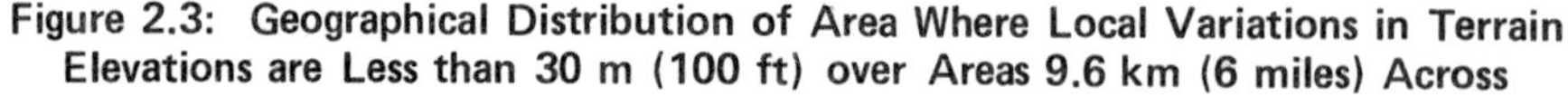

Figure 2.3: Geographical Distribution of Area Where Local Variations in Terrain Elevations are Less than 30 m (100 ft) over Areas 9.6 km (6 miles) Across

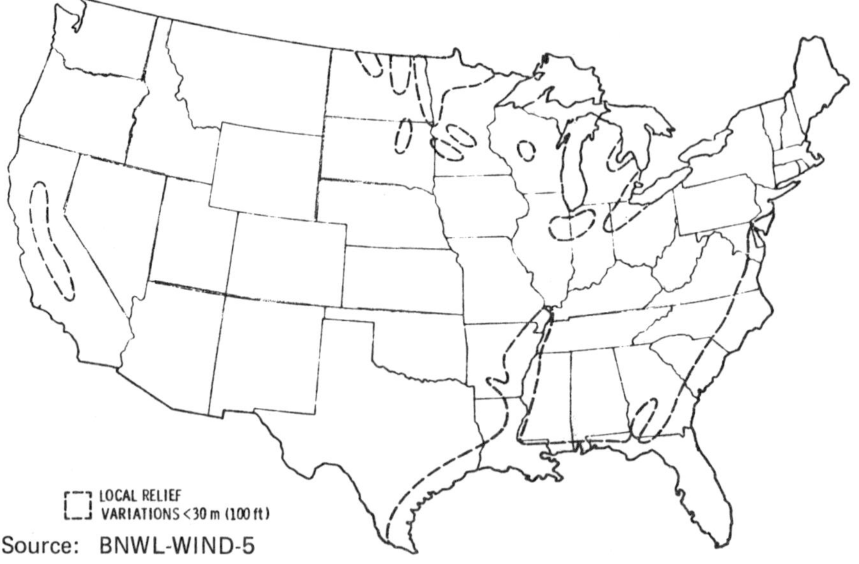

Source: BNWL-WIND-5

Very minor changes in elevation can produce significant differences in wind power. Over regions of gently rolling terrain (defined here as having local relief variations of 30 to 100 m), many stations taking wind measurements are located in the lower areas which are not representative of the higher exposed elevations more suitable for wind turbines. Increases of 1 m/sec in the mean wind speed may give 50% to as much as a 100% increase in available wind power for mean wind speeds of 3 to 7 m/sec (7 to 15 mph). An excellent example is Rochester, Minnesota, located in the rolling terrain of southeast Minnesota.

The Rochester National Weather Service (NWS) station is located on a ridge of about 100 m (300 ft) higher elevation than the city and lower elevations in the local vicinity. The average annual wind speed at the Rochester NWS is 5.7 m/sec (12.7 mph), which is 1 to 2 m/sec (2 to 4 mph) greater than for any other NWS station within a 250 km (150 mi) radius. (These records are based on annual wind speed summaries from 1965 to 1975, for which all anemometer heights have been standardized to approximately 6 m.) Correspondingly, wind power at Rochester is 50 to 100% greater than that at surrounding stations. Thus, Rochester is the only NWS station in this region of gently rolling terrain which appears to have wind power typical of exposed areas representative of probable WECS sites.

Therefore, the elevation of the station with respect to the elevation of the surrounding terrain in the area should be considered to identify those stations which are representative of probable WECS sites. An analysis based only on an average of all the station values within an area will reflect the influences of the relative elevation exposure of the majority of stations.

An alternate method to estimate the wind power at the average height of the exposed areas over regions of rolling and hilly terrain is to utilize low-level rawinsonde data and extrapolate the values to the specified terrain heights. However, there are limitations on the use of rawinsonde data to estimate wind energy, as previously described.

Vertical Extrapolation of Wind

In the surface boundary layer, there are many factors which influence the variations of wind speed with height above ground, such as terrain roughness, atmospheric stability, and wind speed. Two relationships are frequently used to extrapolate the wind speed with height. One is the logarithmic velocity profile

$$\frac{V_2}{V_1} = \frac{\ln(Z_2/Z_0)}{\ln(Z_1/Z_0)}$$

where V_2 and V_1 are the wind speeds at heights Z_2 and Z_1, respectively, and Z_0 is the surface roughness length. The logarithmic profile is suitable for neutral stability conditions and high wind speeds. The second relationship is the power law profile.

$$V_2/V_1 = (Z_2/Z_1)^{\alpha}$$

where α is a dimensionless exponent whose value may depend on the wind speed, atmospheric stability, and surface roughness. Since wind power is related to the cube of wind speed, the exponent becomes 3α for wind power.

The power law profile can also be applied for long-term averages. Reed (1) indicated that a value of $1/7$ for α is reasonably valid for long-term averages. Justus (13) estimated the average value of α as 0.23 ± 0.03, based on an examination of tower data at four sites in different areas of the United States. Justus' and Reed's studies showed that the height variation of wind speed is a function of the mean wind speed and height of the wind measurements, and relationships were developed to estimate α as a function of the wind speed and anemometer height. Basically, these relationships assume that the value of α decreases as the mean wind speed increases; that is, the increase of wind speed with height is not as great for higher wind speeds.

None of these previously mentioned values of α approximate the data very well at all tower sites; e.g., Drexel University/WFIL-TV tower data in Philadelphia, Pennsylvania shows $\alpha = 0.34$ (14). The extreme interstation variability in the surface wind speeds, even over homogeneous regions such as the Midwest and Great Plains, explains much of the variability in the mean values of α. Consequently, it is difficult to justify using one value of α over another for the extrapolation of winds on a national scale. Moreover, values of α for exposed locations representative of probable WECS sites may be considerably different from those evaluated from available tower data. Nevertheless, some assumptions regarding the vertical variation of wind speed (and energy) must be made to extrapolate the wind energy to a wind generator hub height.

Listings of tall instrumented towers have been compiled by Cormier (15) and Wallace (16). An abundant supply of tower wind data also exists for nuclear and coal-fired power plants throughout the country. A report on wind data summaries collected on towers at nuclear power plants in the U.S. has been prepared by Verholek (17). These data sources can be used to gain greater insight into the height variation of wind speed with regard to terrain, climatological, and surface roughness characteristics.

Mountainous Regions

Upon examining the few available studies and data summaries for mountain summit winds, perhaps the best way to describe the relationship between the free air wind speed and mountain summit wind speed is "highly variable and complex." On some mountain summits the mean wind speed may be greater than the free air wind speed, while on others it may be less than 30% of the free air wind speed.

Wahl (18) evaluated the wind speed records for six mountain summits ranging in elevation from about 800 to 3,000 m, all located in Europe. The values for individual summit speeds ranged from less than 30% of the free air wind speed, where the surrounding terrain appeared to induce high frictional effects, to 80% or more for well-exposed summits subject to jet action. Figure 2.4 shows the average typical summit speed versus free air speed for different percentiles.

Figure 2.4: Comparison of Measured Mountain Summit Wind Speeds, V_m, to Estimated Free Air Wind Speeds, V_f*

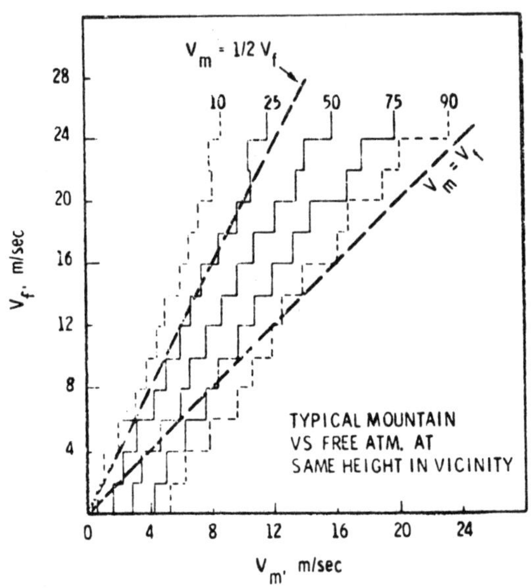

*From rawinsonde data mean frequency distributions for V_f classes are indicated for different percentiles from 10 to 90 (18). Lines for $V_m = V_f$ and $V_m = \frac{1}{2}V_f$ are included.

Source: BNWL-WIND-5

The figure indicates that summit wind speeds are usually less than free air wind speeds at the summit height (except for low wind speeds), but summit wind speeds are less than one-half the free air wind speeds in only about 25% of the cases. Thus it appears that on the average, summit wind speeds are about 60 to 70% of the free air speed for wind speeds greater than 8 m/sec. However, Wahl's analysis also indicates that the relationship between summit speed and free air speed at summit height is reasonably independent of the height of the summit, as relatively low peaks sometimes have greater mean wind speeds than higher peaks.

Pristov (19) has also examined the relationship of free air speed to mountain summit wind speeds. For three mountain summits in the height range of 2,000 to 3,000 m, his results indicate that wind speeds on isolated mountain ridges are not as turbulent or gusty as wind speeds on mountains within mountainous terrain. Wind speed records for a few mountain locations throughout the U.S. also indicate a complex relationship between free air speed and summit speed (2)(3).

Thus, the wind power available on mountain tops appears to depend just as much on the shape and alignment of the mountains and the height of the surrounding terrain as it does on the height of the mountains. In the West, many lower isolated ridges may have as much or more usable wind power as the higher mountain ranges. Gaps or saddles between mountains are often known to have strong winds. Wind tunnel modeling studies have been performed to examine the influence various slope geometries and atmospheric stratifications have upon flow over ridges (20).

Coastal and Offshore Regions

In coastal regions, the mean wind speeds are typically greater over the ocean than over land, primarily due to greater surface friction over land. The wind power just offshore and at exposed coastal sites, such as capes and points, may be a factor of 2 or more greater than the wind power calculated for a coastal town. The wind measurements for most coastal towns are made too far inland (½ km or more) or along sheltered areas, such as bays or inlets, to be representative of the wind speeds for windier coastal and offshore areas. Along coastal regions of rough terrain, such as that of Maine and the Pacific coast states, this decrease of wind speed from the coast inland is substantially greater.

An important factor to consider at a coastal site is whether the prevailing winds come from a direction over the water or over the land. Sites where the prevailing winds come over the water are normally windier than sites where the prevailing winds come over land. For example, the east shore of the Great Lakes should have greater wind power than the west shore, since prevailing winds are generally from a westerly component.

Representative wind data for exposed coastal sites are very limited. Offshore mean wind speed distributions based on ship observations are probably more applicable to exposed coastal sites than are most of the speed distributions from coastal towns, even though the ship observations are averaged over large offshore areas extending 100 km or more seaward. However, ship wind data should be used with caution due to uncertainties in the height of the wind measurements, the locations of the measurements within the area, and the accuracy and frequency of the measurements.

REVIEW OF THE EXISTING NATIONAL WIND ENERGY ASSESSMENTS

Three wind energy assessments have been performed to evaluate the geographical distribution of total wind power over the United States. They are: an assessment by Reed (1) of Sandia Laboratories, an assessment by Lockheed (2), and an assessment by General Electric (GE) (3). The Sandia study was completed before the other studies began and provided a data base for them. A review of these assessments reveals discrepancies and inconsistencies in the geographic distributions of wind power and magnitude of the estimates (Table 2.1). The major problems facing all the assessments were how to deal with the enormous interstation variability in surface wind power estimates and how to estimate the wind power in mountainous and offshore regions and in regions of sparse data. These problems lead to most of the discrepancies and inconsistencies.

Table 2.2 summarized the primary data sets or summaries, the techniques applied, the assumptions made, and the kinds of analyses performed in each assessment. This chapter is broken into two parts—one describing the techniques and data applied to estimate the wind power and one describing the national analyses produced.

Techniques

Surface Wind Data Employed: All three assessments used the same source of surface wind data—the Sandia tape of wind speed frequency distributions. The General Electric and Lockheed assessments were selective, though, in choosing the stations to be utilized. Lockheed eliminated those stations with "limited data," i.e., just three wind speed classes or with fewer than 4,000 total observations in any month. This selection process provided a minimum of six years' observations. (However, wind power estimates for many of the stations with "limited data" that showed relatively high wind power were included on their maps of annual average wind power density.) General Electric used few stations with less than five wind speed classes, and "over the continental U.S., emphasis was given to the use of first order National Weather Service stations." Thus, GE used considerably fewer surface stations than Lockheed, but their assessment was based primarily on rawinsonde data.

Additional wind summaries and data sets are available that were not utilized in the national assessments, including hourly (or 3 hourly) digitized wind data.

Method of Estimating Surface Wind Power: Sandia and GE calculated wind power from wind speed frequency distributions using the midpoint wind speed of each class. General Electric found that wind power values calculated from the median wind speed of each class were consistently larger (from 3.2 to 10.5% for the seven stations evaluated) than values calculated from individual hourly wind speeds. These were all stations with eleven wind speed classes. However, Reed indicated that using the median of wind speed classes does not cause more than a couple percent error in most cases.

Lockheed's study shows that as the number of wind speed classes decreases, the percent error in wind power increases. A total of 128 stations on the Sandia tape have only three wind speed classes and 263 stations have six classes or fewer. The wind power estimated from these summaries is typically 10 to 30% higher when compared to estimates computed from actual hourly observations.

Table 2.2: A Summary of the Criteria Used in the National Wind Energy Assessments by Sandia, General Electric, and Lockheed

Item	Sandia	General Electric	Lockheed	Comments
Surface wind data used	Sandia wind speed frequency distributions	Sandia wind speed frequency distributions	Sandia wind speed frequency distributions	Many additional surface wind summaries and data sets are available
No. of stations used for surface analysis	758 (total available)	Not given; primarily first order stations	478 and selected stations with limited data	–
Method of estimating wind power	V^3, median of wind speed class	V^3, median of wind speed class	V^3, fitted cumulative frequency distribution by spine function and evaluated every decile	Median of class overestimates wind power by 7%, on the average
Accounted for variation in anemometer heights	No	No; heights assumed to be 10 m	Specified where known. Assumed at 10 m where varied over period or not known	Anemometer heights vary from 6 to 60 m. Standard height became 6 m in 1960s
Accounted for density variations with station elevation	No	No	Yes	Use of sea level density overestimates wind power by 8.5 to 9%/km increase in elevation
Low level rawinsonde data, e.g., 150 m level	Not used	Estimated wind power at 150 m from frequency distribution for 65 stations. Annual estimate only for 00Z	Not used	Use of this data underestimates wind power by 30 to 60%
Vertical extrapolation of of wind speed	Suggests 1/7 power law. Different methods discussed	Extrapolated downward from 150 to 50 m using empirical procedure	Extrapolated upward from 10 to 50 and 100 m using 0.2 power law	Extrapolating upward to 50 m using 0.2 power law gives 32% more energy than 1/7 power law
Upper air wind data used	Climatological maps by Crutcher	Climatological maps by Crutcher	Climatological maps by Crutcher	–
Accounted for upper air density	No; sea level density used	Yes	Yes	Density at 3 km is <75% of sea level density
Method of estimating mountain top wind power	850 mb power assumed typical of Eastern mountains; 700 mb power typical of Western mtns.	Assumed $V_m = V_f$ thus $P_m = P_f$ m is mountain, f is free air	Assumed $V_m = ½ V_f$, thus $P_m = 1/8 P_f$, m is mountain, f is free air	Relation of mountain top power to free air power is highly variable, from <1/8 to >1
Coastal and offshore wind data used	Coastal stations on Sandia tape	Ship wind data for Marsden squares and Navy Marine areas	Coastal stations on Sandia tape, and a few additional stations	Many coastal stations are not representative of exposed sites
Types of national analyses made	Surface and upper air isodyns (W/m^2), values smoothed objectively by geometric averaging. Annual and seasonal averages for surface; seasonal for upper air	Isodyns ($kmh/m^2/yr$) for surface, 150 m level, upper air, and 50 m level. Subjective type analyses. Annual averages only.	No isodyns. Power density values plotted on U.S. map for each station. Wind power classes represented by color variations. Annual and seasonal maps of power for 10, 50 and 100 m	Discrepancies are found in the energy patterns and estimates

Source: BNWL-WIND-5

Lockheed did not use the midpoint of wind speed classes to calculate wind power. Instead, the frequencies of wind speed, grouped into classes, were accumulated. This cumulative distribution was fitted by spline functions and evaluated at every decile. Wind power was obtained by cubing the percentiles of wind speed. However, the differences in the wind power estimates based on this technique and the midpoint of classes are usually less than 5%.

Adjustment for Anemometer Height Variations: Only Lockheed conducted a survey of anemometer height data for all the stations used. Wind speeds, where anemometer heights were known, were adjusted to 10 m, and from that level to the 50 and 100 m levels using a power law formula. In their report they state that: "Stations for which anemometer heights were not known, or for which the heights varied during the reported period, an initial height of 10 m was assumed." Prior to the 1960s, typical anemometer heights were mostly 10 to 40 m above ground, as the anemometers were often located on buildings and airport terminals.

The 0.2 power law coefficient for wind speed (used by Lockheed) results in 52% greater wind power at 20 m than that at 10 m. Thus, assuming a 10 m anemometer height for those stations for which the anemometer heights varied over the record period or were not known may result in a significant overestimation of wind power at the 10 m level and, subsequently, at the 50 m level. However, even if the anemometer heights are standardized to 10 m from the actual heights, uncertainties may exist regarding the instrument exposure and effects of the building on the air flow which could cause large errors in the estimates.

With this variability and uncertainty in the heights and exposure of the anemometers, it is difficult to obtain representative estimates of wind speed, and therefore wind power. This explains much of the interstation variability in the wind power estimates. One solution is to consider only wind speed data from the 1960s and 1970s when anemometer heights are constant and instrument exposure is generally good. (During the late 1950s and early 1960s, most airport anemometers were moved to 6 m above the ground and placed away from buildings.) These data have been digitized in 3-hourly intervals and are available on tape from the National Climatic Center. Thus, utilization of this data source would substantially reduce errors caused by uncertainties in anemometer height and exposure.

Adjustments for Atmospheric Density Variations: Atmospheric density variations with station elevations were accounted for in the Lockheed study but not in the GE and Sandia assessments. Neglect of density variation results in an overestimation of wind power by about 9% for every 1 km (3,000 ft) elevation above sea level (Figure 2.1). Thus, the use of sea level density in the GE and Sandia assessments results in station wind power estimates that are 5 to 20% too high over much of the Great Plains and high basins of the West.

Use of 150 m Level Rawinsonde Data: General Electric used 150 m level rawinsonde summaries to obtain wind power estimates at 150 m and extrapolated these values down to 50 m. Their 50 m annual analysis is based only on 00Z rawinsonde observations. However, GE also analyzed monthly values for 20 stations at both 00Z and 12Z and stated in their report that "the pattern of energy distribution shown and the relative magnitude of the maxima and minima are validly represented by the 00Z analyses alone."

A brief comparison of 150 m wind speeds for 00Z and 12Z was conducted and it was found that including the 12Z winds in the analysis would increase the wind power estimate over the southern Great Plains and decrease the wind power slightly over much of the rest of the U.S. Thus, analysis shows that the relative magnitudes are modified, but the large scale patterns of wind power for 00Z and 12Z remain similar, with distinct maxima showing up over the Great Plains just east of the Rockies, along the Northeast coast and south Texas coast. However, because of the large spacing between stations, only gross features can be depicted with these data.

Vertical Extrapolation Techniques: Each assessment used the power law formula to adjust the wind speeds and wind power to different levels. Reed suggests using a $1/7$ power law; Lockheed extrapolated the surface wind speeds for 10 m up to 50 and 100 m using 0.2 power. Wind power estimates for 50 m based on a 0.2 power law are 32% higher than estimates based on a $1/7$ power law. Thus, this is another cause of discrepancies in the wind power estimates. However, as indicated earlier, one cannot justify using one value of the power law over another in a national assessment.

General Electric extrapolated downward to 50 m from the 150 m rawinsonde estimates, stating: "It was felt a more consistent and reliable set of values would result as compared to extrapolation of wind energy up from the surface, where anemometer heights can vary quite radically---." General Electric initially used the $1/7$ law ($3/7$ law for wind power) to extrapolate downward; however, some of the 50 m values were less than the values of surface wind power calculated from the Sandia data tape.

To eliminate this problem, they established an empirical procedure where the ratio of 50 to 150 m wind power used was as a function of the ratio of the 150 m wind power to surface wind power. Nevertheless, it was found that the wind power estimates at 50 m based on this technique are 30 to 60% lower over much of the Great Lakes, Midwest, and Great Plains than average estimates based on extrapolating up from 10 m (see Table 2.1). One explanation is that the twice a day rawinsonde measurements (morning and evening) do not properly account for the diurnal variation in wind speed. Surface winds are typically strongest in the afternoon.

Techniques Applied to Estimate Mountain Top Wind Power: All three national assessments used statistics of Northern Hemisphere wind characteristics compiled by Crutcher (10) to estimate upper air wind power. These values were then applied to estimate typical mountain summit wind power. General Electric assumed that mountain summit wind power is approximately equal to the free air wind power at the summit height. Lockheed assumed that mountain summit wind speed is one-half of the free air wind speed (one-eighth of the free air power), based on a study by Wahl (18), these values were taken to be at 10 m and extrapolated to 50 m using a 0.2 power law. Thus, Lockheed's mountain summit values for 50 m are approximately one-third the free air values of wind power.

Since the relationship of mountain summit wind power to free air wind power at summit appears highly variable, there is no justification that one assumption gives more representative estimates than the other. Observed mean annual wind speeds of various mountain summits throughout the U.S. indicate that some summits or ridges may have greater wind power than the free air. Examples are Rattlesnake Mountain in south-central Washington with over 1000 W/m^2, Sandberg in the southern California coast range with over 500 W/m^2, and Mt. Washington in New Hampshire with over 2000 W/m^2. However, some other mountain summits with wind data indicate considerably less wind power than expected; examples are provided in the Lockheed report.

Techniques Applied to Estimate Offshore Wind Power: General Electric used wind speed distributions from Marsden Squares and Navy Marine Areas, which are based on ship observations. Although there are uncertainties on the heights and accuracy of shipboard measurements, these data are valuable for identifying large scale patterns of wind power over offshore regions. Sandia's and Lockheed's estimates for coastal and offshore regions are primarily based on coastal stations, a few of which are located on offshore islands.

A comparison of the analyses for the coastal regions shows a few discrepancies, e.g., ship wind data indicate there is greater wind power along the central Gulf coast (Louisiana to northwest Florida) than along the Texas coast whereas coastal wind data indicate the reverse. However, in these areas with discrepancies, it is difficult to evaluate which estimates are most representative. If the average shipboard measurements are greater than 10 m, then the values will be overestimated.

Finalized Maps of Wind Power

Each assessment used a different method of analysis in producing their finalized maps of wind power for the U.S. Sandia performed an objective smoothing of the surface estimates by geometric averaging of station values. General Electric did a subjective analysis based on rawinsonde and surface extrapolated estimates; and Lockheed plotted the values calculated for each station location, purposely avoiding isodyns. A brief description of the national analyses made by each assessment is given below.

Sandia: Maps were constructed showing annual and seasonal averages of available wind power at the surface and seasonal averages at the 850 and 700 mb levels over the contiguous U.S. The surface maps reasonably identify regions of relatively high and low wind power. The national analyses of surface wind power potential were based on the unadjusted values computed from the Sandia tape of wind speed frequency distributions. A regional smoothing technique was used to average out the very large interstation variabilities in wind power, as differences were in some cases as much as a factor of 2 to 3 at nearby stations even over fairly homogeneous terrain. However, the regional smoothing technique used causes the analyses to be misrepresentative in some areas, e.g., along coastal regions, the onshore stations depress the offshore values.

General Electric: General Electric's final product was a combined analysis at 50 m for the U.S., based on a subjective analysis of the extrapolated surface, 150 m rawinsonde, offshore ship, and upper air estimates. Instead of wind power, their annual average estimates are in wind energy with units of MWh/m^2/yr. One MWh/m^2/yr is equivalent to wind power of approximately 114 W/m^2. The estimates at 50 m are generally 30 to 60% lower over most of the Great Plains, Midwest, and Great Lakes region than those by Sandia and Lockheed (see Table 2.1). This technique of extrapolating downward from 150 m rawinsonde data appears to significantly underestimate the wind power, as discussed earlier. The wind energy estimates along the central Gulf coast (Louisiana to northwest Florida), where over 4 MWh/m^2 (450 W/m^2) is indicated, may be too high. These estimates are based on ship data for just one Marsden square, and additional offshore data are needed to evaluate if these high estimates are realistic.

The geographical distributions of wind power over mountainous regions show isolated areas of relatively high and relatively low wind power. There are several regions (e.g., Washington and Oregon, and New Mexico and Arizona) where the relative geographical distributions of wind power over mountain regions do not appear consistent with the topography.

Lockheed: Lockheed purposely avoided isodyns and chose to construct maps showing the estimated values of wind power for each station. Maps of wind power were prepared for 10, 50 and 100 m levels. In addition, a separate set of maps was constructed, which designated wind power class intervals by colored dots. Thus, individual stations and areas with relatively high wind power can readily be identified, in addition to the network density of stations within a region. An advantage of this type of analysis is that errors and misrepresentations resulting from drawing isodyns (subjectively or objectively) are eliminated. Interpretation of these values is primarily left up to the user. A disadvantage is that a user may interpret the individual values too literally and be misled.

For example, the highest values may be assumed to be representative estimates for well exposed sites, when really some of these values may be considerably overestimated; likewise, areas with low values may be assumed to have low wind power when really many of these stations may be in sheltered locations. Thus, the user must be aware of the many uncertainties and errors which are inherent and may cause some of the individual estimates to be in error by a factor of 2 or more.

Over mountainous regions, extrapolated values of wind power over mountain peaks, based on the free air wind power at the height of the peak, are given for 1° latitude by 1° longitude squares. Two values are plotted for each 1° square: a value for the highest peak and a value for a representative peak. Areas where these values appear questionable are where the valley estimates are greater than nearby mountain peak estimates.

Summary of Types of Errors and Uncertainties

A summary of factors which lead to errors and/or uncertainties in the wind power estimates is listed below.

(1) Incorrect specification of or neglect of anemometer heights. If the actual height is 20 m, and the height is assumed or specified to be 10 m, then the wind power may be overestimated from 35% (using $1/7$ power) to 52% (using 0.2 power). Conversely, the wind power may be underestimated if the actual height is less than the assumed height.

(2) Number of wind speed classes. Wind power is overestimated by approximately 10 to 30% for stations with just 3 to 6 wind speed classes and by 5 to 15% for stations with 8 to 12 wind speed classes.

(3) Atmospheric density variations with elevation. The use of standard sea level density overestimates the wind power by approximately 9% per km elevation.

(4) Atmospheric density variations with air temperature. Errors from this source vary from region to region but are generally less than 5% annually.

(5) Period of record. Typical year to year variations in mean wind speed are 10 to 20%, which correspond to wind power variations of approximately 30 to 70%. Thus, for stations for which the wind power estimates are based on less than 5 years of data, the estimates may be in considerable error.

(6) Vertical extrapolation techniques. Use of the 0.2 power to extrapolate upward from 10 to 50 m gives 32% greater wind power than the $^1/_7$ power law.

(7) Rawinsonde data at 150 m. Extrapolating rawinsonde-based estimates at 150 m down to 50 m, using a $^1/_7$ power law, generally underestimates the wind power at 50 m by approximately 30 to 60%.

(8) Ship wind measurements. There are uncertainties regarding the accuracy and height of wind speed measurements made by ships.

(9) Station location. Uncertainties may exist on how sheltered or exposed the location is with respect to surrounding terrain.

(10) Instrument exposure. Uncertainties may exist regarding obstructions such as trees, buildings, etc.

(11) Accuracy of the wind speed sensor. Errors of just ½ m/sec (1 mph) in the mean wind speed cause large errors in wind power estimates, e.g., the average wind power for a 6 m/sec wind speed is about 30% greater than that for a 5.5 m/sec wind speed.

(12) Method of observing and recording the wind speeds. The manner in which wind speeds are read and averaged may cause errors (e.g., instantaneous vs 3 minute average vs hourly average).

(13) Analysis and interpolation of the wind power estimates. Different analysis methods may result in significantly different estimates and patterns, even though identical data sets are used.

REFINED ANALYSES OF NATIONAL WIND POWER

Methods Used in the Analyses

Refined analyses of wind power at 50 m above exposed areas over the contiguous U.S. were developed through a synthesis and revision of existing national wind energy assessments and various other sources. This study did not attempt to take the various data sets employed and recompute the wind power. Due to the nature of the wind data used, it is felt that this effort, in itself, would have resulted in little, if any, improvement. Rather, the wind power estimates and analyses in each of the existing national assessments were evaluated subjectively with respect to the techniques applied and the representativeness of the data and the analyses. Over regions where large discrepancies existed, this included identifying the causes of these discrepancies. These criteria have been described throughout previous sections of this report. Differences in the types of wind data used, the assumptions made, the techniques applied in the vertical extrapolation of winds, and the methods of analysis, primarily account for the discrepancies in the wind power estimates and geographical distributions.

In forming the refined analysis, information from the following sources was selectively and discriminately utilized:

(1) surface wind power estimates from the national wind energy assessments,

(2) wind power estimates from low-level rawinsonde wind speed distributions at 150 m (GE study),

(3) offshore wind power estimates calculated for Marsden Square and Navy Marine Areas (GE study),

(4) upper air wind power estimates and mountain summit wind power estimates,

(5) wind data summaries from various other sources, including Hewson (21), Justus (8), Wentink (22), and STAR program and climatological wind summaries from the National Climatic Center,

(6) national maps of topographic relief variation for evaluating the average local height variations of topography over areas of gently rolling and hilly terrain, and

(7) national maps of the physical topographic features, for identifying regions of mountainous terrain.

The representativeness and reliability of the wind power estimates were stressed. Where possible, wind data sets or summaries from other sources (such as those listed above) and from more recent time periods were used for comparison with those in the national wind power assessments. For example, the mean annual speeds given to climatological summaries for recent time periods (for which anemometer heights are constant and instrument exposures are generally good) were compared with those in the Lockheed tabulations (11). In regions of gently rolling and hilly terrain, some of the stations were examined for sheltering effects by local topography (i.e., to determine if the station is located in a relatively low area with respect to the elevation of the surrounding terrain).

Over nonmountainous terrain, the large-scale energy patterns based on 150 m level rawinsonde data are assumed to be generally representative of the relative energy patterns. However, the wind power estimates based on 150 m rawinsonde data are considered to be 30 to 60% too low in most cases (for reasons previously discussed) and were correspondingly adjusted.

Isolated high values for surface stations are not shown in the national wind power analyses. Many of these relatively high values are for stations with "limited" data (i.e., short time periods and 4 or fewer wind speed classes). As discussed earlier, there are several factors or combinations of factors which can cause the wind power estimates for surface stations to be overestimated by as much as a factor of two.

The wind power estimates along coastal regions are based on (1) offshore wind power estimates calculated for Marsden Squares and Navy Marine Areas and (2) estimates for exposed coastal sites. Wind power estimates for the large Navy Marine Areas, which extend seaward 300 km or more, indicate considerably greater wind power (factor of 2) than that for Marsden Squares which are areas located within 100 km of the coast. The relative geographical distribution of wind power along coastal regions is assumed to be similar to the relative distribution for Navy Marine Areas and roughly one-half the magnitude, in most cases.

The offshore wind power values based on ship data and values for coastal stations are applied with discretion in determining the wind power estimates along coastal regions. Only annual average values for offshore areas were calculated in the assessment by General Electric. Thus, in the synthesis, seasonal values for offshore areas were adjusted in proportion to seasonal variations from the annual average at selected coastal stations.

The wind power estimates indicated off the Atlantic, Gulf, and Pacific coasts apply to exposed coastal sites (e.g., capes, points, and islands) and offshore areas. For definition purposes, this offshore zone may be assumed to extend from the shoreline 50 km seaward. (It is suspected that the wind power may increase slightly from the shoreline seaward but there is no conclusive information on this. It is also suspected that the wind power at exposed coastal sites and at 1 km seaward is more typical of the power at 10 to 50 km seaward than at 1 km inland, primarily due to surface roughness effects.) Because of the sharp gradients of wind power along coastal areas, use of isodyns was avoided.

Since wind power over mountain summits appears highly variable and may be greater on relatively "isolated" ridges than on higher summits and ridges in the same region, it was felt that all regions with any significant mountains or ridges, (e.g., 500 m or more higher than the surrounding terrain) should be analyzed with estimates for exposed mountain summits. Thus, regions with scattered or isolated mountain ridges such as the Great Salt Lake basin of northwest Utah are designated by typical estimates for the mountain summits rather than for the lower basin. (An exception was southern Wyoming, where exposed areas in the basin indicate high wind power comparable to that estimated for mountain summits.) These mountainous regions were defined using national relief maps (12), and are the shaded areas on the national wind power maps.

The wind power estimates given for mountain summits are lower limits expected for typical, well exposed sites. Sites where the wind speeds are enhanced by topographic effects may have more than double the amount indicated in the analyses. Examples are: Whiteface Mountain in northern New York with approximately 1500 W/m^2, Rattlesnake Mountain in south-central Washington with approximately 1000 W/m^2, and Sandberg in the southern California Coast Range with approximately 500 W/m^2. Also, gaps and passes may have comparable wind power to nearby mountain summits.

Geographical Distributions

Annual Pattern in the Contiguous U.S.: Regions indicating high wind power (400 W/m^2 or greater), shown in Figure 2.5, are the southern and central Great Plains, offshore and well exposed coastal sites along the north Atlantic coast, north Pacific coast, and south Texas coast, the high plateau area of southern

Wyoming, and areas just east of the Rockies in Montana. The analysis also indicates that most mountainous areas in the East and West should have exposed sites with considerable wind power. (Again, the estimates for mountain summits are lower limits expected for well exposed sites.) As discussed earlier, many lower isolated ridges may also have considerable wind power and be more suitable for wind turbines than higher mountain ranges.

Figure 2.5: Mean Annual Wind Power (W/m^2) Estimated at 50 m Above Exposed Areas*

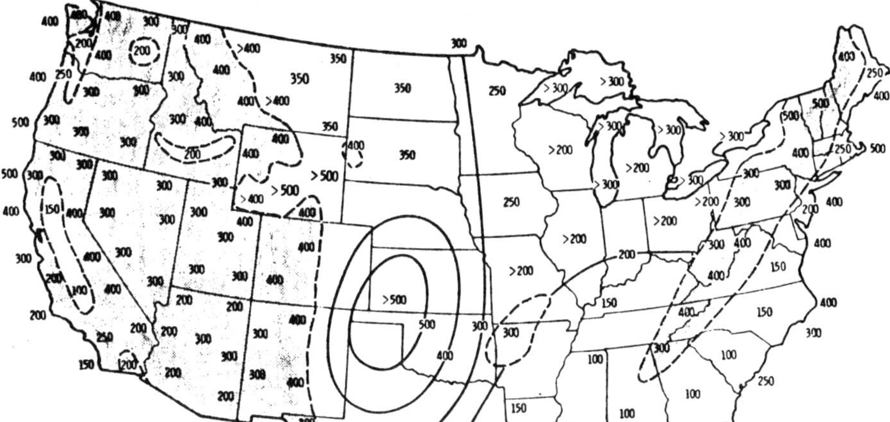

*Over mountainous regions (shaded areas), the estimates are lower limits expected for exposed mountain tops and ridges.

Source: BNWL-WIND-5

Considerable wind power (300 W/m^2 or greater) is also indicated over the northern Great Plains. Some isolated areas in the northern Great Plains indicate over 400 W/m^2 (not shown in the analysis) but the density and representativeness of the data are not sufficient to identify the extent of these areas. More data are needed before additional detail in the analysis can be justified.

Over regions of complex terrain (hilly and mountainous terrain) and coastal areas, where local mesoscale effects can result in higher wind speeds, there is a possibility that a few sites may have as much as 100% greater wind power than the national analyses show.

Very limited data were available for offshore or exposed coastal sites along the Great Lakes. These data indicated that the wind power is substantially greater offshore and at exposed coastal sites than for inland areas.

Thus, the offshore annual wind power over the Great Lakes is estimated to be over 300 W/m^2, which is approximately 100 W/m^2 greater than the average power for nearby inland sites. East shores of the lakes are expected to have greater wind power than the west shores, due to the prevailing westerly winds. Some well exposed sites with limited data indicate over 500 W/m^2 (2) (16).

Seasonal Patterns of Wind Power in the Contiguous U.S.: Figure 2.6 shows the seasons of maximum wind power. Over the eastern one-third of the nation, maximum wind power occurs during the winter and early spring, except for mountain summits where the maximum is in the winter. Spring maxima occur over the Great Plains, the North-Central states, the Texas coast, most nonmountainous regions in the West (e.g., the basins and broad valleys), and offshore areas of central and southern California. Winter maxima occur over all mountainous regions, except for some areas in New Mexico and Arizona where the spring and winter wind power are about equal.

The Northwest coastal region, the high plateau of southern Wyoming, and a strip in Montana which lies just east of the Rocky Mountains, also have winter maxima. The only regions which have summer maxima are coastal and offshore areas of southern Oregon and northern California.

Wind Power in Alaska: A study of the wind power in Alaska and its possible application has been performed by Wentink (22). Most offshore and exposed coastal regions of Alaska have high wind power potential. As in all coastal regions, the wind power typically decreases abruptly onshore, especially along more rugged coastlines. The highest wind power in the Alaska region appears to be along the Aleutian peninsula and island chain between 160° to 172° longitude. Exposed coastal sites and offshore areas of the Aleutians indicate mean annual wind speeds of 7.0 m/sec (16 mph) or greater and annual average wind powers exceeding 500 W/m^2.

Very good wind power (400 to 500 W/m^2) also exists along all other exposed coastal and offshore areas of Alaska. The wind power throughout interior Alaska and sheltered coastal regions (e.g., bays, inlets, and sounds in rugged terrain) appears quite low, except for mountains and isolated ridges. A few interior places show greater than 200 W/m^2; these are primarily canyons and valleys where the winds are enhanced by topography, such as Big Delta (about 80 miles southest of Fairbanks).

Annual mean upper air wind speeds decrease from southwest to northeast Alaska, but the upper winds at 850 mb and 700 mb are sufficiently strong such that well exposed mountain sites throughout Alaska are estimated to have annual average wind powers of at least 300 to 400 W/m^2.

Wind Power in Hawaii: Wind power along exposed coastal and offshore areas of the Hawaiian Islands is estimated to be 200 to 300 W/m^2, but some sites with limited data indicate over 400 W/m^2. Inland areas show much lower wind power.

The variation of wind speed with height in the subtropics is strikingly different from that of the mid-latitudes. Above 850 mb the wind speed generally decreases with height in the subtropics, so that lower mountain ridges in Hawaii are estimated to have greater wind power than the loftier volcanoes (3). Wind power studies for the islands of Hawaii, Oahu, and Maui have been made by the University of Hawaii (23)(24)(25).

Diurnal Variation of Wind Power: Diurnal wind power variations throughout the U.S. have been shown and described in previous national assessments. In the surface layer up to 100 or 200 m, winds tend to be strongest in the afternoon and weakest after midnight for most areas. Over mountain summits, winds are generally stronger at night, based on the limited data available. The survey by Lockheed indicates, however, that the diurnal behavior of wind speed may differ markedly from mountain to mountain.

Figure 2.6: Seasons of Maximum Wind Power*

*Over mountainous regions (shaded areas), the seasons indicated are for exposed mountain sites. Seasons are abbreviated, where necessary, as follows: W = winter, SP = spring.

Source: BNWL-WIND-5

CONCLUSIONS

A synthesis of the existing national wind energy assessments has been performed in an effort to develop an improved, consistent assessment of the available wind power and its geographical distribution over the United States and identify areas of needed research. Maps showing the distribution of annual and seasonal average wind power potential have been developed for the contiguous United States. Some primary results of the synthesized and refined national wind energy assessment are briefly summarized below.

- Wind power estimates for individual stations are subject to various errors and uncertainties in the data and techniques employed. Variations in anemometer heights, and site and instrument exposure can result in nearby wind power values which differ by a factor of 3 even over regions of mostly level terrain. Thus, surface-based wind power estimates should be used with caution.
- Rawinsonde data summarized for the 150 m level above the ground is useful for identifying the large scale energy patterns, but generally underestimates the wind power by 30 to 60%.
- The estimates shown in this assessment are considered to be generally conservative. Over offshore areas and regions of complex topography, there may be well exposed sites with as much as 100% greater wind power than the amount indicated. (This is the total available power, not the extractable wind power.) Over all regions, there may be a few isolated sites or areas with considerably greater (up to 50%) wind power than indicated.
- Over nonmountainous regions, the analysis indicates that annual wind power at 50 m above exposed is high (⩾400 W/m^2) over the central and southern Great Plains, parts of Wyoming, Montana, and offshore and exposed coastal areas of the Northeast, Northwest, Alaska, Hawaii, and south Texas. Moderately high wind power (⩾300 W/m^2) is indicated over the northern Great Plains and the Great Lakes. There is a possibility of isolated good sites throughout the rest of the country.
- All mountainous regions are expected to have some exposed mountains, ridges, or passes with high wind power potential, including regions where the upper air wind power is relatively low (such as the Southwest). The wind power is highly variable over mountainous regions and depends on the shape and orientation of a ridge, as well as the height of a ridge.
- The seasons of maximum wind power are predominately winter over most mountainous regions, late winter and early spring over the eastern third of the nation, and spring over the central U.S. and Great Plains.
- Wind power is generally highest in the afternoon and weakest after midnight for most surface stations. Over mountain summits, winds are typically stronger at night.

RECOMMENDATIONS

As a result of the synthesis of national assessments of wind energy performed in this study, the following recommendations are made.

- Surface values of wind power need to be evaluated using wind data from a recent 5 to 10 year period when the anemometer heights are constant over the period. Hourly, or three-hourly, digitized wind data available from the NCC could be utilized for this task. This would eliminate errors due to variations or uncertainties in the anemometer heights and due to the number of wind speed classes. Also, errors due to poor instrument exposure would be substantially reduced.

- The representativeness of wind power estimates for individual surface stations needs to be examined more thoroughly with regard to the exposure of the station (with respect to surrounding terrain and roughness) and to the exposure of the anemometer (with respect to nearby obstructions).
- The horizontal and vertical structure of wind power along coastal regions and for offshore areas needs to be examined in greater detail. Most coastal stations are located too far inland to be applicable, and there are uncertainties regarding the use of ship data. Techniques for estimating the alongshore and offshore variation of wind power, in the absence of representative data, need to be investigated.
- Summaries of tower wind data exist for numerous sites throughout the U.S. These summaries would be very useful in regional and national assessments of wind power. Wind data summaries from nuclear power plants have recently been compiled by Verholek (17). An effort should also be made to identify and aggregrate other tower wind data summaries, including those for coal-fired power plants.
- Using existing tower data, the variation of wind speed with height should be evaluated with respect to roughness height, local terrain influences, wind speed, seasonal and diurnal variations, and regional and climatological characteristics. Present relationships appear to be based on only a few separate sets of towers widely scattered throughout the country.
- Special wind data summaries compiled by universities, and private and federal agencies could provide additional information needed for improving the national and regional analyses of wind power, especially in regions where data are presently insufficient or questionable. A survey of field programs with mesoscale wind networks has been made for the Wind Energy Conversion Program. Similar surveys should be made to identify other data sources, through literature surveys, telephone contacts, etc.
- Extractable wind power needs to be evaluated and the geographical distribution of extractable wind power described for selected types of wind-powered generators, in a manner similar to that described by Justus et al (4) but using additional data and techniques to refine and improve the analyses.
- An additional effort should be made to evaluate the reliability of wind power estimates derived from rawinsonde and pibal data by comparing with estimates computed from tower winds, provided the sites are close enough to permit such a comparison.
- With the available data sources used in this study, it is not likely that much additional detail can be analyzed into the national assessment of wind power. Using additional data sources, descriptive techniques, and models, regional assessments of wind power should be made to examine the potential wind power and its geographical distribution over various regions in greater detail.

REFERENCES

(1) Reed, J.W., *Wind Power Climatology of the U.S.,* ERDA Contract #AT(29-1)-789, SAND 74-0348, Albuquerque, NM (1975).

(2) "Wind Energy Mission Analysis," Lockheed-California Co., Draft report, ERDA Contract AT(04-3)-1075 (1976).

(3) "Wind Energy Mission Analysis," General Electric Co. (Valley Forge), Draft report, ERDA Contract E(11-1)-2578 (1976).

(4) Justus, C.G., Hargraves, W.R. and Yalcin, A., "Nationwide Assessment of Potential Output from Wind-Powered Generators." *J. Appl. Meteor. 15*(7):673-678 (1976).

(5) Widger, W.K., "Estimating Wind Power Feasibility," *Power Engineering,* (August 1976).

(6) Baker, R.W. and Hennessey, J.P., Jr., "Estimating Wind Power Potential," *Power Engineering,* (March 1977).

(7) Changery, M.J., *Initial Wind Energy Data Assessment Study,* NSF/NOAA, NSF-RA-N-75-020, Asheville, NC (1975).

(8) Justus, C.G., *Wind Energy Statistics for Large Arrays of Wind Turbines (New England and Central U.S. Regions),* Final Report, NSF/RA-76-0191, ERDA/NSF-00547/76/1, Georgia Institute of Technology, Atlanta, GA (1976).

(9) Holzworth, G.C., *Mixing Heights, Wind Speeds, and Potential for Urban Air Pollution Throughout the Contiguous United States,* EPA, Research Triangle Park, NC (1972).

(10) Crutcher, H.L., *Upper Wind Statistics of the Northern Hemisphere,* Vol. 1, NAVAER 50-lC-535, Washington, DC (1959).

(11) Coty, U.A., Court, A. and Reed, J.W., *United States Wind Speed and Wind Power Duration Tables, by Months (Cumulative Distributions),* Scientific Report No. 1 under Contract AT(04-3)-1075, Lockheed-California Co., Burbank, CA (1975).

(12) *The National Atlas of the United States of America,* United States Department of the Interior, Geological Survey, Washington, DC (1970).

(13) Justus, C.G. and Mikhail, A., "Height Variations of Wind Speed and Wind Distributions Statistics," *Geophy. Res. Letters, 3,* 261-264 (1976).

(14) Newstein, H., *An Automated Meteorological Instrumentation and Observing System on a 1000 ft TV Tower,* Drexel Univ., Final Report, ESSA/SDO Contract Cwb-11038 (1976).

(15) Cormier, R.V., *An Annotated Listing—Tall Towers Instrumented for Wind Observations,* Technical Report 73-0179, Air Force Cambridge Research Laboratory, L.G. Hanscom Field, Bedford, MA (1973).

(16) Wallace, J.A., Jr., *An Annotated Bibliography of Meteorological Tower and Mast Studies,* WB/BS-5, U.S. Department of Commerce, ESSA, Environmental Data Service, Washington, DC (1967).

(17) Verholek, M.G., *Summary of Wind Data from Nuclear Power Plant Sites,* BNWL-2220 WIND-4, Battelle, Pacific Northwest Laboratories, Richland, WA (1977).

(18) Wahl, E.W., *Windspeed on Mountains,* Final report, Contract No. AF19(628)-3873, U.S. Air Force, Cambridge Research Laboratories, Bedford, MA (1966).

(19) Pristov, J., "Abweichungen des Windes auf den Alpinen Beabachtungsstationen in Bezug auf die Strömung in der Freien Atmosphäre," *Berichte des Deutschen Wetterdienstes,* Bd. 8, 5, 241-243 (1959).

(20) Meroney, R.N., Sandborn, V.A., Bouwmeester, R.J.B. and Rider, M.A., *Sites for Windpower Installations, Tunnel Simulation of the Influence of Two-Dimensional Ridges on Wind Speed and Turbulence,* NSF/RANN GAER 75-00702 (1976).

(21) Hewson, E.W., Baker, R.W. and Brownlow, R., *Wind Power Potential in Selected Areas of Oregon,* Report No. PUD 76-4, Oregon State University, Corvallis, OR (1976).

(22) Wentink, T., Jr., *Study of Alaskan Wind Power and Its Possible Applications,* Final Report, Contract No. E(45-1)-2229; NSF-G-AER-74-00239A01, Univ. of Alaska, Fairbanks, AK (1976).

(23) Daniels, A., Palmer, B., Tarlpon, T. and Schroeder, T., *A Survey of Wind on the Island of Maui for Potential Wind Power Generation. Part 1,* Mobile Sampling Program, August 7-26, 1976, Dept. of Meteorology, University of Hawaii, UHMET 76-6 (1976).

(24) Schroeder, T., Tarlpon, T. and Daniels, A., *A Survey of Wind on the Island of Hawaii for Potential Wind Power Generation, Part 1,* Mobile Sampling Program, Sept. 3-12, 1976, Dept. of Meteorology, University of Hawaii, UHMET 76-7 (1976).

(25) Ramage, C., Daniels, A., Schroeder, T. and Thompson, N., *Oahu Wind Power Survey, First Report,* Dept. of Meteorology, University of Hawaii, UHMET 77-1 (1977).

ROTOR DEVELOPMENT

Material for this chapter has been based upon reports from the Lewis Research Center (DOE-NASA-1010-77-3); (ERDA/NASA/1004-77/2); and (N77-31614); Martin Marietta Laboratories (COO/2613-2); United Technologies Research Center (COO/2614-76/2); Princeton University (PB-259 898); and various papers presented at the Second Workshop on Wind Energy Conversion Systems (WECS), Washington, D.C., June 9–11, 1975 (NSF-RA-N-75-050).

ERDA/NASA LARGE EXPERIMENTAL WIND TURBINES

The Federal Wind Energy Program (1) includes several projects for the development of large horizontal-axis propeller-type wind turbines for generating electricity. Under the overall program management of ERDA, the NASA-Lewis Research Center (LeRC) provides project management for the large horizontal-axis wind turbines projects and the associated supporting research and technology.

The large wind turbine phase of the Federal Wind Energy Program has several projects; these include:

(1) The 125-foot-diameter rotor 100-kW wind turbine which has been designated Mod-0; this wind turbine became operational on September 4, 1976.

(2) Two 125-foot-diameter-rotor wind turbines of 125 and 200 kW each. These wind turbines are similar to the Mod-0 and have been designated Mod-0A.

(3) One or two nominal 200-foot-diameter rotor wind turbines of 1,500 kW each. These wind turbines have been designated Mod-1.

(4) A nominal 300-foot-diameter rotor wind turbine with an output of 1,000 to 2,000 kW. This wind turbine has been

designated Mod-2 and is being designed to deliver 1,000 kW or more at lower wind speeds than the above wind turbines. This Mod-2 wind turbine is to be operational by the end of 1979.

100 kW Mod-0 Experimental Wind Turbine

The 100 kW experimental wind turbine project has been described in several earlier reports (2)-(5). The objective of this project is to provide engineering data, as early as possible, for use as a base for the entire wind energy program and to serve as a test bed for improved components and subsystems. To meet this objective a 100 kW wind turbine was designed using available technology and catalog items where possible. Information in previously designed large two-bladed wind turbines was obtained from Hutter (6)-(8) and Putnam (9). The design, fabrication, and assembly of the 100 kW Mod-0 wind turbine was completed in approximately 18 months with initial operation beginning on Sept. 4, 1976.

Description: The 100 kW wind turbine has a two-bladed 125-foot-diameter rotor driving a 100 kW synchronous alternator through a step-up gearbox. The rotor is located downwind of the tower and rotates at a constant speed of 40 rpm. The alternator operates at 1,800 rpm and delivers 60 Hz three-phase power. Details of the drive train assembly and yaw system are shown in Figure 3.1. The wind turbine rotor starts rotating at a wind speed of 5 mph and reaches 40 rpm and begins to deliver power at 9.5 mph. Figure 3.2 shows the variation of power output of the wind turbine as a function of wind velocity.

Assembly: The 100 kW wind turbine drive train was assembled at LeRC and tested prior to final assembly at the site. The drive train assembly was tested for 50 hours to make certain all components and subsystems worked satisfactorily. The drive train assembly was then mounted in the yaw drive system and the rotor hub and blades attached. In this condition all systems were checked except for rotation of the blades. The blades were then removed and the system was transported to the site for final checkout and assembly.

Final Site Assembly: The wind turbine was assembled on a test stand near the base of the tower. After all systems were checked out, the wind turbine was lifted on the morning of September 4, 1976, and assembled and placed on the tower by a crane. This operation took approximately 4 hours and went very smoothly.

Test Results: Checkout of the wind turbine systems and preliminary operation, performance, and engineering data were obtained over the next several months. Figure 3.3 is a plot of predicted blade angle at 40 rpm versus wind speed to obtain 0 kW or 100 kW. Shown in the figure are several test points at 0 kW and 100 kW for different wind speeds; these test points agree fairly well with the predicted values. On December 18, 1976, the wind turbine was operated for the first time at the design conditions of 40 rpm and 100 kW. When electric power is generated by the wind turbine, the wind turbine is connected to a variable resistance load bank. To date, the wind turbine is performing as expected except for larger than expected bending moments at the blade roots. During testing at wind speeds of 27 mph these bending moments are nearly twice as high as expected. Figure 3.4 shows the predicted and actual blade bending moments during one revolution of the blade. Zero degrees is with the blade straight down behind the tower. The expected bending moment was calculated from the MOSTAB helicopter rotor loads computer code modified for wind turbines (10). Assuming the tower retarded the wind by 24%, the MOSTAB code predicted a cyclic bending moment that varied from -60,000 ft-lb to about 4,000 ft-lb. The actual test data, however, showed the bending moment varying from -120,000 ft-lb to 7,000 ft-lb. The MOSTAB code nearly predicts this larger bending moment if it is assumed the tower retards the wind by 93% behind the tower.

Figure 3.1: 100 kW Wind Turbine Drive Train Assembly and Yaw System

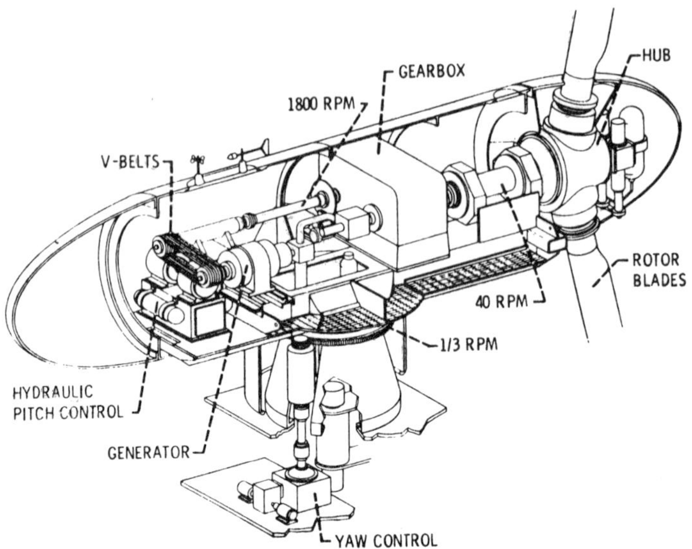

Source: DOE-NASA-1010-77-3

Figure 3.2: Power Output Versus Wind Speed for the 100 kW Wind Turbine

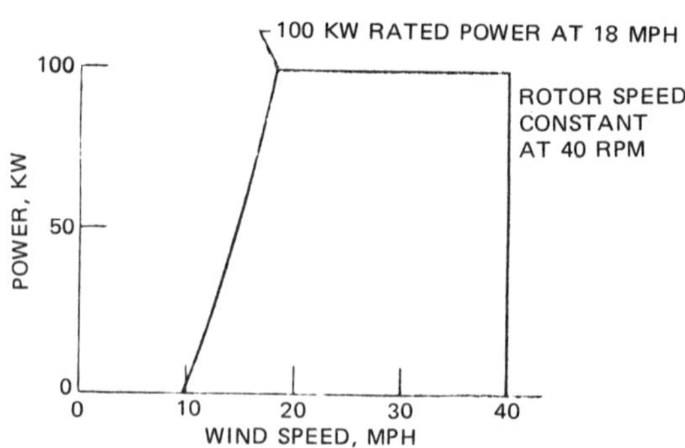

Source: DOE-NASA-1010-77-3

Figure 3.3: Mod-0 Blade Pitch Angle Schedule

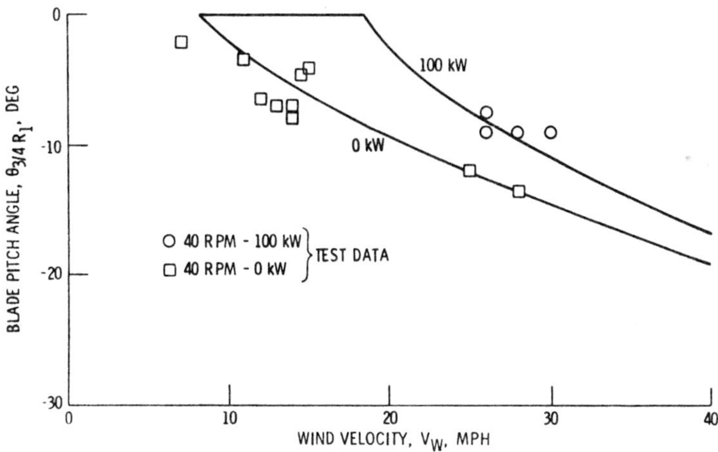

Source: DOE-NASA-1010-77-3

Figure 3.4: Blade Load Comparison for MOSTAB Predictions vs Test Data

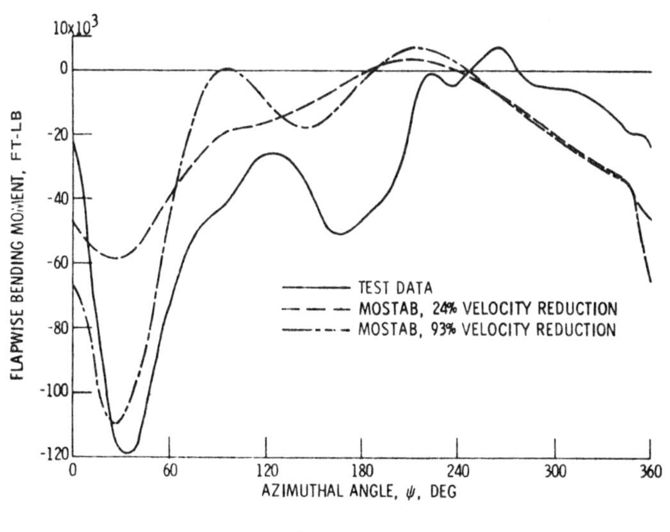

40 rpm, 100 kW, θ_P = - 8°, $V_{WIND} \cong$ 27 mph.

Source: DOE-NASA-1010-77-3

125 to 200 kW Mod-0A Experimental Wind Turbine Project

The objective of the Mod-0A project is to gain early experience with wind turbines operating in utility networks. In particular it was planned to obtain actual utility interface requirements, operations information, and uncover any institutional problems for wind turbines operating in conjunction with a utility network.

To meet this objective, two wind turbines similar to the 100 kW Mod-0 wind turbine but of higher power output were being fabricated. Both wind turbines have been slightly redesigned to increase their power output from 100 kW at 18 mph to 200 kW at 23 mph. However, the first Mod-0A wind turbine is initially rated at 125 kW since it utilizes a spare gearbox from the Mod-0 wind turbine to provide a faster schedule. It was planned to later replace this 125 kW gearbox with a 200 kW gearbox. Both Mod-0A wind turbines use 200 kW alternators. The rotor blades for the Mod-0A will be identical to the metal blades in use on the Mod-0 wind turbine.

The Mod-0A towers are of the steel truss design similar to the Mod-0. The Mod-0A towers, however, are designed to minimize blockage of the wind and to therefore reduce the effect of tower shadow in the blade bending moment. The Mod-0A towers utilize a simple cable-supported electric personnel lift for access to the top with a simple ladder for use in the event of an emergency in place of the Mod-0 tower staircase.

The actual assembly of the wind turbines is under industrial contract as part of an effort to involve and utilize U.S. industry as fully and effectively as possible. The industrial contractor final assembles the wind turbines at the selected utility sites and is responsible for checkout and initial operation.

1500 kW Mod-1 Experimental Wind Turbine Project

The objective of the Mod-1 wind energy project is to involve industry and users in the design and implementation of optimized wind turbines that are capable of supplying electrical power at costs competitive with conventional power sources. Also during the course of this effort, attention will be paid to evaluating the public reaction and/or acceptance of large wind turbines. To meet these objectives the following multiphased project has been initiated:

(1) Design study contracts,
(2) Evaluation of designs and competitive contract preparation,
(3) Detail design, fabrication, assembly, and operation contracts,
(4) Operation and reporting of performance of experimental wind turbines.

The objectives and plans for these phases of the Mod-1 wind energy project have been fully described in Reference (1). This report concentrates on the results of the Mod-1 proejct.

Design Study Contracts—Two parallel 9-month study contracts were awarded in mid-November, 1974, for the selection and design of minimum energy cost wind turbines.

These contracts, at a cost of approximately $500,000 each were awarded to the General Electric Company and the Kaman Aerospace Corporation. The contract scope was limited to the study of horizontal-axis propeller-type wind turbines that could produce electric power acceptable for utility grids. The effort encompassed four major tasks which included conceptual design; parametric analysis; preliminary design and definition of utility interface requirements.

The contractors evaluated a number of candidate configurations for wind turbines. After initial cost studies, both contractors independently concluded that for minimum energy cost the wind turbine should have a variable-pitch constant-speed two-bladed rotor operating downwind of the tower; and the rotor should drive a synchronous or induction alternator through a conventional step-up gearbox. The early results of these studies obtained by Kaman were summarized by Meier in a paper presented at the American Helicopter Society in May, 1975 (11).

During the parametric and optimization task, capital and energy costs were determined for a range of wind turbines optimized for minimum energy cost (¢/kWh). The wind turbines analyzed range from 50 kW to 3,000 kW. NASA furnished wind velocity duration curves which were used for all energy calculations in the design studies. It was assumed that the wind velocity is defined at an elevation of 30 feet.

The results of the parametric and optimization task for the two contracts are shown as solid lines in Figure 3.5 for energy costs (¢/kWh) as a function of median wind speed and rated wind turbine power. These curves represent a wind turbine optimized for lowest energy cost for that specific median wind speed. For comparison, generation cost for power in the U.S. in 1975 generally falls in the 0.1 to 5¢/kWh range. These curves indicate that wind turbines could be competitive with some existing power generation systems.

Both studies show that at each median wind speed the energy costs are higher at the lower power levels over the range studied. The GE study shows that energy costs continue to go down very slowly with increased power levels for all median wind speeds. The Kaman studies show, however, a bottoming effect to the curves with energy costs increasing as power levels go above approximately: (1) 700 kW at 12 mph, (2) 800 kW at 15 mph, and (3) 1,500 kW at 18 mph. The fact that one curve has a minimum while the other continues to slowly decrease is primarily due to the difference in assumptions used for the rotor costs as the diameter goes up.

The energy costs for the parametric studies shown in Figure 3.5 differ by up to a factor of two. These were early studies and are based on the conceptual design results. Later in the program as the preliminary design studies at 1,500 kW and 500 kW were completed, the cost differences between the two studies come closer together with the Kaman energy costs decreasing and the GE costs increasing. The cost estimates resulting from the preliminary designs are shown as data points for each of the designs in Figure 3.5. These later energy costs were calculated using the estimated capital costs, energy capture and the yearly rate of charge for money, operation, maintenance, taxes, etc.

The GE study uses a rate of approximately 0.16/year while Kaman used approximately 0.20/year. This difference also accounts for a major difference between the energy cost results of the earlier parametric studies shown by the solid lines in Figure 3.5. To compare the energy costs from both studies, the 0.16/year rate was used to calculate energy costs for each wind turbine preliminary design and these energy costs are shown as data points on Figure 3.5. The following equation was used for calculating energy cost:

$$\frac{¢}{\text{kWh}} = \frac{(\text{Total capital cost}) \times (0.16/\text{yr})}{(\text{kwh})/\text{y}}$$

Referring to Figure 3.5 it can be seen that the energy cost estimates for the Kaman 500 kW and 1,500 kW design are 5.5¢/kWh and 2.0¢/kWh, respectively, while the energy cost estimates for the GE 500 kW and 1,500 kW designs are 4.2¢/kWh and 1.7¢/kWh, respectively.

Figure 3.5: Wind Turbine Energy Costs as a Function of Median Wind Speed and Rated Power

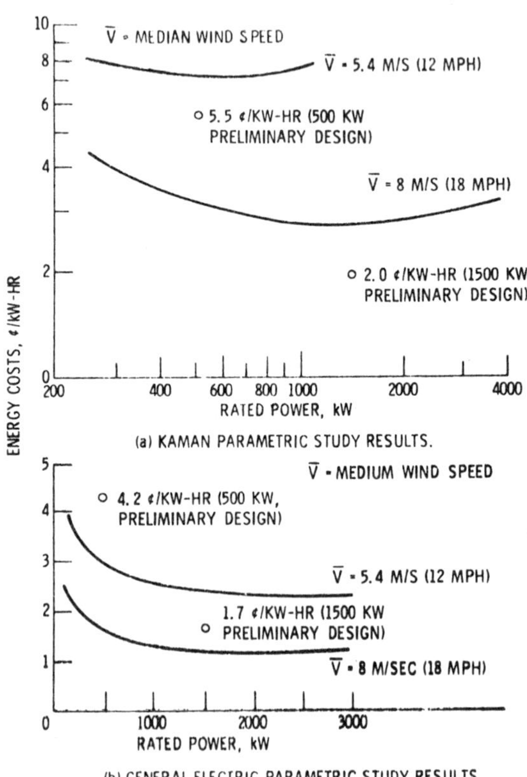

Source: DOE-NASA-1010-77-3

In addition to minimum energy cost there is also the question of the wind turbine plant factor. Plant factor (PF) is defined as the ratio of the energy the wind turbine will deliver over a year for mean wind speed to the amount of energy it would generate in a year if it operated at rated power continuously.

$$PF = \frac{\text{Generated annual-kWh}}{\text{Rated kW x 8,760 h}}$$

Plant factor gives an indication of how effectively a wind turbine is being utilized. High plant factor wind turbines generally have large rotors in comparison to their rated power output or alternator size and thereby operate more hours at rated output. However, wind turbines with high plant factors will not necessarily generate the lowest cost energy. High PF wind turbines generally capture more energy but are also more expensive to build because of the larger rotors.

The selection of a minimum energy cost wind turbine versus a wind turbine with a higher PF is definitely an issue that depends on how the wind turbine is to be used. If the application can use bulk energy whenever it is available, then the minimum energy cost design is preferred. If energy storage is needed, then the higher plant factor wind turbine probably will result in the best overall systems choice. This option is clearly dependent on the resultant user's application.

Before selecting the wind turbine sizes and the mean wind speeds for preliminary design, the following question was addressed: How sensitive to mean wind speed is an optimized design? That is, given an optimized design of, say, 1,500 kW for an 18 mph site, what is the effect on energy costs of putting this design on sites with a mean wind speed of 15 mph or 21 mph. Studies were conducted by both contractors to answer this question. The results of these studies for a 500 kW wind turbine, 12 mph mean wind speed site and a 1,500 kW wind turbine, 18 mph mean wind speed site were plotted.

Significantly, the results show that for less than 10% increase in energy costs the wind turbine designed for 500 kW at 12 mph can be used at sites with 9 to 15 mph and the 1,500 kW design can be used at sites with 15 to 21 mph wind. Thus it appears that the two designs could be used for a relatively wide range of sites with only a small penalty in energy cost.

From the results of the parametric study and optimization tasks, two designs were selected for further definition during the preliminary design task; 500 kW at 12 mph and 1,500 kW at 18 mph. The conceptual, and parametric analysis studies were performed to obtain the minimum energy cost wind turbine designs. However, in the final selection attention was paid to plant factor and the designs selected were slightly off optimum to take advantage of higher plant factors at increase in energy cost of only a few percent. The capital cost of these higher plant factor wind turbines is greater but because more energy is generated, the resulting energy costs are nearly those of the minimum energy-cost design. Table 3.1 summarizes the results of the 500 kW and 1,500 kW wind turbine designs for each of the contractors.

In addition to determining the optimum wind turbine designs, a major task of the design studies was to determine the interface requirements for connecting wind turbines to typical utility grids. These interface requirements included items such as require electrical interface switchgear, automatic protection and controls and required operation and maintenance procedures. These studies show no interface or institutional problems that pose major barriers to the installation of wind turbines on utility grids.

Table 3.1: Results of Preliminary Design Task for 500 kW and 1,500 kW Mod-1 Wind Turbines

	......500 kW......		....1,500 kW.....	
	GE	Kaman	GE	Kaman
Mean wind speed	12	12	18	18
Rated power, kW	500	500	1,500	1,500
Rated wind speed, mph	16.3	20.5	22.5	25
Energy capture, kWh/yr	1.88×10^6	1.3×10^6	6.6×10^6	5.7×10^6
Rotor diameter, ft	183	150	190	180
Rotor solidity, %	3	3	3	3
Rotor speed, rpm	29	32.3	40	34.4
Energy cost, ¢/kWh*	4.2	5.5	1.7	2.0
Capital cost, $/kW	974	901	449	481
Wind turbine cost, $	486,000	450,670	674,000	720,800
Plant factor	0.42	0.29	0.51	0.43

*Assumed 16% of capital costs per year for interest, operations, maintenance, taxes, etc.

Source: DOE-NASA-1010-77-3

1000 to 2000 kW Mod-2 Experimental Wind Turbine Project

The object of the Mod-2 project is to develop large wind turbines that can generate low-cost electricity when operating at sites that have median wind speeds near 12 mph. Preliminary designs of these wind turbines indicate that rotors of approximately 300 feet in diameter are required for optimized wind turbines at the 12 mph median wind sites. The federal wind energy plan calls for the operation of the first of these large wind turbines to occur near the end of 1979.

Supporting Research and Technology

In parallel with the large wind turbine projects, a supporting research and technology project (SR&T) has been implemented. The objective of this project is to reduce the costs of wind turbine components and subsystems, simplify the overall designs, and to make the wind turbines more reliable.

Rotor Systems: Low-Cost Rotor Blades—The major emphasis of the SR&T project is directed toward reducing the costs of the rotor systems. The Mod-0 and Mod-1 projects have clearly shown that the rotor systems offer the greatest potential for reducing costs.

Several tasks have been initiated to explore low-cost technologies for manufacturing rotor blades, these include: (1) the design and fabrication of filament wound composite blades for the Mod-0 wind turbine; (2) the investigation of prestressed materials such as concrete and the use of variable density urethanes; and (3) low-cost methods of fabricating all metal blades similar to the present Mod-0 blades.

Large Rotor Blades—Another major effort of the SR&T project is directed toward the fabrication of large blades, nominally 150 ft, for the future large Mod-2 wind turbines.

Rotor Hubs and Blade Pitch Controls—The SR&T project also contains tasks for the investigation of low-cost, reliable hubs. These tasks include the analysis, design, fabrication, and testing of selected hubs in the Mod-0 wind turbine. One hub under consideration is a teetered hub that allows the two blades to move in an art-of-the-plane of rotation. Such a svstem is expected to reduce the blade bending moments and result in lower-cost and/or longer blade lives. The advantages of a teetered hub have been reported by Spera (12). A three-bladed hub is also under consideration as well as other hubs and methods of blade attachment for possible tests on the Mod-0 wind turbine.

The purpose of the three-bladed tests would be to experimentally quantify the difference in loads between two-bladed and three-bladed wind turbines. Such test results may be important as the wind turbine program is directed toward progressively larger wind turbines to take advantage of the many available lower wind speed sites.

Towers: The Mod-0 100 kW wind turbine has a steel truss tower similar to many of the utility transmission towers. The Mod-1 design studies investigated steel shell and concrete towers in addition to the steel truss towers. The design studies showed that the steel shell and concrete towers could compete with the truss tower economically if they were produced in sufficient quantity and if favorable soil conditions existed. These alternate towers appear to most people to be more esthetically pleasing than the truss towers and will be investigated as well as other designs that show promise in further detail under the SR&T project.

Other Components: In addition to the rotor systems and the towers, other major wind turbine components and subsystems will also be investigated under the SR&T project. These include the gearboxes, the alternators, the electric control systems, and the yaw systems for orienting the rotor into the wind. As particular components and subsystems show promise for simpler and/or lower cost wind turbines, they will be tested in the field on the Mod-0 wind turbine. The Mod-0 wind turbine is designed to serve the wind energy program as a test bed for obtaining the necessary engineering and operation data for evaluation of promising components and subsystems.

100 kW Hingeless Metal Wind Turbine Blade Design

At the Second Workshop on WECS, R.E. Donham of the Lockheed-California Company, division of Lockheed Aircraft Corporation described the work done on designing and building the blades for the Mod-0 rotor: The NASA-LeRC wind energy program included the design and operation of a 100 kW experimental wind turbine generator (WTG) to be used to provide an engineering data base and to evaluate candidate components and subsystems. The Lockheed-California Company had the contract to design and build blades for the 100 kW WTG rotor. This contract provided for design, analysis, fabrication, instrumentation, and shake testing of three blades (set of 2, one spare) which meet the following NASA-LeRC requirements:

- Rotor diameter of 125 feet
- Operating speed of 40 rpm
- Collecting pitch range of blades, -3.5° to 92.5° at 0.75 radius

- 33.8° nonlinear twist on each blade
- NACA 23,000 series airfoils
- All aluminum construction

Background: The 100 kW experimental WTG includes a two-bladed, 125-foot-diameter, horizontal-axis, hingeless rotor system which drives a synchronous electric generator at 1,800 rpm through a step-up gearbox. The blades are designed to provide 133 kW of shaft power when rotating at 40 rpm in an 18 mph wind. The use of spars, ribs, stringers, the heavy nose skins, the light trailing-edge skins (thickness determined by avoidance of hail damage), the root-end fitting and the two heavy machined ribs is derived from conventional aircraft design/fabrication technology.

Figure 3.6 shows general structural dimensions for each blade. A D-spar structural concept was used which provides strength, rigidity, balance and low weight; these qualities are also consistent with rotary wing designs. It is also interesting to note that the rotationally generated centrifugal force field at the tip of most propellers is about 900 *g*, at the tip of most helicopter main rotors, 550 *g*, and for the wind turbine rotor, 34 *g*. These comparisons emphasize the similarity of the design/fabrication of WTG rotor blades to airplane wings.

Figure 3.6: 100 kW Wind Turbine Blade

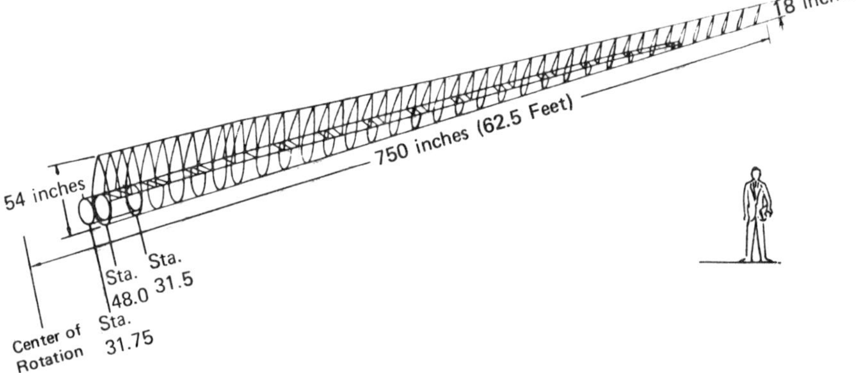

Source: NSF-RA-N-75-050

Blade design criteria were specified by NASA-LeRC:

(1) Thickness, chord distribution, and setting angle distribution shall be as defined in Figure 3.7.

(2) Spanwise aerodynamic loads were specified.

(3) The weight of each blade shall not exceed 2,000 lb.

(4) The center of mass for any blade section, outboard of radial station 301, shall be located forward of both the elastic axis and the center of pressure.

(5) Accessible locations will be provided to add weights to balance each blade.

(6) The maximum rate at which the power unit will be

rotated with changing wind direction is 0.33 rpm.

(7) The blade shall be designed to the following structural criteria:

The design goal is a life of 50,000 hours when operating at the design condition.

The blades shall be capable of operating in gusty weather.

The blades shall be capable of being feathered at design speed without structural failure.

(8) The blades shall be free of flutter.

(9) The blade shall be operable year-round during exposure to the elements customarily found at Sandusky, Ohio.

Figure 3.7: Blade Geometry

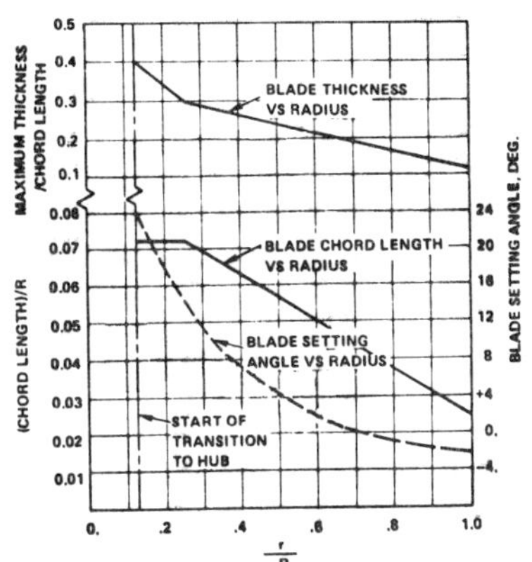

Source: NSF-RA-N-75-050

Lockheed's aeroelastic experience in the design, analysis, and test of a full-scale stopped-rotor model (13)(14)(15) showed that wind speeds which the blade can withstand when stopped are strongly dependent on the blade torsional rigidity as well as the bending rigidity when a quartering wind is experienced (loading the blade 0.75 chord). Selection of the D-spar concept aided in attaining the required torsional rigidity.

Hub design allows only the pitch change degree of freedom, as in the case of variable-pitch propeller systems. Hydraulic actuators provide the primary pitch control function with a pneumatic backup system. The stiffness of the primary control circuit is computed to be 170,000 in-lb/degree per blade.

Dynamics: In the design of a rotor system, detailed consideration must be given to dynamic characteristics. Items that generally require attention are frequency placement, flutter including wake effects, torsion-flap-inplane stability, stall flutter, gust response, and effect of tower support and control system characteristics on coupled rotor-tower dynamic stability, response, clearance, and transient loads associated with system response during starting and stopping.

The cantilevered blade frequency spectrum from 0 to 80 rpm is presented in Figure 3.8. These results are for the fully coupled flapping-inplane bending and torsion description. The results of a quasi-steady flutter analysis of the cantilevered blade at wind speeds of 18 mph and 50 mph up to a 100% shaft overspeed, which include a g = 0.02 structural damping level, show that the blade system is free from flutter. Nonrotating blade flutter analyses were made and showed no tendency to flutter or diverge.

Figure 3.8: Wind Turbine Cantilever Blade Bending-Torsion Frequencies vs Rotor Speed

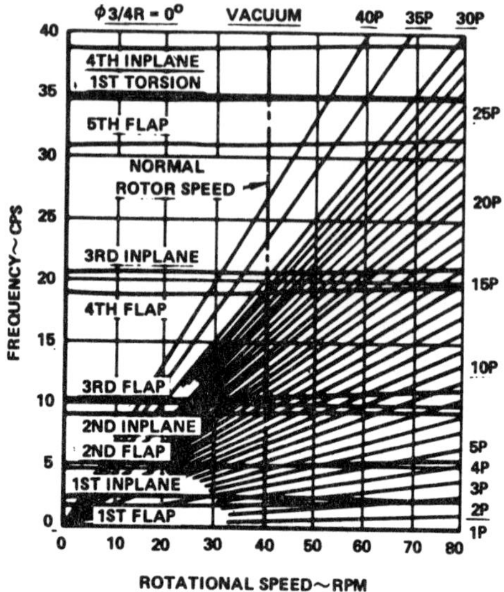

Source: NSF-RA-N-75-050

The tower was designed by the NASA-LeRC Facilities Engineering Division. In recognition of the importance of this experimental facility to the testing of future systems, the possibility of dynamic interaction between the rotor and tower has been minimized by designing a high stiffness and high natural frequency tower. A NASTRAN model of this system has been formulated and the results of the dynamic analysis show that the first bending natural frequency of the tower/rotor is 2.5 Hz (3.75 P at 40 rpm).

The elastic characteristics of the support shaft, bedplate, and tower have been coupled to the rotor system, and blade spectrum frequencies as well as flutter stability solutions obtained. Since the characteristics of the tower support are asymmetric and rotor is two-bladed, the system cannot be treated rigorously by stationary nonrotating system coordinates with constant coefficients, i.e., time variant equations must be solved. The influence of the elastic support on the rotating blade frequency spectrum and flutter stability margins was investigated using averaged stiffness characteristics. These properties were largely due to shaft flexibility. Cyclic and collective blade flutter eigenvalues were obtained with quasi-steady aerodynamics including tower, shafting, and bedplate averaged characteristics. The system is stable.

Comparison of Internal Blade Load Predictions: The Modular Stability Derivative Program (MOSTAB), developed by Mechanics Research, Inc. in 1970 under contract with the U.S. Army to analyze rotorcraft, was modified to MOSTAB-WT (wind turbine), (16). The MOSTAB-WT is being used by NASA-LeRC to predict loads for the 100 kW metal blades. MOSTAB-WT only uses the first mode flapwise bending of the rotor to predict internal blade loads. The Lockheed Rotor Loads Prediction Model was also modified to analyze wind turbines and the modified version is called WINTUR.

It is used by Lockheed to predict internal loads and displacements for the blades. The WINTUR program accounts for coupled flapwise and chordwise bending of the rotor blades. Therefore, the WINTUR program provides a more rigorous solution than MOSTAB-WT, but results from both programs compare well. All forces and bending moments have a 1 P frequency. The shear forces are small compared to the average centrifugal force of 22,700 lb, acting at the hub blade interface. The torsion moment at this blade location is small compared with the chord and flap bending moments. Tower shadow effects and vertical wind shear are included.

In addition to the conditions specified as minimal in the contract, other conditions were investigated which include the effects of:

- Rotor yaw angle
- Rotor yaw rates
- Rotor speed variation
- Blade pitch variation
- Wind ground shear gradient
- Windmill tower interference
- Wind velocity variations
- Start-stop cycles

Some of these conditions were treated grossly and, hopefully, conservatively. More analytic work on some of these should be done to enhance confidence. However, because of a conservative design condition with remote possibility of occurrence the blade is designed to have appreciably more structural capability than the 100 kW loadings associated with routine operation. It appears that this rating could be easily increased to 500 kW.

Rotor/Tower Interference: The static elastic deformation of the rotor tip-path plane with aerodynamic loading (power absorption) moves the rotor away from the tower structure, since the rotor system is downwind of the tower. The elastic deformation of the rotor plane due to gusts, sudden wind shifts, or control failures does, however, require consideration of the tower/rotor elastic deformations to permit assessment of clearance margins. The tower geometry is shown in Figure 3.9 for both the widest and narrowest aspects of the 100 kW truss tower. The static elastic equilibrium positions of the rotor blade including the 7-degree cone for the nominal 18 mph and 40 rpm operating conditions over the blade pitch-angle range -6 to +6 degrees are shown in Figure 3.10. These deflection gradients, resulting from angle change when combined with transient-load amplification considerations, indicate that this system has substantial rotor/tower interference margins. In general, feather rate and wind velocity must be known to determine whether this condition is critical for blade strike.

Figure 3.9: Tower Geometry and Clearance of Wind Turbine System

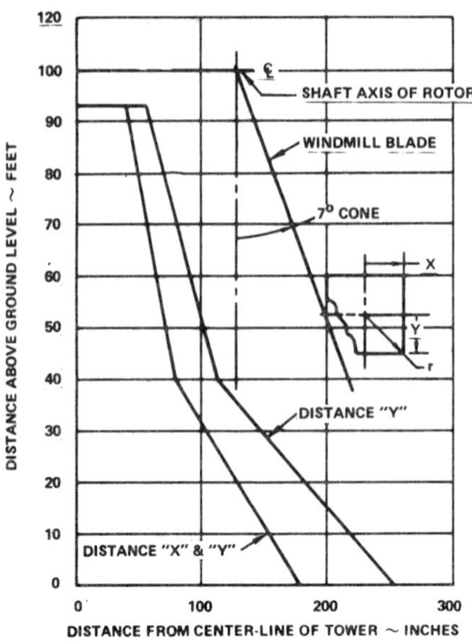

Figure 3.10: Wind Turbine Steady Blade Flap (Cone and Elastic) Displacement

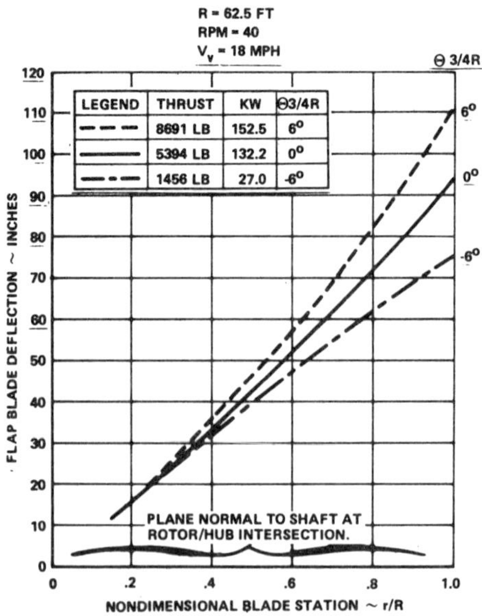

Source: NSF-RA-N-75-050

Several alternatives or tradeoffs exist for alleviating potential tower/blade interference:

- Increase blade stiffness
- Change blade twist
- Increase hub precone
- Increase rotor overhang from tower
- Pitch rotor pylon assembly away from the tower
- Precone blade away from feather axis in direction of hub precone

A preliminary investigation was conducted to determine the influence of geometric blade twist on blade-deflection characteristics for two rotor power output settings using the WINTUR program. The analysis was based on the mass and stiffness data for a composite nonmetallic blade and the blade planform and taper of the 100 kW metal windmill blade design. The data show a shift in the mean displacements between the two twist configurations for the power settings selected and a dramatic increase in the overall blade tip deflection for the untwisted blade. These data show that there is a greater probability of tower strike when the root bending principal axes are not twisted. Preliminary analysis indicates that the flapping stiffness to preclude blade-tower strike from occurring will require frequency above 2 P.

Fabrication: The fabrication of the metal blades derives from conventional aircraft technology. Lockheed's experience with airplanes and helicopter rotors was directly applied. The blade consists of thick aluminum skins and a large number of formed ribs, and conforms to a prescribed aerodynamic shape which has a large amount of twist and taper.

The taper and twist are such that straight-line elements along the blade span are not located at constant percent chord locations along the span. Identification of straight-line elements was essential to the design (locations of longerons) and to the procedure for forming curvatures on the thick leading-edge skins.

Structural Tests: Nonrotating structural tests have been conducted on the first metal wind turbine blade mounted on a test fixture. This support fixture simulated the hub/spindle stiffness of the 100 kW experimental wind turbine generator. The test/analysis frequency summary comparison presented in Table 3.2 shows that excellent correlation was obtained. The frequencies of the flapping modes and the inplane modes are both slightly higher than calculated which is attributed largely to higher blade bending stiffness levels being obtained. The first torsion mode was slightly lower than predicted, but for a 2,000 cpm mode this is good correlation. It is concluded that the mass and stiffness distribution properties calculated by computer graphics were an unqualified success.

During the calibration test loading of the strain gage instrumentation, deflection measurements were taken concurrent with loading applications. Structural influence coefficients determined from these measurements were compared with theoretical prediction. These results show excellent correlation with the flapping and inplane measurements. The flapping is 1.06 times stiffer than theory and the inplane is 1.10 times stiffer than theory at blade station 301. These values when compared with the nonrotating first blade mode frequencies would explain the higher frequency results obtained between test and theory on Table 3.2.

Table 3.2: Metal Windmill Blade Frequency Spectrum Determined by Nonrotating Tests

Mode Shape	. . . Test/Analysis Frequency Summary (cpm) . . . Full Tip Weight* 7.4 lb/Blade (test)	. . . No Tip Weight* . . . (test)	(analysis)
1st flapping	97.8	103.8	98.7
2nd flapping	280.8	299.4	286
3rd flapping	600.0	622.8	610
1st inplane	141.0	159.6	143
2nd inplane	567.6	588.0	558
1st torsion	not tested	1,968	2,040

*Metal blade was primed but not through final paint; surface area weight would increase by 25 lb/blade.

Source: NSF-RA-N-75-050

The first metal windmill blade was weighed at 1,981.3 lb on April 22, 1975, just prior to trailer shipment to NASA-LeRC. This compares well with 1,955.5 lb predicted by analysis. The weight increment due to final painting is believed to account for the 25.8 lb above the prediction. The spanwise and chordwise location of the blades c.g. measured at Station 270.18 versus Station 269.97 (calc) and 1.26 inches aft of 0.25 chord versus 1.13 inches (calc) is also in good agreement with the theory (5).

Conclusions and Recommendations: The metal wind turbine blade-frequency spectrum provided good separation between natural vibrating frequencies and excitations from nP higher harmonic aerodynamic forces as well as the 1P gravity field excitation. Flutter stability analysis results showed margins for the rotating blade in excess of 100% shaft overspeed for both cantilever and elastic (tower) rotor support boundary conditions.

The stability results of bending-torsion divergence (including quartering wind conditions) and stopped-rotor flutter analytic studies to 140 mph indicated that the aeroelastic properties of this design provide wide latitudes with respect to its intended use and expected environment.

Loads correlation between MOSTAB/WINTUR programs indicated that the blade loadings can be reasonably predicted for known flow field.

Ground vibration tests as well as carefully controlled powered whirl tests should be conducted to verify system characteristics and stability within the operating limits.

Real-time correlation of the test results with analysis should be emphasized to further ensure an orderly, safe development of this first 100 kilowatt unit.

Dynamic Blade Loading Study

The blades of large wind turbines are continually collecting, amplifying, and transmitting dynamic loads. These loads may be aerodynamic, gravitational, or inertial in origin. Whatever their source, dynamic blade loads influence the weight and cost of almost all structural components in a wind turbine. Therefore, it is important to develop and improve methods of predicting and controlling dynamic loads in wind power systems. Some progress has been made in understanding dynamic loads in large, two-bladed horizontal-axis wind turbines.

Two ERDA/NASA wind power systems were investigated: (1) the 100 kW Mod-0 system, an experimental machine in operation at NASA's Plum Brook Station near Sandusky, Ohio, and (2) the 200 kW Mod-0A system, being assembled for installation as part of the municipal power plant of Clayton, New Mexico. The two wind turbines are almost identical in outward appearance. Both have two-bladed downwind rotors 125-feet in diameter and are mounted on truss towers with shaft elevations of 100 feet.

The design, fabrication, installation, and initial testing of the Mod-0 wind turbine are described in references (3)-(5). The Mod-0 wind power system develops its rated power of 100 kW at a wind speed of 18 mph, while the Mod-0A system is designed to produce 200 kW of power at a wind speed of 24 mph. Both machines have rotor speeds of 40 rpm.

There are several structural differences between the Mod-0 and Mod-0A wind turbines. Mod-0A blades are stronger and somewhat heavier than Mod-0 blades (2,300 lb versus 2,000 lb). The Mod-0A tower is an all-pipe truss while only the legs of the Mod-0 tower are pipes. The Mod-0A gearbox, high speed shaft and generator are all larger than the same equipment in the Mod-0 power system.

The Mod-0 wind turbine first achieved its rated speed and power in December of 1975. While its performance was generally as predicted, bending loads measured near the blade roots were very high. At wind speeds of 35 mph, bending moments were measured which were two to three times as large as expected. This was true for both flatwise (out-of-plane) and edgewise (in-plane) moment loads. Continued operation at these high levels of load would have resulted in early fatigue failure of the blades. An intensive study of the data recorded during this brief period of operation in December, 1975, was undertaken to (1) determine the causes of these excessive blade loads and (2) recommend modifications to reduce loads.

Results of Vibratory Loads Study: Flatwise Blade Loading—The dynamic loading of the Mod-0 blade in the flatwise direction is dominated by the impulse applied to the blade each time it passes through the tower's wake. It was concluded that the Mod-0 tower was blocking the airflow to a much greater degree than expected and that this was the cause of the excessive flatwise blade loads.

It was recommended that the service stairs and elevator rails originally installed in the tower be removed. Wind tunnel tests of scale models of the tower with and without stairs confirmed that the blockage was very high with stairs and was greatly reduced when they were removed.

The average velocity reduction $\Delta\overline{V}$ over the width of the tower was determined for various tower orientations and elevations. Tower blockage was then calculated as the ratio of this average velocity reduction to the free stream velocity V_0. Removing the stairs reduced the blockage from 0.64 to 0.35. As a result flatwise blade loads would be reduced by about one-third. To further reduce blockage, the Mod-0A tower should be fabricated using structural pipes for all truss elements. The use of round sections throughout the truss produces a small but significant additional reduction in tower blockage.

Edgewise Blade Loads—Examination of the harmonic content of the excessive edgewise blade loads led to the conclusion that these loads were caused by nacelle yawing motion. As originally constructed, the nacelle was connected torsionally to the tower by means of a single yaw drive shaft. It was determined that the spring constant of this shaft in torsion produced a yawing resonance which resulted in large lateral motions of the rotor hub. These motions in turn were assumed to cause edgewise blade loads in a manner analogous to that known for airplane propellers.

To reduce edgewise loads it was recommended that the single yaw drive be replaced by a dual drive system. Nacelle motions would then be reduced, because of three factors: (1) avoidance of a resonance, which was of greatest significance, (2) stiffening the nacelle-to-tower connection, and (3) eliminating the free play and nonlinearity present in the single yaw drive system. A pretorque of 50,000 lb-in was recommended for the dual yaw system. In this way the rotation caused by a nominal yaw moment of 80,000 lb-ft would be reduced by 60% from 0.42° to 0.16°.

Effects of the Mod-0 Structural Modifications: Removal of the service stairs and elevator rails and installation of the dual yaw drive were completed in April, 1977.

Flatwise Moment Loads—Reductions in load are substantial, varying between one-third and one-half. In addition, the predicted load levels as a result of structural modification were verified.

Edgewise Moment Loads—The structural modifications decreased both the mean loads and their variations dramatically. This is particularly evident when the gravity cyclic load is recognized as the minimum possible value. Predictions of edgewise load were found to be somewhat conservative.

Mod-0A Blade Loads: The load calculation methods verified by the data obtained above were used to predict blade loads for the Mod-0A wind turbine. Accordingly, the strengthened Mod-0A blade should be free of fatigue damage from flatwise loads over almost all of its operating range. The same is true for edgewise cyclic loads, according to the predictions.

Investigation of Excitation Control for Generator Stability

As discussed in the previous section, high-speed horizontal-axis two-bladed wind turbine generators (WTG) with blades on the downwind side of the support tower are being developed by NASA/ERDA. This design has advantages of high aerodynamic efficiency, high output power, and using materials efficiently. It does, however, require special design considerations to handle disturbances introduced by the flow wake behind the tower. Aerodynamic forces that change as the blades travel through the wake can result in high mechanical stresses and electrical power fluctuations. The problem has been noted in the literature (17) and observed during certain operating conditions of the NASA-Lewis Mod-0 WTG.

Several schemes for adding damping to the drive train system and/or reducing the band width of the drive train system in order to suppress the effects of input torque disturbances have been proposed. This report investigates the possibility of adding damping to the system by controlling the generator exciter. Effective damping will result if torque can be applied to the generator rotor with a phase relationship such that the torque opposes changes in the generator's power angle. Because the effective time constants are small in excitation control loops, it was assumed that a large control effort could be expended with relatively small input of control energy (18). This technique was encouraged by successes obtained in the 1960s that improved power system stability of large utiiity networks.

Experiments and analyses of the NASA Mod-0 WTG system have been made to determine benefits that might be obtained by using the generator exciter to provide system damping.

Two conclusions were obtained from the investigation. The mathematical analysis disclosed that the first-mode amplification factor (first resonance) of the Mod-0 WTG can be reduced but not eliminated by adding damping to the generator. Secondly, it was clearly demonstrated by response tests that any advantages for excitation control to obtain improvement in system dynamics would be completely cancelled by the accompanying excursions in reactive power. Thus it was determined that excitation control is not a suitable method for suppressing Mod-0 WTG power oscillations caused by the flow wake behind the tower.

SEGMENTED AND SELF-ADJUSTING ROTORS

This report by P.F. Jordan and R.L. Goldman of Martin Marietta Laboratories (COO/2613-2) examines the feasibility of utilizing an aeroelastic approach for providing a wind turbine generator (WTG) with a means of automatically adjusting to changes in wind speed. Operation of a WTG at nearly constant rpm is sought in order to facilitate the incorporation of the generator into a conventional electric power network.

The study was directed at developing a design concept for a WTG rotor blade that eliminates the need for the mechanical pitch change mechanisms previously required for large wind turbines. The specific task was to investigate the possibility of utilizing aeroelastic effects to control the blade forces acting on the rotor at varying wind speeds.

The key difference between a standard WTG rotor blade and an aeroelastically-active blade is that the former is essentially rigid in torsion and has a root pitch angle that is either fixed or mechanically controlled whereas the aeroelastic blade is flexible in torsion such that the pitch angle of each blade element responds to the aerodynamic moment acting about the blade's elastic axis.

A major facet of the study was an exploration into how the rotor response might be beneficially influenced by providing an appropriate aeroelastic feedback, that is, an optimum form of pitch response to a given aerodynamic moment.

Initial phases of the investigation involved the preparation of a number of rotor performance analytical programs. In these the Glauert (19) blade element theory was used to evaluate the aerodynamic forces. Exploratory studies examined rotor performance under various pitch-determining laws. It became evident early on that rotor blades having a flexible spar of conventional structural design would not meet the study objectives.

Specifically, torsional deformations of such blades would not be large enough nor would they yield the desired pitch distribution along the blade axis. From these considerations, a concept of a local pitch compensation mechanism evolved. Essentially, this concept assumes a straight blade formed by a spar, which is substantially rigid in torsion. Although basically free to rotate about the spar, each individual blade element is subject to a pitching moment applied to the blade by a compensating mechanical spring. The actual pitch angle of the element will then be the angle that provides an equilibrium between the external aerodynamic moment and the compensating internal mechanical spring moment.

Parametric studies and performance analyses were conducted to evaluate the pitch compensation concept. Initially a fictitious model of a smooth, continuous blade with a continuous distribution of mechanical springs was studied, followed by an investigation of a more realistic model using a segmented rotor blade. The segmented rotor is one in which the blade shape is formed by a few light but rigid shell-type, airfoil-shaped segments arranged along the blade axis. Each segment is connected to the blade spar by bearings and each contains its own compensating spring mechanism.

In addition to performance evaluations for a number of such blade models, preliminary vibration and flutter analyses were carried out to examine the stability and response characteristics of representative configurations.

Within the constraints of Glauert's blade element theory and under certain qualifying analytical assumptions, the theoretical performance characteristics of WTG rotor blades with unique aeroelastic properties have been examined. In particular, a parametric study of a segmented rotor blade design with internal variable-moment pitch compensation has been conducted. This concept shows the promise of obtaining highly desirable performance characteristics for a WTG system operating in a varying wind field. The design concept investigated envisioned a rotor blade in which no mechanical pitch changing mechanism would be required.

The analytical studies indicate the feasibility of developing a segmented rotor blade whose torque is nearly invariant with wind speed and whose drag decreases markedly with increase in wind speed. Vibration and flutter analyses show that the individual blade segments are well damped in pitch, a characteristic that would help a segmented rotor to respond smoothly to wind speed variations. Although flapping flutter was shown initially to be a possibility, this adverse characteristic was eliminated by judicious design changes to the tip segment.

This study, although limited in scope, demonstrates, theoretically, the feasibility of utilizing aeroelastic blade properties to provide rotational speed control for a WTG. The potential operational advantages which might accrue to this concept would indicate that additional design effort leading to a modest experimental verification program would be an appropriate objective for a next phase program. A typical experiment, for example, could involve a scaled, elastic rotor operating in a controlled, variable wind environment.

Although the object of this investigation was not to design a rotor system, but rather to explore conceptually the basic concept of an aeroelastically self-adjusting rotor blade, it seems appropriate to comment on some aspects of the design. Both this concept and the more conventional one have features which complicate practical implementation.

The conventional rotor has a single long twisted and tapered blade with an active root pitch changing mechanism. The self-adjusting blade would be made of several straight untapered sections and would have no root pitch mechanism. On the other hand it has a set of bearings and springs in each section to allow the blade to adjust and the various sections must be fitted together to adequate tolerance. The relative costs of manufacture and maintenance in the two cases are not obvious and require detailed analysis.

COMPOSITE BEARINGLESS ROTOR CONTROL

Much of the work in recent years in the field of wind energy conversion systems has been related to investigations of total system concepts and the general energy producing capability of rotors designed around current state-of-the-art aerodynamics. The general theme which one obtains from these investigations is that wind energy systems must be developed, installed, and operated at costs that are competitive with alternate forms of energy conversion. To achieve this, all high cost aspects of such systems must be identified and dealt with as efficiently as possible before proceeding to full-scale development. It is believed that significant progress can be made in this regard in the area of the principal component of a wind energy conversion system, the rotor itself.

A major activity in rotary wing research in recent years has been directed at the task of reducing costs. Significant progress has been made to achieve lower costs through the application of new rotor concepts, improved manufacturing and construction techniques, and the use of advanced composite materials. The experience gained in these areas can now be applied to other rotor or propeller systems, such as wind energy conversion systems, to achieve significant improvements in life cycle costs over that which would be expected for systems based on technology of a few years ago.

In addition to cost reductions available through general advancements in rotary wing state-of-the-art, potential improvements specifically related to wind energy conversion systems become apparent. These are related to the control of such systems to maximize efficiency over the full range of operating conditions, which requires control of rotor axis alignment, rotational speed, and blade pitch. Further controls are desirable to guard against blade or tower failures resulting from gusts and excessive wind velocities. It is proposed that most or conceivably all of these controls can be achieved automatically (aerodynamically or mechanically) without the need for exotic sensing mechanisms or energy absorbing subsystems which would increase costs and reduce reliability. The specific rotor system proposed is the composite bearingless rotor (CBR) which has been under development at the United Technologies Research Center (UTRC) and the Sikorsky Aircraft Division. Full-scale main and tail rotors have been fabricated and tested and 30% reductions in weight have been achieved and cost reductions up to 30% have been projected for production designs. The composite bearingless rotor particularly lends itself to innovative concepts in automatic control by virtue of its aeroelastic characteristics.

According to a report given at the Second Workshop on WECS by M.C. Cheney, Jr. of United Technologies Research Center, the objective of their research is to evaluate the potential benefits the composite bearingless rotor can provide in minimizing costs of wind energy conversion, particularly as related to self-regulating controls designed to maximize efficiency under all wind conditions and to protect from dangerous overloading. The research program would consist of a preliminary design of a full-scale rotor, stress and loads analysis of this design, fabrication of a dynamically scaled model incorporating the capability to examine various automatic control concepts, and wind tunnel tests of this model to demonstrate the basic feasibility of the composite bearingless rotor as a wind energy rotor as well as to evaluate various control concepts.

The sketches shown in Figure 3.11, presented as an assembly sequence, depict the rotor in its basic components: spar, airfoil, control torque tube, and hub. The spar assembly shown in Figure 3.11a consists of fiber reinforced composite material with the fibers aligned unidirectionally, which provides relatively high bending stiffness but low torsional stiffness. A variety of fibers are commercially available for this application including glass, graphite, boron, and Kevlar.

A four-bladed rotor consists of two spars stacked one on the other as shown in Figure 3.11. The airfoil surfaces are shown in Figure 3.11b where composite materials are also employed, however here the fibers are embodied in a $\pm 45°$ arrangement to provide maximum torsional rigidity. The volume within the airfoil, between the cover and the spar, can be filled with lightweight foam or honeycomb. This serves to transmit the thrust and centrifugal shear loads from the cover to the spar. At this stage in the assembly, the blades are comprised of torsionally rigid outboard sections supported by torsionally flexible inboard "flexbeam" sections. Alternate configurations, which would eliminate the problem of fabricating extremely long spars and not effect this required stiffness distribution, would incorporate a simple joint between the outboard end of the flexbeam and the blade proper.

Pitch control of the blade is achieved by applying a moment at the outer end of the flexbeam, which is accomplished by extending the torsionally rigid cover inboard forming a control torque tube as shown in Figure 3.11c. Deflections are then applied at the root end in a conventional manner. Finally, the blades are held together with a hub that simply consists of two plates bolted together as shown in Figure 3.11d. The hub is not required to react with the centrifugal force of the blades since the carry-through feature of the flexbeams allows opposite blades to resist each other. The hub merely supports the blade thrust loads and transmits the torque to the energy receiving system.

Figure 3.11: Composite Bearingless Rotor

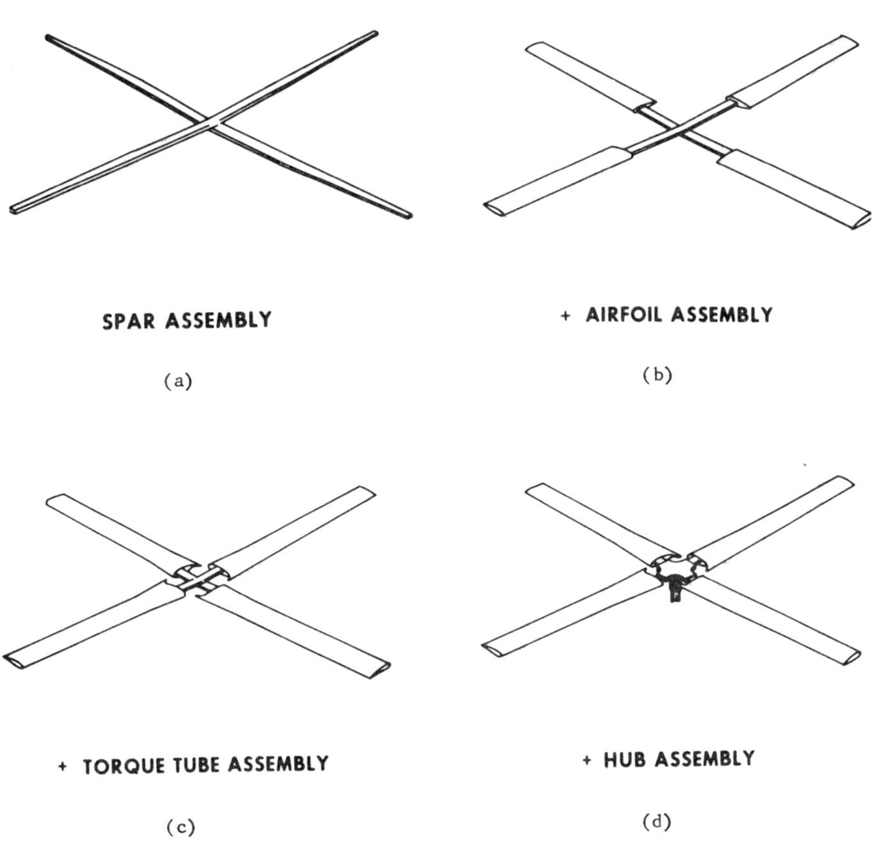

Source: NSF-RA-N-75-050

Rotor Controls

A key element of the composite bearingless rotor (CBR) concept is the inboard flexbeam region which represents a torsion spring against which blade pitch changes are imposed. Centrifugal sensing mechanisms to produce pitch changes can be employed in such a system without providing external springs. Two such devices will be examined to determine their feasibility in providing pitch and rotational speed control to (1) achieve minimum variations in rpm over the nominal wind speed range and (2) to provide rpm and pitch adjustments with variations in wind speed to attempt to achieve optimum performance. A further characteristic of the composite bearingless rotor is its capability to provide certain pitch-flap coupling phenomena so that blade bending responses caused by changes in wind speed can be reduced by directly coupling with pitch. This coupling phenomenon can be produced by utilizing the relative motion between the flexbeam and torque tube (similar to δ_3 coupling of a conventional articulated helicopter blade).

The basic rotor axis control is achieved by virtue of the natural tendency of a hingeless rotor to pitch away from an angle of attack change. The support tower to be fabricated for the model test will consist of a free pivot to allow the shaft axis to respond to rotor moments. A diagram of the rotor and support are shown in Figure 3.12. The rotor would continue to respond until its axis and the wind direction are parallel. Without cyclic pitch control, a hingeless rotor cannot experience an inplane component of flow without producing rotor moments. These moments will exist until the inplane velocity goes to zero which is the point at which the rotor axis is parallel to the wind.

The moment results from the fact that blades advancing into the inplane component of wind experience a higher dynamic pressure and therefore produce higher lift. The rotor hub senses this lift increment at some finite time later due to the dynamic lag of the blade. The net result is a moment which will rotate the axis until it is parallel to the free stream. The rotor can also be viewed as a gyroscope which precesses about an axis perpendicular to the moment axis. The rate of axis adjustment would be relatively low due to the low rotation speed and high inertia, and therefore rapid turn rates, due to gusty conditions, would be avoided. An additional moment is also produced in a direction perpendicular to the pivot axis and is resisted by the tower.

This moment also disappears when final axis alignment is achieved. The model will be designed to position the rotor downstream of the tower since this would represent the most stable configuration.

Figure 3.12: Rotor Axis Control

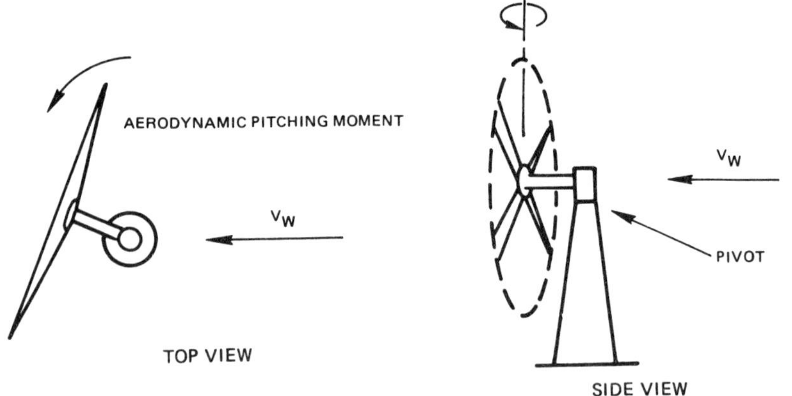

Source: NSF-RA-N-75-050

The model rotor will be designed to allow a variety of pitch control concepts to be examined. Two methods of sensing centrifugal force and coupling this with the blade will be tested. Figure 3.13 shows two tentative approaches. The upper configuration incorporates a pendulum supported at the hub where the weight is out of the plane of the rotor under static conditions.

As rpm increases, the weight responds to the centrifugal force and rotates against the flexbeam spring towards the plane of rotation and in so doing, increases blade pitch. The lower figure shows a centrifugal control system where the masses are built rigidly into the blade torque tube. They also respond to centrifugal loads and increase pitch in a similar manner. Such devices would be used to provide a pitch schedule versus rpm in an attempt to achieve maximum average power for all wind speeds, or used to achieve minimum rpm excursions.

The model blades shall be capable of freely feathering by providing a joint at the root or outboard end of the flexbeam which can be manually decoupled to free the blade of pitch restraint. Such a device could be incorporated to withstand severe wind conditions and could conceivably be activated by a specified increase in rpm or a remote triggering signal.

Figure 3.13: RPM-Pitch Coupling Mechanisms

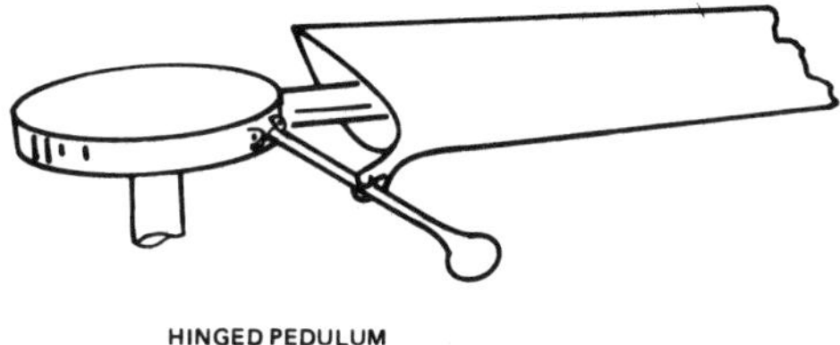

HINGED PEDULUM

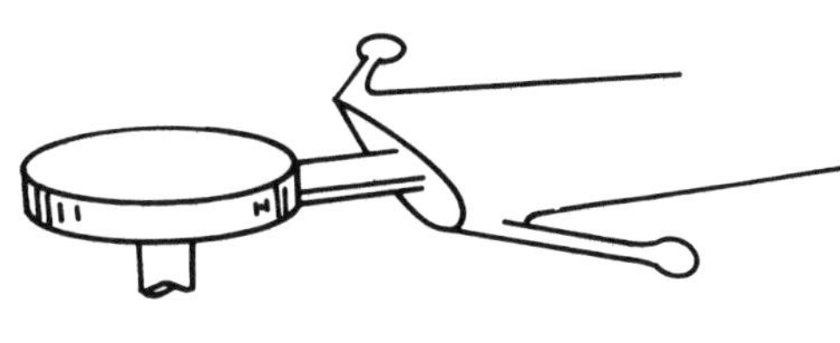

FIXED PENDULUM

Source: NSF-RA-N-75-050

Experimental Program

Model Design: Before a dynamically scaled model could be designed, a typical full-scale wind turbine had to be selected so that its stiffness characteristics could be realistically scaled. The design selected was similar to the 125-ft Mod-0 turbine constructed by NASA. The important nondimensional frequency and mass parameters were maintained for the model design to ensure that the response characteristics would be equivalent.

Full-scale tip speeds and wind speeds were retained; however, gravity effects could not be accurately simulated.

In keeping with the size restrictions of the low speed wind tunnel scheduled for this experiment, the model diameter was set at 4.5 ft. Four blades were used and consisted of a carbon/epoxy spar and inboard flexbeam with the aerodynamic portion constructed of balsa wood. Tip speeds up to 300 ft/sec were anticipated.

Model Tests: The model tests consisted of four basic control configurations: (1) fixed pitch without pitch-flap coupling, (2) fixed pitch with pitch-flap coupling, (3) control pendulum fixed to the outboard end of the flexbeam, and (4) control pendulum supported at the hub.

The fixed pitch configurations produce performance and stress results typical of conventional designs and were tested primarily to obtain baseline data. The blade supported pendulum configuration resulted in a low torsional frequency, approximately equal to the rotational frequency, and therefore was easily excited by once-per-revolution excitation such as that produced by the tower wake. The tests of this configuration indicated questionable stability and therefore were discontinued. The final configuration, the hub mounted pendulum, achieved the self-starting and self-regulating features desired, and it is only this configuration, which will be discussed.

In the pendulum control mechanism, the weights are mounted so as to form a pendulum with a natural frequency of approximately 1.5 times the rotational frequency. The flexbeam torsion spring and the spring due to pendulum flapping are continuously in equilibrium via a steel connector strap. The connector can flex to absorb the changes in distance between the pendulum and the end of the flexbeam due to coning angle and rpm. When the pendulum is mounted statically on the downwind side of the turbine disk, a negative sign for $d\theta/d\Omega$ results, producing an rpm-governor type mechanism. With the pendulum located at the upstream side of the turbine disk, a positive $d\theta/d\Omega$, or self-starting configuration is obtained.

The effects of both types of coupling on the performance of the wind turbine are shown in Figure 3.14. In the rpm-governor configuration of the pendulum, the tip speed variations with wind speed are less, as expected, than those for a fixed pitch condition. In contrast, the tip speed for the self-starting configuration increases more rapidly with wind speed. Figure 3.15 shows the time history of the self-starting configuration at a wind speed of 25.8 mph where the static pitch angle was set at -31°. The turbine is initially yawed -90°. It is then released at t = 0, and initially the turbine weathervanes to the downwind position in a second order fashion.

The tip speed then increases, and after about 2 seconds the pitch angle, under the action of the pendulum motion, begins to increase. This causes the continued acceleration through the startup. The torsional stresses change proportionally with the pitch angle. The edgewise and flatwise stresses show, at 14 seconds after release, a slight amplification at 6 P which is caused by the coalescence of the first edgewise and the second flatwise natural frequencies. However, the amplification is relatively mild and is quickly passed through.

Figure 3.14: Performance of Hub-Mounted Pendulum Configuration

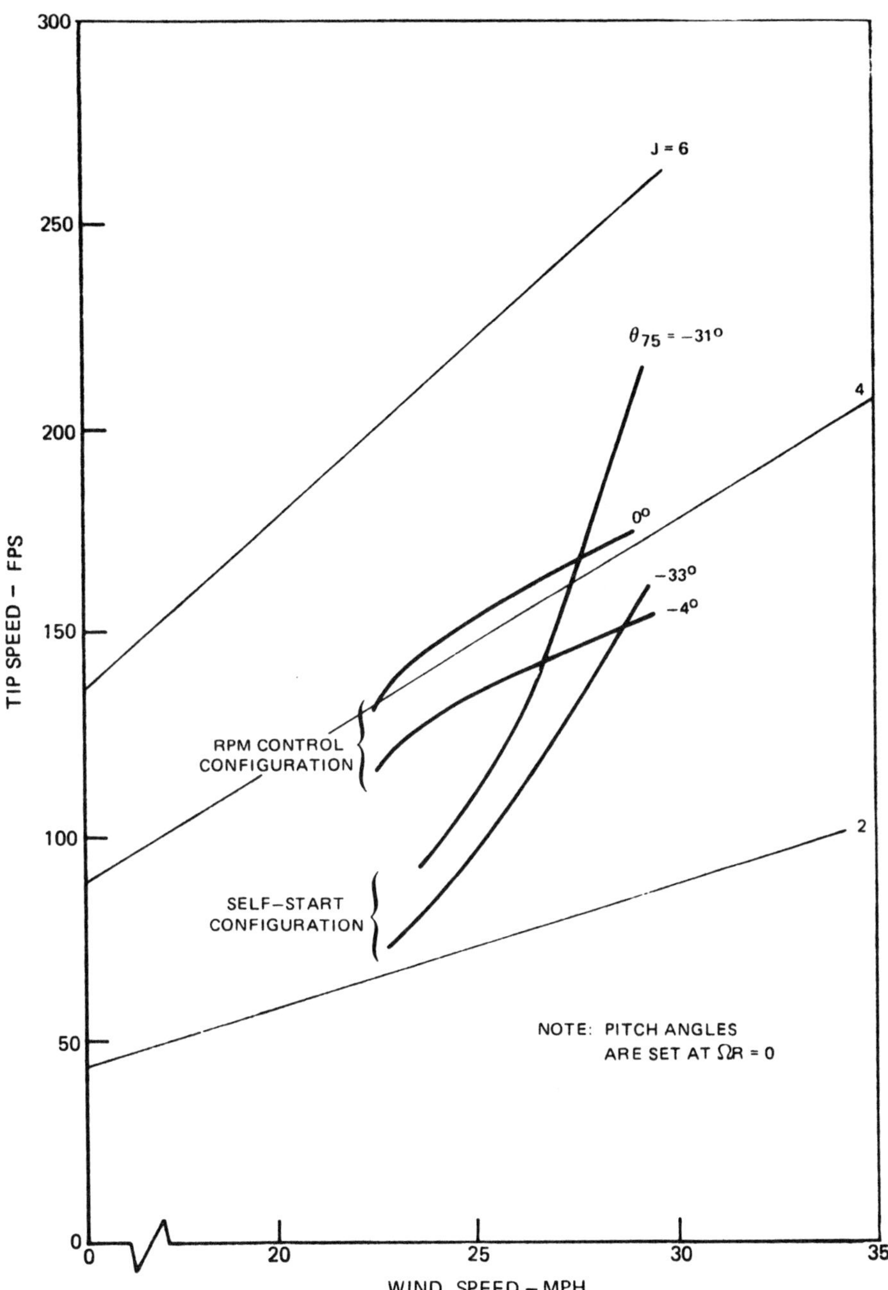

Source: COO/2614-76/2

Figure 3.15: Time History of Startup in Self-Starting Configuration

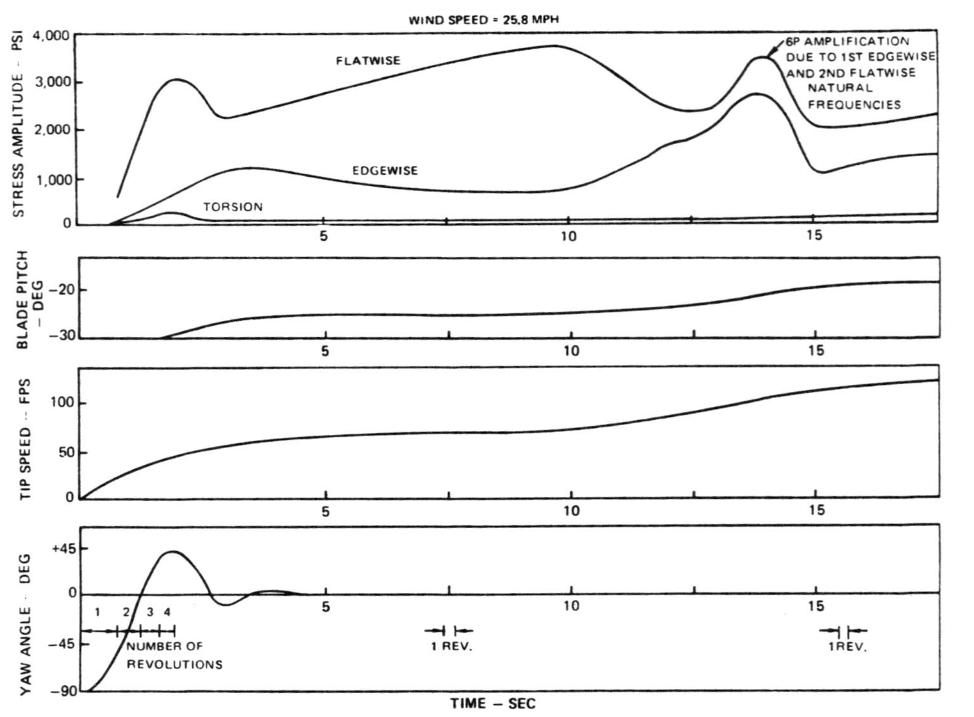

Source: COO/2614-76/2

The capability of the turbine in the self-starting configuration to respond to sudden wind direction changes is shown in Figure 3.16. The turbine, rotating at its high rotational speed equilibrium is released from a preset yaw angle. In a first order fashion the turbine tends to realign with the wind. The initial flatwise stress amplitude is due to the large helicopter-type cyclic input because of the initial yaw angle. The results demonstrate that the CBR wind turbine in its self-starting configuration is self-aligning to the wind, self-regulating in rpm by means of automatic pitch changes, and has low vibratory stress levels during normal maneuvers of startup and alignment.

Figure 3.16: Response of Wind Turbine to Simulated Abrupt Wind Change

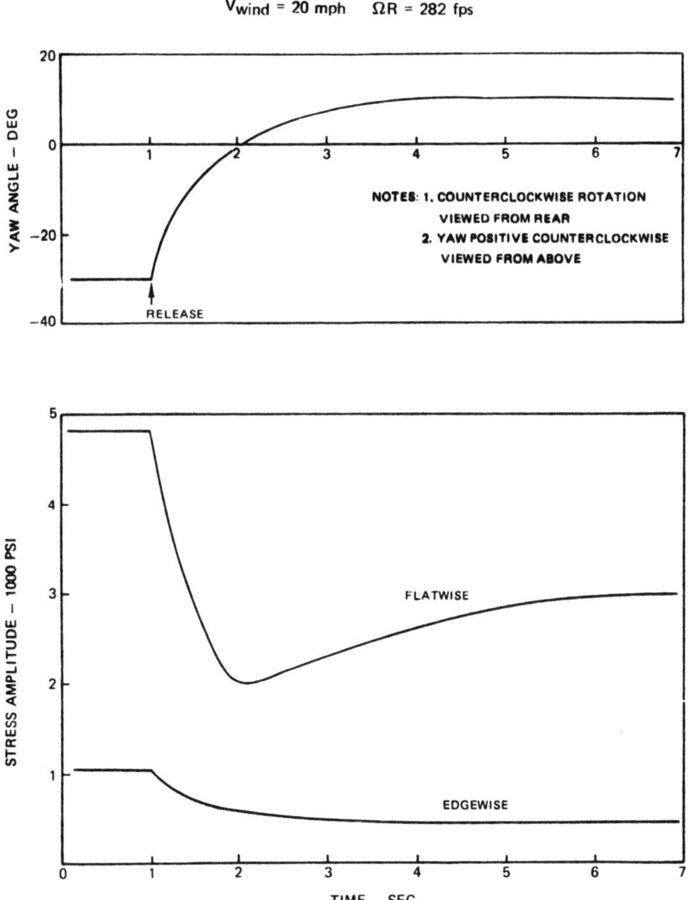

Source: COO/2614-76/2

Conclusions: The conclusions set forth below were arrived at as a direct result of the model test program conducted. Since the model was dynamically scaled it is expected that the conclusions are applicable as well to full-scale designs. However, some variations in the full-scale characteristics might be expected as a result of the Reynolds number and Froude number difference, but the conclusions reached should generally be applicable. Reference to normal wind speeds means speeds up to approximately 30 mph.

(1) The composite bearingless rotor (CBR) can be used in conjunction with a hub-mounted pendulum to produce a self-regulating stable wind turbine. Self-regulating is defined as self-starting with an automatic pitch-rpm schedule to provide efficient power output at all reasonable wind speeds.

(2) Wind turbines having hingeless blades of conventional stiffness and free to yaw, have inherent dynamic moments which tend to maintain alignment of the turbine axis with the wind.

(3) The CBR wind turbine in its self-regulating configuration operates well below stress allowables for normal wind speeds. This characteristic is independent of wind direction, initial yaw position, and initial rpm, and encompasses transient as well as steady operation.

(4) For a free yawing wind turbine designed for operating downstream of the support tower, the rotor drag when nonrotating is normally sufficient to align the turbine to the wind for any initial wind direction.

(5) The equilibrium yaw position for a free yawing wind turbine operating downstream of its support tower is skewed from the downwind position. This is due to unloading of the downward blade caused by the reduction in wind velocity behind the tower.

(6) The CBR wind turbine with fixed pitch performs as a conventional wind turbine.

(7) Blade stress levels are well below allowables when fully feathered and nonrotating at wind speeds up to 60 mph.

(8) A CBR wind turbine with a self-regulating pitch control comprised of a pendulum fixed to the blade which produces a torsional frequency near the rotor speed (1 P), is unstable.

OPTIMIZATION AND CHARACTERISTICS OF A SAILWING WINDMILL ROTOR

The Princeton Jeep Windmill Testing Facility

Early in the efforts of the windmill research program it became apparent that atmospheric testing was unsuitable for the optimization and rapid development of rotor performance that was the goal of the Princeton undertaking. While such techniques are suitable, and even preferred, in the field testing of wind turbines, where such factors as the machine's dynamic response to wind velocity fluctuations can have a dramatic effect on the net power generation when integrated over a long period of time, such techniques are almost useless when an accurate determination of the effect of a single configuration change on rotor performance is desired in a relatively short time period. Thus, the development of a moving-vehicle windmill test-bed, a full-scale wind machine mounted on the front of a standard World War II Army Jeep, was begun. The gradual evolution of this facility has now reached a point of sophistication where a more detailed description is justified. Due to both the reasonably low scatter and the high repeatability of the data collected using the Jeep windmill testing facility in its current stage of development, it is doubted whether any significant improvements that are within a reasonable cost are possible.

Description of the Test Facility: Central to the performance determination of any wind turbine is an accurate measurement of the undisturbed free-stream wind velocity. In the case of the Princeton Jeep facility, a number of unsatisfactory anemometry systems were tried before the successful type was employed. These attempts included both mechanical vane and hot-wire types until the "home-made" cup-type anemometer, which was ultimately successful, was installed. The overall problem in this application is that in addition to a fairly high degree of sensitivity and accuracy, a reasonably steady meter reading is required such that it can be easily tracked by the driver to match the speed of the vehicle to the wind velocity desired.

Both the mechanical and the hot-wire types proved to be too susceptible to road-bumps to allow accurate speed control. The cup-type, on the other hand, is insensitive to bumps and gives an extremely steady reading, yet, due to its lightweight, halved ping-pong ball type construction, responds quickly to changes in velocity.

In an attempt to remove the anemometer as much as possible from the disturbing influence on the flow-field caused by the presence of the Jeep, the anemometer is mounted a considerable distance in front of the vehicle. The cup portion of the instrument is used to drive a small generator whose signal is transmitted to a milliammeter mounted on the Jeep's steering column within easy view of the driver. The initial calibration of this system was accomplished in the wind-tunnel and, in order to prevent errors, is verified before each series of tests by recording the time required for the Jeep to travel a given distance and comparing the resulting calculated velocity to that indicated by the instrument. This procedure has the additional purpose of indicating whether or not the natural winds are within the limit at which meaningful data can be obtained.

Throughout the development of the Jeep windmill testing facility, several methods of loading the windmill were tried, the last and most successful being that of a hydraulically actuated motorcycle disk-brake. The master cylinder and actuating mechanism for this system are mounted at the base of the supporting tower drag brace and operated by a person sitting in the bed of the Jeep. The brake is applied by turning down the turnbuckle. The springs allow a more steady load to be applied in that they have replaced a displacement signal to the actuator-arm with a constant force signal.

The amount of torque that is generated by the rotor when the brake is applied is measured by a strain-gage bridge mounted on the horizontal portion of the supporting structure. The signal from the strain-gage is monitored on the microammeter. Thus, by watching the meter, a desired amount of braking torque can be administered to the windmill by turning down the turnbuckle an appropriate amount.

The braking system is calibrated by locking the disk and noting the meter response to a known applied torque as obtained by hanging weights at a given radius on one of the rotor blades.

The final testing parameter that is required in the determination of windmill performance is rotor rpm. This is easily measured by means of a small tachometer mounted at the rotor hub, the signal from which is sent to a voltmeter, and read by the crew member riding on the right side of the Jeep. Calibration of this unit is performed by rotating the hub by means of an electric drill and correlating a strobe light rpm measurement with the given voltmeter deflection.

Data Collection: The most important requirement for the collection of data using the Jeep windmill testing facility is that of very light or, preferably, no natural winds. It has been found that a light crosswind of possibly two and certainly three mph is sufficient to invalidate a data sample. This requirement for absolute calm test conditions is only met with any regularity in the early dawn, and, thus, most of the Princeton data have been obtained during the time between first daylight and sunrise.

The actual collection of data using the moving test-bed requires a minimum of three researchers. The driver's duty is to maintain the Jeep's speed constant, according to a predetermined anemometer reading, for each pass down the 3,000 ft Princeton University runway. The man on the right is responsible for commanding a given braking torque to be applied, reading the rpm meter, and for recording all of the information required at each datum point on a worksheet. This man has the further responsibility of determining when all of the readings have stabilized sufficiently for a point to be taken. Finally, the man sitting in the bed of the Jeep maintains the torque setting required for a given point by watching the torquemeter and adjusting the brake turnbuckle accordingly.

Finally, it is important that the individuals operating the testing vehicle become experienced at their posts and familiar with the various subtleties associated with the experimental facility. The fact that the quality of data samples improved significantly as the experience levels of operators increased has been notably evident throughout the Princeton program.

Reduction of Data: In order to ascertain the performance of a windmill, the data that are obtained using the Jeep test facility are generally reduced and presented in the standard plot of power coefficient, Pc, as a function of tip speed ratio, μ. This nondimensional form of data presentation makes it possible to rapidly evaluate a particular windmill or configuration and compare its operation to that of other designs. For optimization research, such as that performed by the Princeton group, this capability to immediately determine the effect on performance of a single configuration change is obviously of great benefit.

Although ideally all data collection utilizing any moving vehicle test-bed should be conducted in absolute calm conditions, it can be reasonably argued that if the wind is blowing parallel to the test-strip, the tests should still be valid in that the data is reduced on the basis of the indicated air speed of the vehicle and should be the same velocity as that seen by the windmill. However, after some experimentation, it was found that because of a vertical wind velocity gradient above the test-strip (boundary layer), along with the fact that the windmill and the anemometer were not mounted at the same elevation above the ground, such was not the case. Thus, in an effort to extend somewhat the conditions under which data could be obtained, a method of correction was developed for those occasions when a light wind was blowing parallel, or nearly parallel, to the test-strip. All in all, the best data is still that collected in calm air; however, this correction technique has added a little flexibility to the Princeton test program and has prevented a number of early morning efforts from being entirely wasted.

Performance Build-Up of the Princeton Sailwing

This phase of the Princeton windmill research program is concerned with the systematic performance build-up of a 12-foot-diameter, 2-bladed sailwing rotor. The experiments were designed to determine which configuration elements are necessary in obtaining reasonable performance and which elements can be eliminated with the benefit of lowering the overall cost. In addition, unlike the preceding systematic build-up of a 3-bladed Princeton design (20), this particular rotor was constructed with the capability of easily optimizing each configuration in both pitch and twist. Thus, it was hoped that such capability would add significant insight toward generalized blade design technology.

As a further important point, it should be noted, that a two-bladed rotor was chosen for the basic design because of the much greater ease with which configuration changes could be accomplished. In the carefully controlled test environment in which this machine was to be utilized, the dynamic problems of a two-bladed rotor when a rapid change about the azimuth occurs could be effectively eliminated; however, even though such problems are minimal because of the low inertia associated with the lightweight sailwing blades, the Princeton group does recognize the advantages of and strongly recommends the use of three-bladed machines in nearly any practical field application.

Data Presentation and Discussion: The basic sailwing structure from which all the subsequent configurations were evolved is shown in Figure 3.17a. The windmill itself consists of a 12-foot-diameter, two-bladed rotor which is constructed using a tapered tubular leading-edge spar and fixed root (manually rotatable) and tip members. The blade trailing edge consists of a cable in tension to which is firmly attached the trailing edge seam of a Dacron sailcloth envelope. The pronounced catenary arc of the trailing edge produces chordwise tension in the sail itself which can, when adequately high in tension, eliminate sail luffing. The remaining configurations tested in this program are summarized in Figure 3.17, and involve the addition of a leading-edge fairing (b), blade-tips (c), center-body disks (d) and (g), and various types of trailing edge stiffeners (e)-(g).

During the actual testing of each configuration, the blade pitch angle was adjusted until the performance was found to drop off on either side of an optimum. Furthermore, each configuration was tested at free-stream wind velocities of 19.6 ft/sec (13.3 mph) and 31.3 ft/sec (21.3 mph). Thus, the nondimensionality of the windmill performance parameters is tested and the rate of performance degradation with increasing wind speeds can be evaluated.

The data for each configuration and wind speed that was evaluated was reduced and used to generate a conventional windmill performance curve that plots power coefficient, Cp, against tip-speed ratio, μ. These curves summarized the overall performance of each configuration tested and permit a rapid comparison of one configuration with another.

Figure 3.17: The Several Blade Planforms Tested

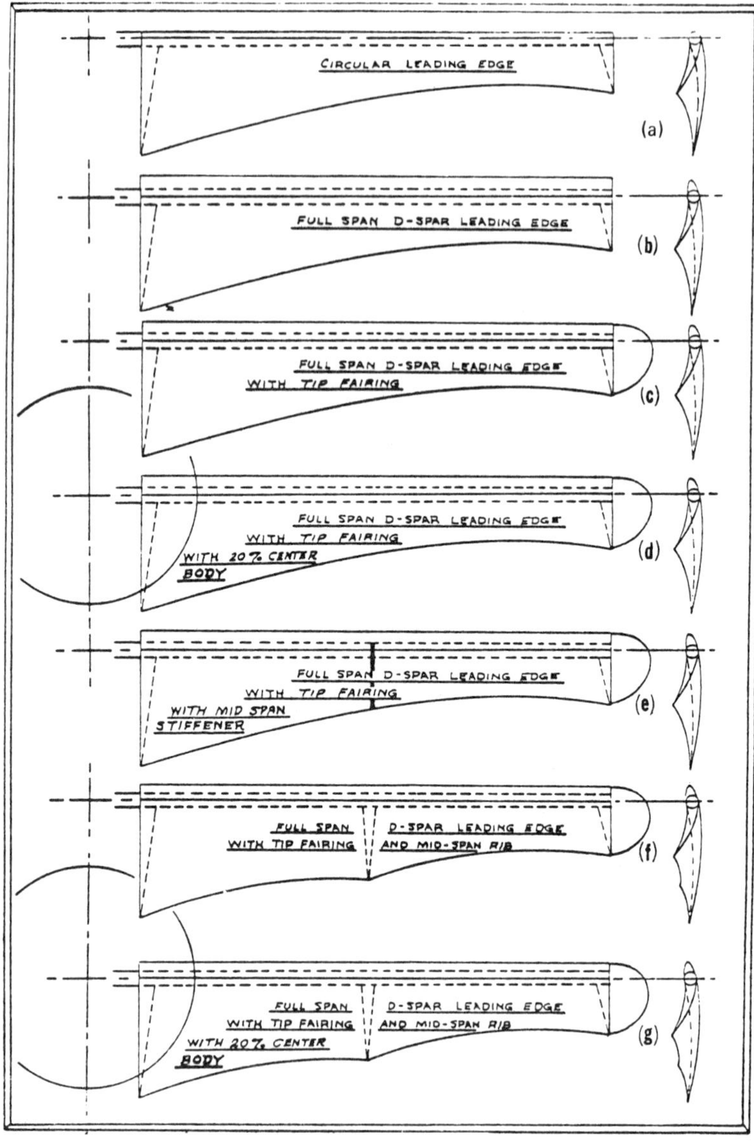

Source: PB 259 898

The performance of the basic windmill of Figure 3.17a is shown in Figure 3.18a. It will, of course, be noted that the maximum power coefficient leaves much to be desired. Although this was somewhat expected from previous sailwing experience, it was felt that use of tubular leading-edge should be documented to answer one of the most often posed questions concerning the sailwing blade design. Shown in Figure 3.18a is the blade tip cross section of this configuration which was the only one with a circular leading-edge tested.

Figure 3.18: Blade Tip Cross Section

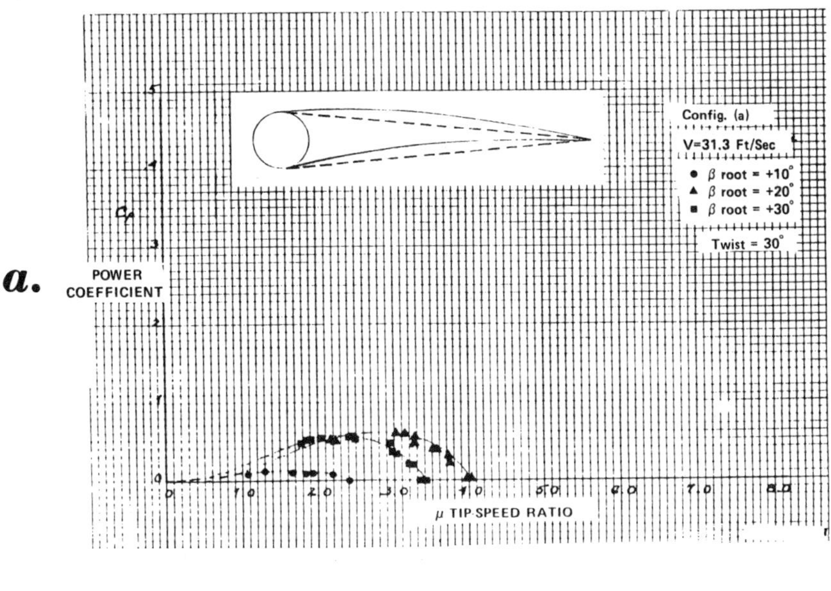

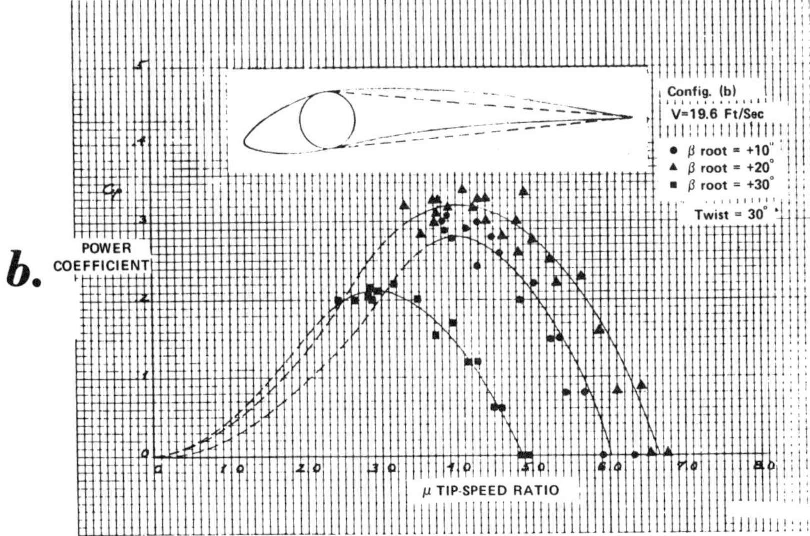

Source: PB 259 898

The performance of the configuration shown in Figure 3.17b is shown in Figure 3.18b. By the simple addition of a full-span D-tube fairing to the leading-edge, the power coefficient has been increased over 5 times. This is thought to be due to optimizing the blade sectional shape by improving the fluid flow accelerations around the leading-edge, by moving the point of maximum thickness aft, by increasing the sectional chamber by introducing approximately seven degrees of leading edge droop, and, finally, by significantly reducing the relatively large thickness ratio of each section along the blade span. Also shown in Figure 3.18b is the blade tip profile that was used on all of the remaining configurations.

With regard to the effect of the geometric addition of smoothly faired blade tips as shown in Figure 3.17c, although the overall power generated by this configuration was increased over that of the tipless version, it was more than balanced by the increased rotor radius used to nondimensionalize the power in the form of a power coefficient. The fact that the power coefficient itself did not improve is somewhat surprising but might be explained by a slight deterioration from the optimum loading distribution along the blade span.

Motivated by previous center-body fairing research performed at Princeton (20) (21), the addition of a center-body disk having a radius 20% that of the overall blade radius, Figure 3.17d, resulted in a small improvement in the overall performance.

A significant improvement of nearly 20% in rotor power coefficient resulted from the addition of the mid-radius trailing-edge stiffener, Figure 3.17e. This brace between the leading-edge spar and the trailing-edge cable firmly fixed the trailing-edge cable relative to the primary structure at that point but did not affect the basic catenary arc of the trailing-edge. The addition of the 20% center-body disk to this configuration produced very little change in rotor performance.

An increase in the geometrical solidity of the rotor is achieved by the configuration shown in Figure 3.17f but resulted in little noticeable change in performance. However, by dividing the original single-catenary arc trailing-edge into two smaller catenaries, the unsupported trailing-edge span is halved with the result of significantly raising the critical velocity threshold at which sail-luffing occurs (20).

Finally, a 20% center-body disk was employed to obtain the configuration shown in Figure 3.17g. This modification resulted in the largest power coefficient that was obtained, Cp = 0.40, for the lower wind speed. A summary of the significant parameters is given in Table 3.3

Table 3.3: Summary of the Significant Parameters Obtained in the Performance Build-Up of the 12-ft-Diameter, Two-Bladed Sailwing Windmill Rotor

	..... V = 19.6 fps			...... V = 31.3 fps		
Configuration*	μ_{max}	$C_{P_{max}}$	μ (opt)	μ_{max}	$C_{P_{max}}$	μ (opt)
a	4.0	0.06	2.8	–	–	–
b	6.7	0.32	4.0	6.0	0.31	3.4
c	7.0	0.31	4.1	6.6	0.30	3.9
d	7.6	0.34	4.8	6.6	0.31	4.2

(continued)

Table 3.3: (continued)

Configuration*	V = 19.6 fps μ_{max}	$C_{P_{max}}$	μ (opt)	V = 31.3 fps μ_{max}	$C_{P_{max}}$	μ (opt)
e	8.0	0.37	4.9	7.0	0.34	4.1
e**	7.6	0.39	4.4	7.0	0.33	4.2
f	7.3	0.37	4.3	–	–	–
g	8.0	0.40	4.3	7.1	0.36	4.1

*As per Figure 3.17.
**Plus center-body

Source: PB 259 898

Conclusions: In summarizing the findings of these experiments, it can be fairly stated that a circular leading-edge does not produce a good windmill rotor; however, by the simple addition of a drooped leading-edge fairing, the performance of the sailwing rotor becomes reasonably good. It can further be stated that the addition of wing-tip fairings on this configuration did not improve the overall power coefficient obtained. The installation of a 20% center-body disk did modestly improve the rotor performance while a significant gain in rotor efficiency was obtained by halving the unsupported span of the trailing-edge cable.

Over all of the configurations tested, the effect of increasing the wind velocity from 19.6 ft/sec to 31.3 ft/sec resulted in an average deterioration of approximately eight percent in the maximum power coefficient obtainable. While this loss in performance is not considered crucial, some attention must be given to the fact that this condition will become more pronounced as the velocity approaches that of the critical value, for a given trailing-edge tension coefficient, at which sail luffing occurs (20); however, lest the impression be given that this effect must be entirely detrimental, it has been suggested that by careful control of the pertinent parameters, the performance degradation with increasing wind speed can be used to aid in the prevention of rotor self-destruction in high winds.

Finally, in comparing the final results of the two-bladed sailwing rotor build-up with those of the three-bladed rotor, it was expected that the lower overall solidity ratio of the two-bladed design would allow it to ultimately achieve larger maximum values of power coefficient. Thus, it was not only disappointing but somewhat confounding as well when, in spite of higher corresponding tip-speed ratios, the two-bladed machine achieved a maximum power coefficient of 0.4 and required considerably more effort than the three-bladed design needed to obtain 0.44 (20).

While the explanation of this dilemma is obviously tied up in the fact that the three-bladed design has an overall more optimum induced velocity distribution over the disk area, this has yet to be determined rigorously. Thus, even though both rotors are considered excellent when compared to other modern machines, one can appreciate the fact that there remains a great deal to be learned concerning the aerodynamics of windmill design.

Multistage Windmill to Obtain Center-Body Fairing Benefits

Based solely on the intuitive notion that the improved windmill performance that was realized with the addition of a 20% center-body disk on the 12-foot-diameter, two-bladed sailwing rotor could also be successfully achieved by the large pressure drop generated at the rotor-disk center by a small, first-stage rotor mounted coaxially in front of the main rotor, a multistage rotor assembly was built and tested. As there had been virtually no previous experience with such a device, the research was to be of a purely exploratory nature and the prototype assembly was designed to permit a number of operational modes as well as easily incorporated modifications. Specifically, as one could only hypothesize as to the most suitable mode of operation for such a device, the symmetrically sectioned, untwisted blades of the small, upstream rotor were constructed in such a manner that the pitch could be adjusted to allow rotation in the same direction or in the opposite direction to that of the main rotor.

Furthermore, it was possible in the course of testing to apply a variable load to the first-stage rotor and, therefore, control its operational tip-speed ratio independently from that of the main rotor. This capability was included because it was not intuitively clear whether the flow field would be most beneficially disturbed with the small rotor free-wheeling, extracting additional energy from the stream tube or, although unlikely, drawing energy from the main rotor in order to put energy into the center portions of the stream tube so that substantially more working fluid would be diverted to the outboard regions. Obviously, a very broad exploratory program would be required to determine the most suitable operating mode and only then could that configuration be optimized and the overall potential of the concept evaluated.

Data Presentation and Summary: The basic rotor configuration with which the multistage windmill experiments were conducted is that of Figure 3.17f but without blade tips. The baseline performance curve of the test-bed windmill without the first-stage rotor assembly mounted was plotted and it was by comparison with this curve that the relative merits of the various operating modes of the coaxial rotor were determined. A summation of the important points taken from the performance plots is tabulated in Table 3.4.

Table 3.4: Summary of the Significant Parameters Obtained in the Multistage Windmill Rotor Experiments

Configuration	μ_{max}	$C_{p_{max}}$	μ (opt)
Baseline windmill	7.3	0.36	4.1
Coaxial rotor: rotations synch., free-wheeling, $\beta = 20°$	7.0	0.36	4.1
Coaxial rotor: rotations synch., intermediate braking, $\beta = 20°$	6.8	0.33	3.9
Coaxial rotor: rotations synch., brake locked, $\beta = 20°$	6.8	0.33	4.0
Coaxial rotor: counterrotating free-wheeling, $\beta = 20°$	7.2	0.36	3.8
Coaxial rotor: counterrotating free-wheeling, $\beta = 5°$	7.2	0.35	3.9

Source: PB 259 898

Conclusions: In examining the data presented in Table 3.4, one can only come to the conclusion that all of the coaxial windmill operating conditions yielded performances that were either the same or slightly worse than that of the baseline windmill.

In any case, the upstream rotor's effect appeared to be so little that further research efforts were ceased. The only explanation as to the reason for these disappointing results is that while the upstream rotor may very well produce a fluid-flow disturbance much like the center-body disk, it also produced an in-plane interference due to the drag of its blades. Thus, any beneficial interference or additional rotor torque is nullified and the net rotor performance remains essentially unchanged. In any case, on the basis of these experiments, an overall evaluation of the concept would have to conclude that a simple center-body disk is not only more effective, but less costly as well.

Wind Tunnel Investigation of Various Sailwing and Sailvane Cross-Sectional Profiles

The motivation for exploring the aerodynamic characteristics of various sailwing and sailvane cross sections was to determine if simplification or modification of the conventional double-membraned sailwing could yield any economical benefits without causing excessive performance penalties. Thus, a wind tunnel program was undertaken and structured in such a manner as to ascertain the relative magnitudes of the penalties associated with using commercially streamlined sailboat mast and circular cross-sectional tubing in place of the sailwing's D-section leading-edge. Furthermore, the importance of the full double-membrane was tested by including several sections not utilizing the lower membrane as well as several having only a partial lower membrane. In total, eight wings, identical in all respects except for the section utilized, were tested.

Model Description: The tests of the eight different wing profiles, shown in Figure 3.19 were performed in the Princeton University 4' x 5' wind tunnel. The wing planform that was utilized is shown in Figure 3.20 and characterized by a span, b, of 37.8 inches, a mean aerodynamic chord, $\bar{c}_A$, of 4.5 inches, and a total area, S, of 168 square inches. The resulting aspect ratio of the planform, $AR = b^2/b\bar{c}$, is equal to 8.4. Relative to the mean aerodynamic chord length, the sectional thickness ratio, t, is 11.5%. The trailing-edge cable tension in the models could be varied and was adjusted to 9.5 lb and 36 lb for the experimental results discussed herein. Respectively, these values yield a trailing-edge cable tension coefficient, C_t of 0.07 and 0.28.

Test Conditions and Data Reduction: All of the data collected in the series of experiments included in this report were obtained with the tunnel speed adjusted to yield a dynamic pressure, q, of 13.0 lb/ft^2. Although the corresponding Reynolds number, based on the mean aerodynamic chord, can be calculated to be approximately 250,000, because of the turbulence level in the tunnel, the aerodynamic data collected is more representative of a higher Reynolds number on the order of 750,000.

The mounting arrangement of a test model permits the wing angle-of-attack to be adjusted while the tunnel is in operation to any value between -12 and +24 degrees. Thus, force balance data for lift, drag, and pitching moment were obtained at each 2° angle increment between the limits. These data were then reduced to the standard coefficient form and plotted as a function of the wing angle-of-attack as referenced to the unloaded (no wind) orientation of the mean aerodynamic chord. In addition, the efficiency of each of the wings was summarized in a plot of the lift-to-drag ratio, L/D, as a function of wing angle-of-attack.

Figure 3.19: Sailvane and Sailwing Cross Sections Tested in the Wind Tunnel

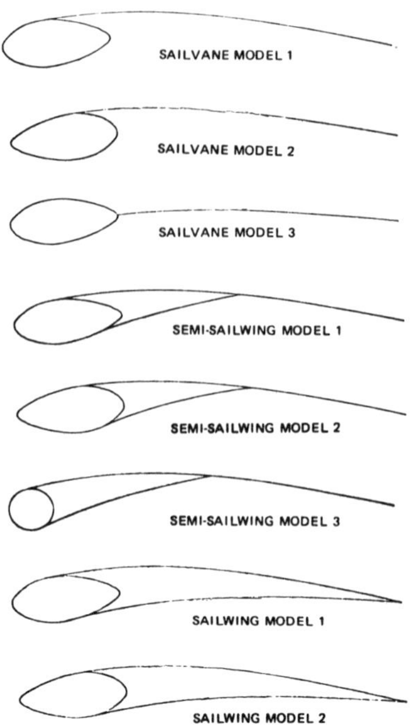

Figure 3.20: Sailvane and Sailwing Planforms

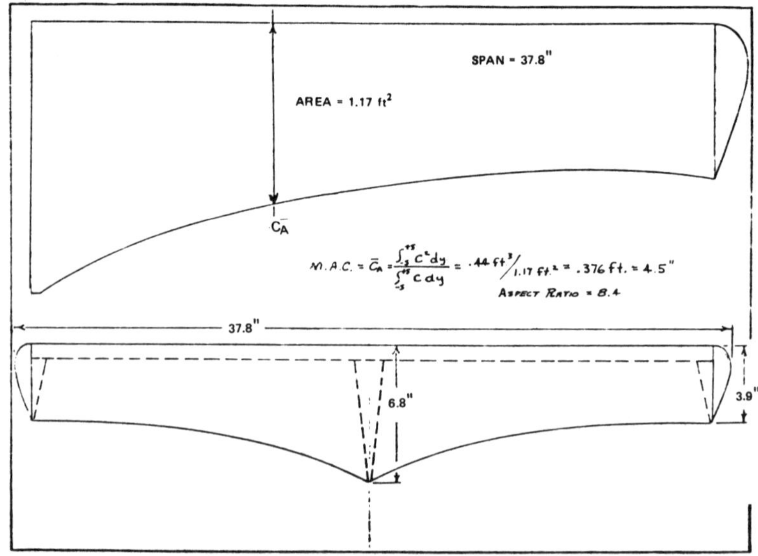

Source: PB 259 898

One difficulty with the experimental technique that should be pointed out is concerned with the fact that all of the drag measurements are obtained by subtracting a very large tare drag reading from a very large overall drag reading to obtain the very small drag contribution of the model. Although this is the common procedure when a limited amount of time and funding are available, it undoubtedly allows small errors in the drag measurement to enter into the experiment. This is not to say that the procedure yields data which are unsuitable for design purposes and it is certainly acceptable for comparative purposes; however, the results obtained in this manner should not be considered absolute.

Characteristics of the Sailwing: Because of the flexible nature of the sailwing, it possesses several features that cause it to differ greatly from a conventional rigid wing and it is therefore appropriate to discuss some of these characteristics in relation to the sailwing's operation. For example, when the sailwing is at rest (no-wind), the cloth membrane is held taut by the trailing-edge cable and is essentially, except for the leading-edge, a symmetrical section as the upper and lower surfaces experience the same pressure. As the wing becomes operational in a lifting orientation (wind-on), the asymmetrical pressure distribution that is established between the upper and lower surfaces causes the membrane (or membranes) to deform away from the high pressure regions (underside) and toward the lower pressure regions (upperside).

Thus, when the wing is lifting upward, the airfoil section assumes a positive camber distribution that fairs in smoothly with the airfoil's leading-edge shape. It should be noted that the actual shape of the sailwing section is a function of the wind velocity, the wing angle-of-attack, the no wind airfoil shape, and the amount of tension in the trailing-edge cable. Thus, as the angle-of-attack is increased, thereby increasing the amount of lift that is generated (up until the wing stalls), the resulting increased pressure differential between the upper and lower surfaces causes the amount of camber in the section to increase.

This not only causes the maximum value of wing efficiency, lift-to-drag ratio, to occur at fairly high angles-of-attack, but also delays the impending stall. At this point, the importance of maintaining the desired trailing-edge cable tension should be noted. As might be expected, relaxing the cable tension allows for a greater amount of camber and therefore a higher maximum lift coefficient; however, it simultaneously decreases the maximum lift-to-drag ratio obtainable as well as the threshold of critical velocity at which detrimental sail-luffing occurs. Thus, the amount of tension in the sailwing's trailing-edge cable controls the important trade-off between C_{Lmax} and $(L/D)_{max}$. As an upper limit, it might be considered that as the cable tension becomes hiqher and higher, the sailwing's behavior becomes more and more like that of a rigid wing.

Another interesting characteristic of the lifting sailwing is the upward deformation of the trailing-edge in the unsupported mid-span regions of each wing panel. A result of this action is a reduced angle-of-attack in these regions and one would expect a local reduction in lift; however, it is generally the case that this effect is more than offset by the increased amount of camber that occurs and results in a locally increased lift generation. In fact, because of this effect, the lift distributions that occur over some sailwings are often very close to that of the elliptical optimum.

The constant chordwise tension that is a result of the trailing-edge cable and the catenary arc sail cut is responsible for many of the desirable features of the sailwing over other flexible designs. One such feature is that relatively low drags are present at low angles-of-attack and lift coefficients. Furthermore, the sailwing has the ability to pass through the zero-lift angle-of-attack without flapping or luffing. Below the zero-lift angle-of-attack, the asymmetrical pressure distribution on the wing is reversed resulting in the section having negative camber. All in all, through many years of extensive research, the sailwing has been found to provide a simple, lightweight and low-cost alternative to the conventional rigid wing while not suffering any performance penalties throughout most low-speed applications.

Test Results: For the eight sailwing and sailvane models tested, the lift coefficient, drag coefficient, pitching moment coefficient about the quarter-chord point, and lift-to-drag ratio, are plotted as functions of the angle-of-attack of the unloaded mean aerodynamic chord. The significant parameters that have been obtained from these data are summarized in Table 3.5.

It is important to note that a direct comparison of these data to those of a conventional rigid wing is complicated a great deal by the flexible nature of the sailwing. For example, the sailwing data is likened to that of a rigid wing in which a movable flap is deflected additionally for each increasing angle-of-attack increment. This characteristic is responsible for the fact that it is generally impossible in the case of a sailwing to linearize the drag polar or obtain a meaningful span-efficiency factor as is done from wind tunnel test data for a conventional wing.

Similarly, it should further be noted that at lower angles-of-attack (up to approximately 5°), most sailwings have a lift-curve slope which exceeds the theoretical maximum for rigid wings (2π per radian = 0.11 per degree). This occurs because the section is continually varying camber over the angle-of-attack range. At higher angles-of-attack, the section is unable to deform as much as when it is less loaded. Therefore, the lift-curve becomes increasingly more like that of a rigid wing when the angle-of-attack is increased to higher values.

Table 3.5: Three-Dimensional Sailwing and Sailvane Aerodynamic Parameters, Aspect Ratio = 8.4

Section	$dC_L/d\alpha$ (per degree)	α_{L_o} (degree)	$C_{L_{max}}$	$(L/D)_{max}$	$(L/D)_{C_L = 1.0}$
Sailvane Model 1	0.125	-1.5	1.30	16.0	16.0
Sailvane Model 2	0.104	-3.0	1.41	18.0	18.0
Sailvane Model 3	0.112	-1.5	0.92	12.0	3.6
Semisailwing Model 1	0.110	-1.5	1.18	21.0	17.0
Semisailwing Model 2	0.098	-2.8	1.47	18.0	17.7
Semisailwing Model 3	0.108	-1.0	1.12	13.4	9.0
Sailwing Model 1	0.118	-0.8	1.30	29.0	27.8
Sailwing Model 2	0.118	-2.1	1.49	29.5	27.0

Source: PB 259 898

As previously mentioned, the model tests were conducted using two different trailing-edge cable tension coefficients; however, the data presented is for $C_t = 0.28$ as the results varied only slightly, in accordance with the expected trends, for $C_t = 0.07$. It is therefore concluded that for the dynamic pressure at which the tests were performed, both of these values are sufficiently high to allow only minimal increased sail deformation (camber), even at the lower value of C_t.

For the purpose of design, it is often the established procedure in wind tunnel testing of rigid wings to generate two-dimensional sectional data from three-dimensional data such as that presented for the sailwing; however, the flexibility of the sailwing airfoil makes this extremely difficult and the Princeton group does not know of any reasonable method to determine the exact breakdown of total drag into its three-dimensional induced contribution and its two-dimensional profile contribution as this would require that the drag polar be linearized to some degree such that a span efficiency factor could be obtained. As previously mentioned, this is usually not possible in the case of a sailwing.

However, by carefully comparing the information presented in Table 3.3 with that of a rigid wing utilizing a similar profile, for example the NACA 4412, it can be realized that the data for the double-membraned sailwing resembles that of the rigid section to a large extent and it is not at all unreasonable to assume that the sailwing's two-dimensional characteristics are such that the maximum sectional lift-to-drag ratios are of the same order. Obviously, this approximation leaves a lot to be desired quantitatively; however, the argument is being presented only to give some indication that the sectional efficiency of the sailwing does not differ significantly from that of a conventional wing section often utilized in windmill-rotor blade design.

Conclusions: In examining the data presented in Table 3.5, it becomes clearly evident that the performance penalties paid for deviating from the conventional sailwing sections are fairly high. In fact, although some reasonably good rotors have been constructed using the single-membraned sailwing, it seems that some particularly convincing arguments concerning reduced costs and increased simplicity would have to be presented to justify using anything other than the double-membraned cross sections.

The most outstanding characteristic in comparing the effect of the leading-edge shape are the larger values of lift coefficient, along with the associated more abrupt stalling characteristics, that occur with the smaller radius leading-edge. In addition, the smaller radius leading-edge gives rise to a slight shift in the entire lift curve such that a lower angle-of-attack is required to obtain a given lift coefficient.

In considering the application of these sections, the data in Table 3.5 demonstrate that the sailwing's three-dimensional performance is quite competitive with most rigid wings of the same aspect ratio. Thus, the use of the sailwing should allow the benefits of simpler construction and lower costs to be realized without paying any significant price in performance. In fact, some consideration should be given to the fact that, unlike many of its rigid counterparts, the cambering characteristics found in the sailwing cause its three-dimensional lift-to-drag ratios to be very near maximum at lift coefficients of close to 1.0. This situation is probably very close to that which is ideally desired for optimum windmill rotor performance (22).

Summary

(1) The Princeton moving-vehicle windmill testing has reached a level of sophistication where it is possible to obtain windmill performance data having both a high degree of repeatability and an acceptable range of experimental scatter with a minimum of effort.

(2) The step-by-step geometrical build-up of the 12-ft-diameter two-bladed sailwing rotor resulted in the value of the maximum power coefficient being increased nearly seven times. In addition, a significant amount of information was collected which should prove to be of considerable benefit in furthering the understanding of windmill blade technology.

(3) An exploratory research effort to evaluate the concept of a first-stage, coaxial rotor to obtain beneficial aerodynamic interference has found that such a device has either a degrading effect, or no effect whatsoever, on windmill performance. Thus, it is concluded that future efforts would be more wisely directed toward attempting to optimize the utilization of a center-body assembly.

(4) Finally, a comparison based on wind tunnel studies of various simplified versions of the sailwing have found the aerodynamic penalties for deviating from the standard cross section to be quite high. Although it might be possible to justify the use of such sections in certain limited applications (for example, where a low-cost sail reefing capability is a mandatory requirement), in the general case it is more likely that any cost savings would be insignificant when compared to the overall loss in performance.

REFERENCES

(1) *Federal Wind Energy Program,* Energy Research and Development Administration Report ERDA-84, (1975).
(2) Thomas, R., et al, *Plans and Status of the NASA-Lewis Research Center Wind Energy Project,* NASA TM X-71701, (1975).
(3) Puthoff, R.L., and Sirocky, P.J., *Preliminary Design of a 100 kW Wind Turbine Generator,* NASA TM X-71585, (1974).
(4) Puthoff, R.L., and Sirocky, P.J., *Status Report of 100 kW Experimental Wind Turbine Generator Project,* NASA TM X-71758, (1975).
(5) Donham, R.E., Schmidt, J., and Linscott, B.S., "100 kW Hingeless Metal Wind Turbine Blade Design, Analysis and Fabrication," American Helicopter Society, Preprint No. S-998, (1975).
(6) Hutter, U., *Operating Experience Obtained with a 100 kW Wind Power Plant,* NASA TT-F-15, 068, (1973).
(7) Hutter, U., *Influence of Wind Frequency on Rotational Speed Adjustments of Windmill Generators,* NASA TT-F-15, 184, (1973).
(8) Hutter, U., *The Development of Wind Power Installations for Electrical Power Generation in Germany,* NASA TT-F-15, 050, (1973).
(9) Putnam, P.C., *Power from the Wind,* D. Van Nostrand Co., Inc., (1948).
(10) Hoffman, G.A., "Basic Methodology for the Modular Stability Derivative Program (C-version)," Mechanics Research Inc. Report 4343-2, (1970).

(11) Meier, R.C., "Concept Selection and Analysis of Large Wind Generator Systems," presented at the 31st Annual National Forum of the American Helicopter Society, Washington, D.C., (May, 1975).

(12) Spera, D.A., *Structural Analysis of Wind Turbines Rotors for NSF-NASA Mod-0 Wind Power Systems,* NASA TM X-3198, (1975).

(13) Donham, R.E., "Analysis of Stowed Rotor Aerodynamic/Aeroelastic Characteristics," *AGARD Conference Proceedings,* 22, (September, 1967), Goettingen, Germany.

(14) Cardinale, S., and Donham, R.E., "Full-Scale Wind Tunnel Tests and Analysis of a Stopped/Folded Rotor," Lockheed Report, LR 21016, Naval Air Systems Command, (June, 1968).

(15) Donham, R.E., and Harvick, W.P., "Analysis of Stowed Rotor Aeroelastic Characteristics," *Journal of the American Helicopter Society,* 12, No. 2 (January, 1967), pp 21-44.

(16) Hoffman, J.A., and Holchin, B.W., "Modifications to MOSTAB for Wind Turbine Applications," Mechanics Research Inc., Report 2711 (July, 1974).

(17) Johnson, C.C., and Smith, R.T., "Dynamics of Wind Generators on Electric Utility Networks," *IEEE Trans. Aerosp. Electron. Syst.* Vol. AES-12, No. 4, (July, 1976), pp. 483-493.

(18) Anderson, P.M., and Fouad, A.A., *Power System Control and Stability,* Iowa State Univ. Press, (1977).

(19) Glauert, H., "Windmills and Fans," Vol. IV, Div. LXI, *Aerodynamic Theory,* W.F. Durand, Editor, Springer (1935).

(20) Sweeney, T.E., et al., *Sailwing Windmill Characteristics and Related Topics,* Princeton AMS Report No. 1240, (Spring, 1975).

(21) Blaha, R., *The Effect of a Center-Body on Axial Flow Windmill Performance,* Princeton AMS Report No. 1266, (March, 1976).

(22) Rohrbach, C., *Experimental and Analytical Research on the Aerodynamics of Wind Turbines,* Mid-Term Technical Report, ERDA Contract E (11)-2615, (February, 1976).

INNOVATIVE WIND TURBINES

Material for this chapter has been based upon reports by Sandia Laboratories (SAND 76-5683, SAND 76-0650 and SAND 76-5535); Grumman Aerospace Corporation [COO/2616-2(Pt. 1)]; Polytechnic Institute of New York (PB 263 749); the University of Dayton (COO/4130-77/1); and West Virginia University (NSF-RA-N-75-050).

DARRIEUS TURBINES

Status of the ERDA/SANDIA 17-Meter Darrieus Turbine Design

Although patent applications for the Darrieus turbine were filed in France and the United States in 1925 and 1926, respectively, the technical data base was essentially nonexistent until the mid-1960s when the National Research Council of Canada (NRC) began their program. Even today, the development of Darrieus turbine technology is in its infancy when compared to conventional horizontal-axis wind turbines. The general objective of the Sandia Laboratories' work is to develop a sound technology base in order to obtain a more valid comparison of the relative energy costs between Darrieus and horizontal-axis wind turbines. This technology base is being developed through combined analytical, experimental, and hardware activities. Particular emphasis is placed on the synchronous grid application which will probably be the first large-scale (MW size) application in the United States of wind energy systems.

Efforts at Sandia Laboratories have been directed toward the design and fabrication of a 17-meter diameter Darrieus turbine. This size turbine was chosen because it represented a reasonable extrapolation from the technology available from a number of 5-meter turbine designs, thus should not be a high risk design. Figure 4.1 presents an artist's concept of the 17-meter experimental turbine. The basic rotor configuration consists of three NACA 0012 blades, each blade being supported by two struts that extend from the rotating shaft to the blades. (The tower design will also be capable of handling a two-bladed rotor configuration.) These struts are considered necessary to keep the blade's natural frequency above the design specification of 4.5 Hz; however, the system design will allow operation without struts provided vibration problems are not significant. The rotating shaft will be of the tube type instead of the truss type because of a combination of stiffness and size requirements. For a given stiffness, a tube type shaft will have a considerably smaller diameter than the corresponding truss type shaft. The rotating shaft, which will be fabricated from steel, has an outer diameter of 0.508 m (20-inch) with a 2.54 cm (1-inch) wall thickness.

Figure 4.1: Artist's Concept of the ERDA/Sandia 17-Meter Darrieus Turbine Design

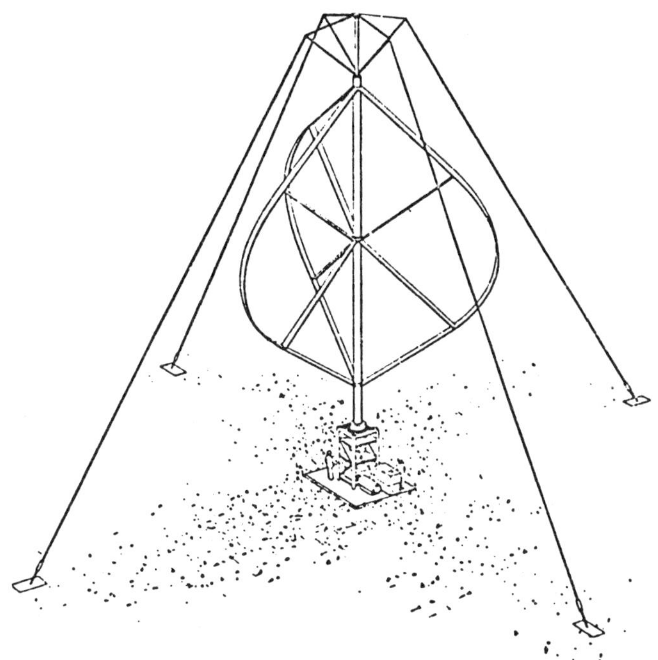

Source: SAND 76-5683

The turbine blades are being fabricated by Kaman Aerospace Corporation, Bloomfield, Connecticut. Figure 4.2 presents a representative cross section of the NACA 0012 blades. The leading section is an extruded aluminum "D" section while the trailing section is paper honeycomb with aluminum spline for the trailing edge. Fiber glass covers the trailing section. Blade specifications include 0.533 m chord, center of gravity of the cross section will be within 3% of the chord length forward of the quarter-chord point, and the center of twist of the cross section will be within 3% of the chord length aft of the quarter-chord point. The specifications on center of gravity and center of twist were made to reduce potential coupling between bending and torsional motions of the blade.

The blades are also required to operate up to 75 rpm in a 26.8 m/s (60 mph) wind without permanent deformation and to withstand 53.6 m/s (120 mph) wind at zero rpm.

Figure 4.2: Blade Design by Kaman Aerospace Corporation

Source: SAND 76-5683

The turbine base contains the means for axial support of the turbine, the power train (consisting of a speed increaser, synchronous generator, and an inductive starter), and a braking system. The right-angle drive permits installing the generator/starter system on a horizontal axis. This feature, which is admittedly more complex and less efficient than a straight drive, does offer important advantages when used in this prototype. For example, the horizontal components can be easily removed and changed due to their accessibility; and the constraint of compactness (to avoid excess base height) is relieved by having two rotational axes. This will simplify making any modification to the drive train once it is installed. These advantages are believed to be much less important for production machines, and a production-oriented design would probably have a straight vertical axis drive.

The speed increaser is a commercially available planetary gear unit with a gear ratio of 42.9:1. For a generator speed of 1,800 rpm, the turbine speed would be approximately 42 rpm. By including a variable gear ratio, more hours of operation at a given site can be achieved. This variable gear ratio will be obtained by timing belt drive with standard sprockets with 24, 26, 30, and 34 teeth. By using appropriate combinations of these sprockets, the following discrete turbine operating rpms are obtained: 29.6, 32.1, 33.6, 36.4, 37.0, 38.7, 42.0, 45.4, 47.5, 48.4, and 52.5.

The electrical generator will be of the brushless synchronous type, 1,800 rpm, 60 kW, 3-phase, 60 Hz, and 277/480-volt output. The synchronous generator will also be wired to act as a starter motor. In the event that problems are encountered with this type of starter system, a separate induction starter motor is included in the system design. A hydraulically actuated caliper brake system will be utilized as a safety brake in the event that synchronization is lost. Table 4.1 summarizes some of the general specifications of the 17 m Darrieus turbine design.

Table 4.1: 17-Meter Darrieus General Specifications

Diameter, m (ft)	17 (55)
Number of blades	1, 2 or 3
Blade section	NACA 0012
Blade shape	Straight-circular-straight
Chord length, m (in)	0.533 (21)
Turbine swept area, m^2 (ft^2)	180 (1,937)

(continued)

Table 4.1: (continued)

Power:	
Turbine power, kW	70
Generator output, kW	60
Generator, rpm	1,800
Rated wind speed, m/s (mph)	14.3 (32)
Cut-in wind speed, m/s (mph)	4.5 (10)
Cut-out wind speed, m/s (mph)	26.8 (60)
Turbine at rated power, rpm	45.4
Direction of rotation (looking down)	CCW

Source: SAND 76-5683

The anticipated power vs wind speed curve for the 17-meter turbine design is shown in Figure 4.3. The design rotational speed is 45.4 rpm and maximum power is 70 kW. Due to turbine component inefficiencies, the electrical generator will reach approximately 60 kW, which is its rated capacity. The reference wind speed for Figure 4.3 is at a height of 3.66 m (12 ft). The power vs wind speed curve for the minimum turbine rotational speed (29.6 rpm) is also shown. It is anticipated that turbine checkout will begin with the minimum rotational speed and progress toward the design condition of 45.4 rpm.

Figure 4.3: Calculated Power vs Wind Speed Curve for the 17-m Darrieus Turbine

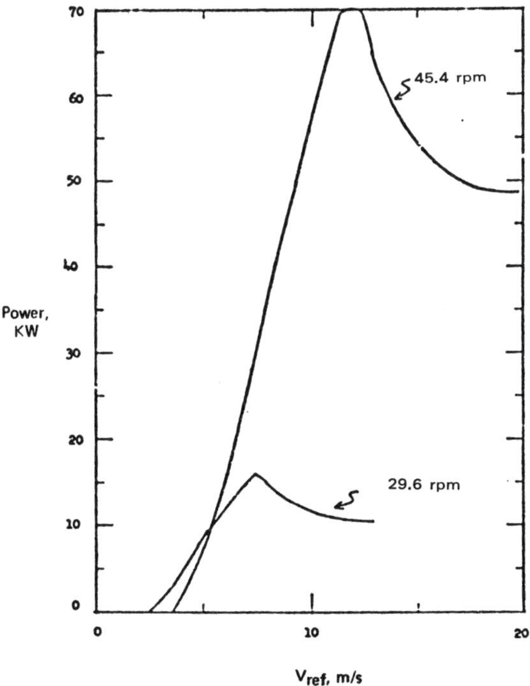

Source: SAND 76-5683

Structural Studies

Initial efforts at Darrieus turbine design have led to the use of relatively slender, high-aspect-ratio structural elements, such as the blades and supporting tower. This trend, which is shared by other modern wind turbine designs (such as the NASA Mod 0 horizontal-axis turbine), has apparently been governed by the aerodynamic incentive to use low-solidity elements. These design trends have accentuated the importance of developing a relatively complete and accurate structural understanding of the Darrieus turbine. This is because it is not possible, particularly for the blades, to cover our lack of experience and understanding by arbitrarily modifying structural sections to achieve large factors of safety.

Although the design of Darrieus turbines built to date has been primarily constrained through aerodynamic requirements, future "optimized" turbines designed for low cost should consider the tradeoffs that are certain to exist among aerodynamics, structures, and systems requirements. For example, a simpler, less costly structure that compromises aerodynamic performance may be preferred, provided that the net cost per unit of energy produced can be reduced. The ability to carry out such tradeoff studies clearly requires the capability for the quantitative evaluation of a variety of turbine structures.

This section reviews some of the techniques, both analytical and experimental, which have been used for Darrieus turbine design. The discussion is conveniently divided to consider two major turbine subsystems: the turbine blades and the turbine support system. The latter consists of the tower, base, and tiedowns. Where coupling between these two subsystems is believed to be significant, the combined structural behavior is discussed. The section concludes with some comments on design guidelines that are pertinent to acceptable structural performance.

The major conclusion of this review is that considerable testing and experience and analysis tools have evolved to support the structural design of Darrieus turbines. The tests and analyses, however, do not include every conceivable phenomenon that may affect turbine structural performance. The greatest amount of uncertainty appears to be in blade dynamics, where quantitative understanding is relatively incomplete. There is a need to continue collecting experimental data related to operational turbines so that the base of experience can be expanded to guide the improvement and revision of existing structural design tools.

Turbine Blades: Blade Statics — The response of the turbine to quasi-statically applied loads has received considerable attention because acceptable static performance is, at the least, a necessary condition for suitable blade design. Static analysis can also be useful as a design tool in special cases, provided that the lowest natural blade frequencies are well above the aerodynamic load excitation frequencies.

One of the first published static analyses (1) considered the flatwise response of the Sandia 5-m-diameter rotor to centrifugal loads. A general-purpose, nonlinear finite element program was used. This analysis indicated two important results. First, even a slight deviation of the blades from a troposkein shape can lead to significant bending stresses. (Troposkein shape means the blades are curved in the shape that a perfectly flexible member would assume under the action of centrifugal forces.) Second, the tension in the blades due to centrifugal forces stiffens the structure considerably. The latter effect is shown in Figure 4.4, where the bending stress evidently increases with roughly the first power of the rotation speed, Ω. Without the centrifugal stiffening effect, the stresses would be proportional to the applied centrifugal load, which increases as Ω^2. Since the Darrieus rotor is curved, a proper accounting of centrifugal stiffening is aided by the use of nonlinear finite-element programs, where the load is incrementally applied and the stiffness matrix is adjusted after each load increment.

Figure 4.4: Calculated Maximum Blade Bending Stress for the Sandia 5-m Turbine (1) (1 psi, 6.9 x 10³ Pa)

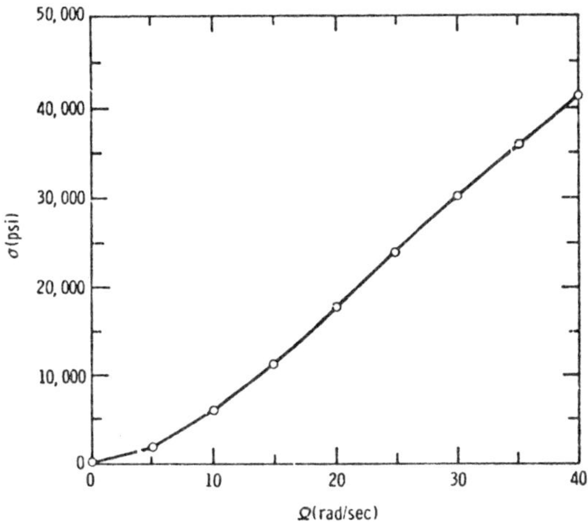

Source: SAND 76-0650

More recent analyses have considered the effect of aerodynamic loads on the blades. The aerodynamic loads are significant because they vary cyclically during turbine rotation and govern the fatigue life and dynamic behavior of the turbine. The aerodynamic loads also introduce chordwise (lead/lag) forces on the blades. These chordwise loads on the curved blade produce coupled twist and bending blade deformations, necessitating the inclusion of additional degrees of freedom in the analytical models.

A number of aerodynamic load models have been proposed (2)(3)(4) for use in structural analysis. All were based on the single (5) or multiple streamtube (6)(7) aerodynamic models. An example of the cyclic normal blade load variation is indicated by the net radial aerodynamic load resultant at the blade root (see Figure 4.5). These load resultants, predicted by the single streamtube model for a NACA 0012 airfoil, are normalized to unit peak magnitude and are primarily a function of the turbine speed ratio. (The normalized loads also depend, though less strongly, on such other turbine parameters as solidity, turbine shape, and blade Reynolds number.) The erratic behavior of the loads for speed ratios below 3 is due to periodic blade stall. The peak magnitudes of the normal loads at the turbine equator for the Sandia 17-m turbine now under construction are indicated in Figure 4.6, where it is evident that the most substantial loads occur with the combination of high turbine rpm and high wind speeds. It is noteworthy that for this numerical example the aerodynamic normal blade loads can exceed the centrifugal loads under certain conditions.

Figure 4.5: Variation with Azimuthal Angle of the Net Radial Blade Root Resultant Due to Aerodynamic Normal Blade Loading*

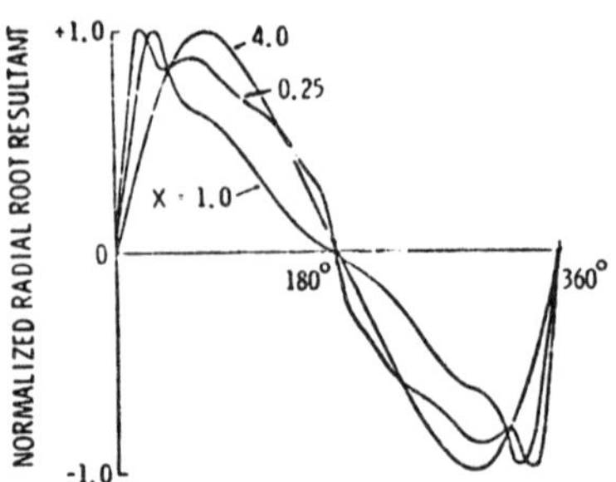

*Results are normalized to unit magnitude and show the effect of the speed ratio.

Source: SAND 76-0650

Figure 4.6: Calculated Centrifugal and Peak Aerodynamic Loads at the 17-m Turbine Equator*

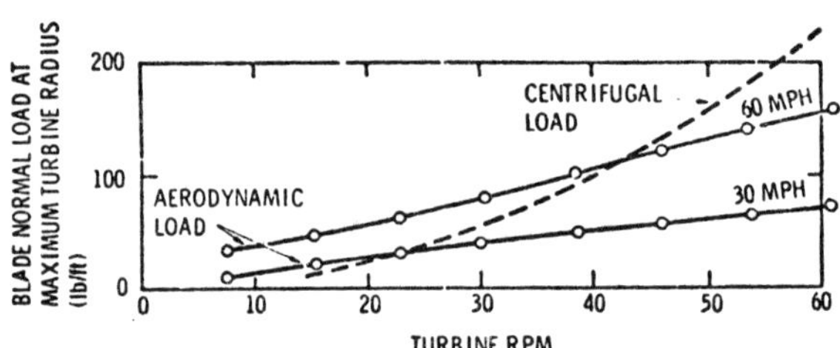

*Blade specific gravity is 0.5. (1 mph = 0.44 m/s, 1 lbf/ft = 14.6 N/m)

Source: SAND 76-0650

The existing load models are, of course, limited by the simplifications implicit in the aerodynamic strip theories used in their derivation. For example, the effects of tower or blade shadows on loads are not included in the models. Such effects are probably not so significant for these models as for propeller-type turbines because of the relatively large spacing between the blades and the tower structure. However, experimental and analytical investigations are needed to clarify this issue.

The aerodynamic load and structural response models have been used primarily to establish blade performance during turbine operation. Experimental work at the National Aeronautical Establishment of the National Research Council of Canada (NAE/NRC) has indicated that the resistance of the parked turbine to gravitational or aerodynamic buckling in gale-force winds may also be a critical design condition. Experimental results in dimensionless form from wind tunnel model tests of blade buckling are available (8). Analyses of the parked rotor response to gravitational and aerodynamic loads, using simple beam theory (9) as well as finite element models (10), have also been published.

Blade Dynamics — There are several features of the structural evaluation of wind turbine blades that could be considered dynamic analyses. These include the calculation of blade frequencies and mode shapes, evaluation of dynamic loading effects, life expectancy calculations, and flutter analysis.

Finite-element structural programs capable of solving eigenvalue problems are well suited for frequency and mode shape analyses because of the analytically cumbersome shape of the blades. The numerical programs must, however, be capable of accounting for the presence of blade tensions that are induced by the centrifugal forces acting during rotation. Figure 4.7 illustrates how the calculated natural frequencies of the first and second normal blade bending modes depend on turbine speed. These results for the Sandia 5-m turbine show the centrifugal stiffening effect on normal mode frequencies that is due to the blade tensions.

Figure 4.7: Natural Frequency, ω (First Two Normal Blade Modes) vs Turbine Angular Velocity for the Sandia 5-m Turbine (1)

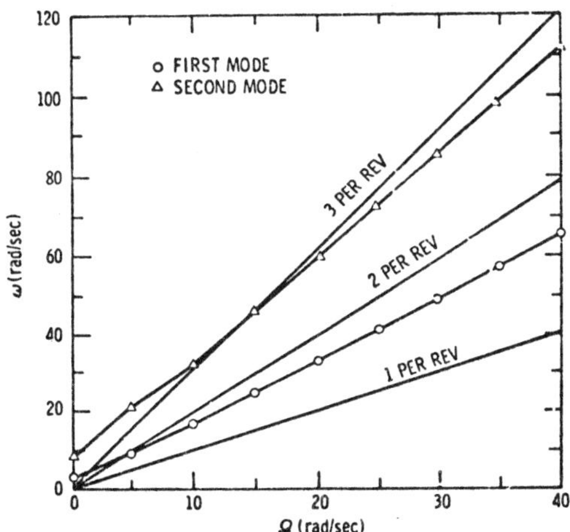

Source: SAND 76-0650

Also shown in Figure 4.7 are constant slope rays, sometimes called fans, which represent the loci of points where the natural frequency is some constant multiple of the turbine rpm. Since multiple-bladed rotors are capable of reacting to oscillatory forces that occur an integral number of times in one revolution of the turbine, the intersections of the fans with natural frequency curves are of particular interest. For example, at speed ratios above ~3, normal aerodynamic forces on one blade vary through one cycle in each turbine revolution (2). A turbine with three blades, each loaded by this 1/rev excitation at 120° phase angles, will experience a 3/rev excitation which, in turn, will be transmitted back to the blades because of the elastic nature of the support structure. This turbine will experience edgewise bending resonant conditions if operated at rotational speeds, which are given by the projections onto the ordinates in Figure 4.7, of the intersections of the 1/rev and 3/rev fans with the normal-mode frequencies. This illustrates the need for careful turbine design and integration with systems phases where turbine operating speed ranges are selected.

A conservative approach in design is to raise blade natural frequencies sufficiently to preclude resonances through the operating range of the turbine. Blade resonant frequencies can be raised only a limited amount through the structural design of the blade cross section because of the aerodynamic constraints on the section envelope. The use of auxiliary support struts, such as those shown in Figure 4.1, can substantially increase resonant frequencies (10). The fact that these struts can also improve turbine response to gravitational loads may be important for large (30-m-diameter or greater) turbines. Although struts do have these positive structural effects, there are economic disadvantages to their use, and future efforts should attempt to establish the conditions for which struts are necessary.

In calculating blade frequencies, one does not necessarily treat turbine components independently. For example, the elastic nature of the turbine blades and their central support tower suggests that motion in one will induce motion in the others. The relative motion between the attachment points of each blade will be affected by the presence of the other blades and the elastic characteristics of the central support tower. Although independent evaluation of turbine components is useful and informative for initial understanding and design, coupling effects among appropriate components can change component resonant frequencies and introduce additional modes.

Because turbine rotational frequency tends to be the same order of magnitude as blade natural frequencies, only the first few modes are ever within range of being actively excited. These low-frequency modes are relatively easy to visualize (see Figure 4.8). Here, the first and second normal blade modes with auxiliary blade support struts are shown. Additional information can be found in Reference (11). Notice that the first normal blade mode is dominated by strut motion; the second, by curved blade section motion.

It is clear that operation of the turbine at a speed which significantly excites a turbine mode is undesirable because dynamic stresses can be high, leading to a rapid consumption of fatigue life. Given that a blade resonance does occur, there have been efforts to calculate response amplitudes (9) using modal analysis. This analysis effects a Fourier decomposition of the aerodynamic loads to evaluate the response amplitude of each turbine mode. Magnitudes of the resulting calculations of dynamic stresses are uncertain because of the need for a realistic representation of the damping characteristics of the blade and of the failure to account for changes in aerodynamic loads resulting from blade deformation.

Figure 4.8: Isometric Views of Two Normal Blade Bending Modes for the Sandia 17-m Turbine

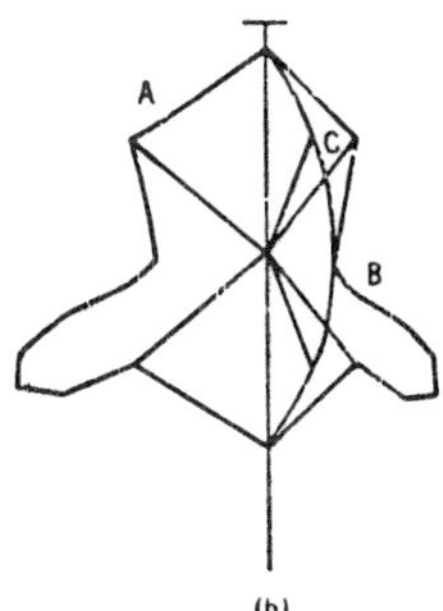

Source: SAND 76-0650

The difficulties in establishing dynamic response amplitudes and their significance with regard to fatigue life emphasizes the desirability of collecting experimental data and operating experience with Darrieus turbines. NAE/NRC (8) has published structural data from an instrumented wind tunnel scale model of their 200-kW Magdalen Island turbine. These data, which show that the dynamic stresses can be maintained at low levels with proper blade design, have not yet been used as a basis for evaluating and improving analytical techniques.

A potential blade flutter problem was identified while a 2-m turbine was being operated by Sandia in a low-speed wind tunnel (12) for the purpose of gathering aerodynamic performance data. Both solid aluminum and plywood blades with various chord lengths were tested in two- and three-blade configurations. When the plywood blades were tested in a three-blade configuration, an operational turbine speed was found where all blades underwent large vibrational deformations in both the wind-off and the wind-on condition. The deformations indicate strong coupling among normal blade bending, lead/lag twist, and bending displacements. The deformation resembled what one might classify as blade flutter instability with angle-of-attack changes, which occurred during the motion, that produce destabilizing aerodynamic lift forces.

The same effect was not observed in the structurally stiffer aluminum blade turbine. Because of the destructive potential in this type of motion, it was clear that its characteristics and properties would have to be understood before turbine designs could be advanced with confidence that flutter would not occur. To this end, an approximate modal analysis of the flutter condition was undertaken (13). When an incomplete elastic ring was used as an analytical model for the unstrutted turbine blades, the normal blade bending and the coupled torsion/chordwise bending equations of motion were obtained. Coriolis force effects were included in both equations of motion, and aerodynamic lift force effects were added to the normal blade bending equation. Unsteady chordwise aerodynamic forces were ignored. The resulting set of equations represented the coupled flutter equations for the turbine rotating in still air.

A stability analysis of these equations indicated that predicted flutter speeds are sensitive to a geometric coupling coefficient that relates chordwise and torsional stiffnesses. When theoretical coupling coefficient values were used, predicted flutter speeds for the plywood blades were in fair agreement with observations. The predicted speeds are lower than those found during the wind tunnel tests. The analysis also showed that increasing the cross section torsional stiffness tends to raise the flutter speed. This is consistent with the fact that flutter difficulties were absent when the aluminum blades were tested.

When auxiliary blade support struts were added to the analytical model, it was found that flutter frequencies were raised considerably. Numerical evaluation of the blades for the Sandia 17-m turbine indicated flutter speeds well in excess of operating speeds, despite the apparent conservatism of the analysis.

The effect of incident wind on aerodynamic loading of the blades was estimated by adding normal blade and chordwise loads to the flutter equations. This change was found to have negligible effect on elastic blade motion. This conclusion is corroborated by the fact that flutter during the wind tunnel tests occurred both with wind-on and with wind-off.

Possibly one of the most interesting and far-reaching results that came out of this study was the conclusion that the degree of cross section mass balancing has little effect on the location of flutter boundaries. Blade twist due to chordwise Coriolis forces is the dominant driving force for flutter in the Darrieus turbine and, therefore, does not depend on the location of the blade center of mass along the chordline. Removal of the constraint to mass-balance blade cross sections could lead to less expensive blades by allowing new manufacturing methods.

The NAE/NRC has collected the most comprehensive experimental data on flutter speeds (8). The results are presented in dimensionless form and are therefore applicable for predicting flutter speeds with geometrically similar turbines over a range of sizes and operating conditions. The qualitative observations made by NAE/NRC on the variables affecting flutter are in excellent agreement with the trends predicted by the above flutter analysis. There has not yet been an attempt to predict the Canadian results quantitatively through the analysis of the turbine geometry used in the NAE test program.

Turbine Support Structure: Virtually all the Darrieus turbines built to date have been supported by a guyed central tower connected to either a pinned or a cantilevered support at the turbine base. The principal advantages of a tiedown-type support system are the relative ease of obtaining support stiffness, a substantial reduction of tower bending moments due to aerodynamic loading, and simplicity of design. The major disadvantages are difficulties in obtaining blade-to-tiedown clearance, increased land usage, the introduction of additional axial loads in the tower and support bearings, and the potential for cable vibration problems.

The most severe problem, that of obtaining adequate blade clearance, is not so difficult to overcome with the curved-blade Darrieus turbine as for propeller-type or straight-blade vertical-axis turbines. The various advantages of tiedowns, however, have apparently led to the predominant use of these tiedowns on operational Darrieus turbines. Nevertheless, a number of innovative alternative support schemes using enveloping structures have been proposed (14). These alternatives will undoubtedly receive more attention as the development of the Darrieus turbine continues.

For a number of reasons, the structural design of the turbine support system may proceed along more conventional lines than for the blades. The most significant reason for this is the relaxation of aerodynamic constraints. It is desirable to have the support system occupy only a small fraction of the aerodynamic area of the turbine; nevertheless, this constraint permits considerable latitude in the choice of structural sections. Although the tower rotates, its rotating mass is located close to the axis of rotation. This eliminates the need to account for centrifugal stiffening and minimizes dynamic effects due to rotation (15). These features have led to the use of linear beam theories (16)(17), linear finite element analyses (16), and conventional guy cable analyses as the principal tools in tower, base, and tiedown design.

The design criteria for the turbine support system have considered resonant frequencies; stress due to torsional, axial, and transverse loads at the blade attachment points; and resistance to axial tower buckling. An example of design calculations, in this case the Sandia 17-m turbine, has been published (17). This example indicates that a simple tubular tower can meet all the design requirements with substantial factors of safety. The same study suggests that relatively lightweight truss-type towers can also meet the design requirements; however, these towers have large diameters and occupy more of the turbine central area.

Concluding Remarks on Structural Studies: The successful operation of a number of Darrieus turbines has shown that it is possible to construct turbines that can survive the foregoing structural problems. Through the experience gained in constructing prototypes, as well as deductions based on analytical models, a number of principles pertinent to successful structural design have evolved. These principles, speculative in nature, are guidelines reflecting the current state-of-the-art, with changes anticipated as the technology develops. Briefly, the principles are as follows:

(1) Turbine Scaling – The uniform geometric scaling up of turbine components in size has a surprisingly small effect on structural performance, except that gravitational stresses will increase in proportion to the scale. This rule follows from dimensional arguments (8). It assumes that the scaled-up turbine has similar aerodynamic performance and will be operated at the same tip speed in a similar wind environment.

(2) Blade Section Properties – The blade strength (moments of inertia) relative to the applied loads increases with the ratio of blade chord to turbine radius. Turbines with blade-to-chord ratios as low as 0.05 have been successfully operated. The minimum possible chord-to-radius ratio has not been established, but it will depend on the specific nature of the blade cross section and turbine geometry. However, for a given design, increasing the chord-to-radius ratio will tend to reduce structural problems.

(3) Blade Mass Balancing – As discussed above, coincidence of the elastic, aerodynamic, and mass centers of the blade is not required for precluding blade flutter.

(4) Blade Weight – Increasing the blade specific gravity produces some favorable structural effects. Specifically, both centrifugal stiffening and the ratio of steady centrifugal loads to cyclic aerodynamic loads are increased. This tends to reduce vibratory stresses seen by the blades, thereby increasing fatigue life. An increase in blade weight is limited by the tendency for gravitational stresses to increase. However, this limit is quite high for typical small turbines less than 10 m in diameter.

(5) Turbine Struts – The use of support struts improves turbine response to aerodynamic operating loads, flutter, gravitational loads, and parked buckling. Thus, struts can increase the conservatism of a design; however, operating experience with smaller turbines suggests that struts are not an essential feature of design.

Low-Cost Blade Design Considerations

Sandia Laboratories has carried out a study to identify and upgrade promising blade designs for the Darrieus wind turbine. The goal of this effort is to establish several low-cost blade designs.

Existing Designs: A new design should, at some time during the process, include a consideration of what others have already accomplished in similar or related projects. Various designs for small turbines (5 m or less in diameter) have been produced. Many of the features of these designs are refreshingly innovative. Selected features of these smaller blade designs are also applicable to the design of much larger blades. Furthermore, there are strategic advantages in advancing a particular design if certain of its features have already been practically demonstrated on a smaller scale system. Many of the blades known to have been produced are shown in Figures 4.9a and 4.9b.

Figure 4.9: Existing Designs

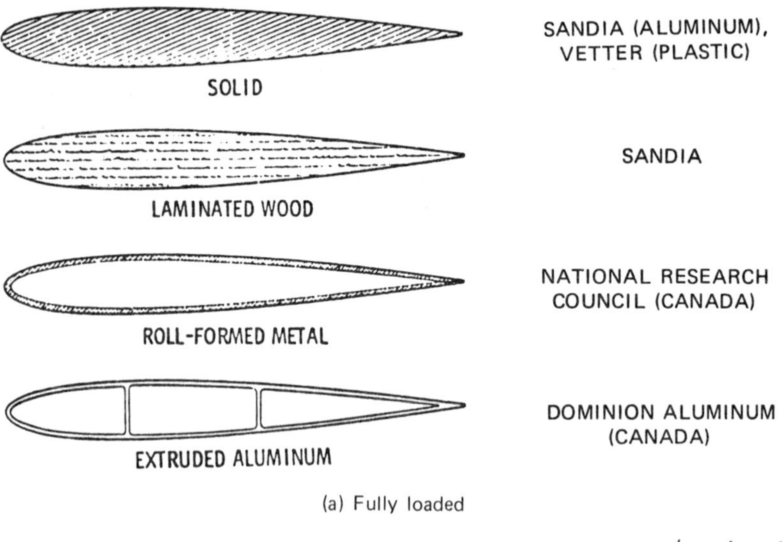

(a) Fully loaded

(continued)

Figure 4.9: (continued)

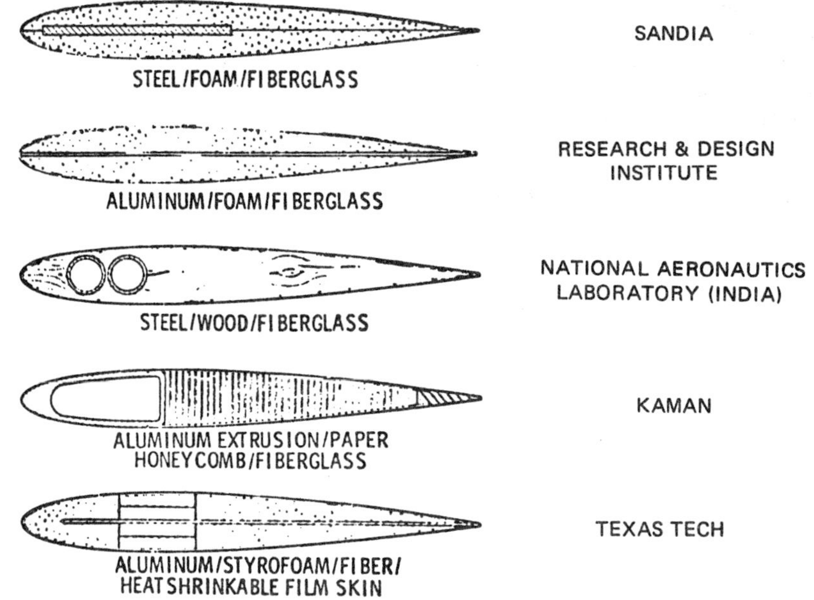

(b) Backbone loaded

Source: SAND 76-5535

The designs can be categorized by the manner in which the blade structure is loaded. Figure 4.9a presents those designs in which all of the blade material is load-carrying. Figure 4.9b presents those designs wherein the structure required for establishing the aerodynamic shape is, in varying degrees, parasitic on the load-carrying structure.

Proposed 17-m Blade Designs: Three provisional blade designs proposed for the 17-m Darrieus wind turbine are shown in Figures 4.10, 4.11 and 4.12. These blades have identical external geometries. It is also presumed that each design can be constructed to accommodate the structural requirements. Descriptions of these designs are accompanied by examples of how the weighting of selected values of the input criteria, especially the production quantity anticipated, can affect the selection of a particular design. The proposed construction methods and materials of each design, along with the manufacturing technology level involved as related to the expected production quantities, will be emphasized. Also, features of prior blade designs that have contributed directly to the 17-m blade designs will be acknowledged.

The design in Figure 4.10 is an all-steel, welded monocoque structure. The shapes are produced (in straight sections) by roll-forming from continuous strip sheet metal stock. The roll-forming element of this design was accomplished on a small scale in the NRC design. The NRC researchers, in fact, produced their own roll-forming tooling to accomplish this task in their wind tunnel maintenance machine shop. After being roll-formed to contour, each individual component is then stretch formed into the troposkein blade curve. The components are then joined by automatic seam welding into an integral blade assembly, and the end attachments are added by welding. There are no cross chord structure components (ribs) except the surface skin. The hinge end fixtures are of weldable investment-cast steel, e.g., AISI 4130 alloy. The investment casting process will provide for substantial savings both in machining and in material costs. The joint (end attachment) structural efficiency can be made very high by welding.

Figure 4.10: Proposed Low-Cost Design No. 1

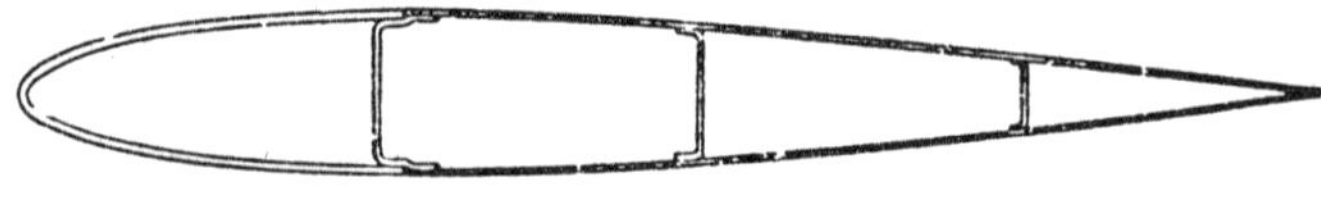

FEATURES:

- Roll-formed components
- Automatic seam welds
- Low-alloy steel
- Stretch-formed to curve
- Investment-cast end fixtures
- High technology - high volume

Source: SAND 76-5535

Each of the sheet metal components of this design is a low-alloy weldable steel that is also resistant to corrosion. The entire structure is considered to be load-carrying. The finished blade would be given extended service life by the addition of a protective coating to further retard any corrosion tendencies. The production of raw steel is considered to be relatively energy efficient when compared to competing structural materials. Use of primarily machine processes will afford excellent reliability and reproducibility.

Costs of blades produced by this approach can be projected with acceptable accuracy. If the anticipated high costs of tooling could be ignored, this design would be projected to be the least costly of all the designs so far considered. Of course, tooling costs cannot be ignored; therefore, this relatively large expense must be diluted (cost per blade consideration) by the production of a very large number of blades.

It appears that production from this design will require processing by four separate suppliers. The manufacturability was established by contacting individually the roll-forming and stretch-forming industry specialists. The welding of the components into the finished assembly and the investment casting of the end attachments are more widely understood to be state-of-the-art processes.

The second design, shown in Figure 4.11, employs a different approach to the structural loading. It provides an acceptable design for the situation where very large quantities of blades cannot be anticipated. Tooling costs will be minimal, but the labor costs will be quite high. Material costs will probably be higher than for an all-steel blade design.

Figure 4.11: Proposed Low-Cost Design No. 2

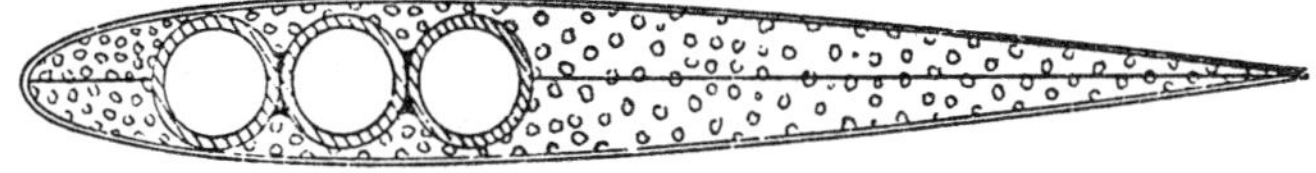

FEATURES:
- Commercial tubing backbone
- Backbone bent to shape
- Molded or formed foam airfoil segments
- Fiber glass ribs interdigitated
- Fiber glass skin - hand layup
- Welded end fixtures
- Low technology - low volume
- Low tooling costs

Source: SAND 76-5535

The technology level required for the production of this blade is lower than that required for the metal monocoque blade. If large production quantities can be projected, the bottom-line cost for this design is anticipated to be higher than for the monocoque design because this one is inherently labor intensive. Conversely, for low production the cost per blade will be significantly lower because the tooling costs for this design can be minimal.

The blade loads are accommodated primarily by a steel backbone similar to the earlier Sandia and NAL designs for smaller turbines. The aerodynamic shape is established by the rigid-foam airfoil halves, which are bonded to the steel backbone. A smooth fiber glass aerodynamic skin also provides impact resistance; e.g., for hail. With regard to structural load capacity, both the foam and the skin are largely parasitic on the steel backbone.

The manufacturing sequence is fairly simple. The backbone is constructed from inexpensive and easily obtainable mild-steel pipe or tube. The pipe elements would be welded to each other as shown in Figure 4.11, forming a structural "strap" or "ribbon." This strap would then be bent to the desired blade curvature, and end attachments similar to those of the first design would be attached by welding. Lengths of rigid polyurethane foam would then be bonded to the strap. These foam airfoil segments could either be formed to shape from slabs or be molded to the desired shape. Flat, foam airfoil shapes could probably be used because the anticlastic tendency of rigid foam is very low. (Anticlastic behavior is the tendency for a material to curve about an axis that is transverse to the axis about which it is being bent. The degree of severity of this behavior is indicated by the value of the Poisson ratio.) The foam segments would be split along the chordal line (airfoil centerline) to facilitate assembly.

If deemed necessary for providing additional torsional rigidity, structural fiber glass ribs could be interdigitated between the foam segments along the blade. Finally, a fiber glass skin would be bonded to the structure either by use of preformed skin segments or by the wet layup technique. The finished assembly would be smoothed to the aerodynamic finish desired, and an abrasion-resistant finish would be applied.

Individual blades might need to be selected (or modified) for the blade set in order to achieve a tolerable dynamic mass balance. The reproducibility of the blades (accuracy of form, weight, etc.) will not be so good for this design as for the monocoque design but it may still meet the operational requirements of the turbine. The anticipated departure from an "ideal" geometry will probably be acceptable, inasmuch as preliminary analytic studies have indicated that blade angles of up to 2° will not significantly degrade the turbine performance.

In the two preceding designs are presented the possibilities of using what are considered to be practical upper and lower boundaries of the existing manufacturing technology. The final design, shown in Figure 4.12, is included as an illustration of how features from both the high-technology and low-technology approaches can be combined into a single design. The pictorial representation of the components included in this third design is not intended to be so specific as the details of the first two designs but, rather, to present a design approach involving various technology levels.

Figure 4.12: Proposed Low-Cost Design No. 3

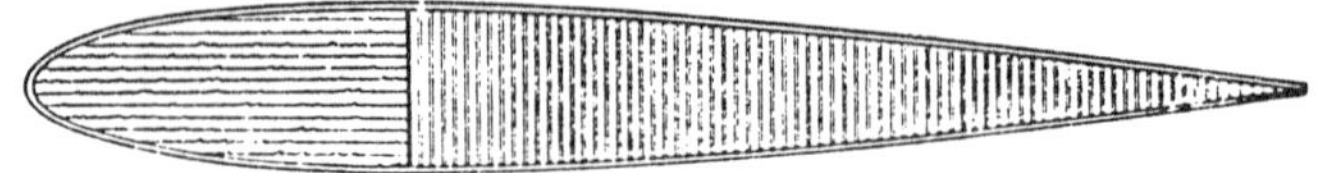

FEATURES:

- Wood laminated
- Honeycomb aft section
- Fiber glass ribs
- Fiber glass skin
- Bonded end attachments
- Technology mix (by components)
- Medium volume design

Source: SAND 76-5535

In this approach, certain high-level technology components would be furnished to an assembly team which might possess considerably lower fabricating skills. This team would construct a finished blade structure by assembling the components, using only elementary techniques. This particular approach would permit either the more critical blade components or the entire blade being furnished to the assemblers in kit form.

The precision of a blade could be assured by the machine processing of critical components and by the supplying (on a loan basis) of such special fixturing as deemed necessary for accurate assembly. Utilizing a trained blade fabrication technician to assist in the assembly might also be advisable. The assembly labor would be drawn from the location of the user. Although Figure 4.12 shows specific materials, laminated wood for the leading edge, paper honeycomb for the aft filler, and fiber glass skin, the individual components would be selected on a merit basis. The same approach might yield a design with an extruded aluminum D-spar leading edge, polyurethane rigid foam filler, and bonded-metal skins. Many of the features of the various blade designs for small turbines could be considered applicable to this design philosophy.

Obviously, no complete appraisal of each blade that might be produced by such an iteration of the various components exists. Such an approach may be prescribed, however, if the anticipated production quantity is too low or too high for the first and second design approaches presented, respectively. This design must have a production that is high enough to introduce such high-technology components as the laminated-wood leading edge or an aluminum extrusion D-section. If very high production is anticipated, a design similar to that of Figure 4.10 might prove to be the optimum approach.

This third design approach, like that of Figure 4.11, has the advantage of allowing assembly at or near the turbine site. This would minimize such shipping concerns as the size and bulk of the blades and the protection to be furnished enroute. Also, it would employ the local labor force, a consideration which might foster the acceptance of wind machines by the community.

Conclusions: The design of the turbine blades is crucial to the performance and economic success of the Darrieus wind turbine endeavor. The determination of many of the input criteria required for accomplishing this task has not yet been completed and the information turnover in this area is considerable. The design approaches and the design team assemblage are strongly influenced by the procedure and timing of the Darrieus turbine technical data bank accumulation. Several contractual approaches appear to have merit for accomplishing low-cost blade designs.

Although many of the design inputs are not yet rigidly established, it is considered that reasonable values may be selected so that the establishment of the hardware definitions may proceed. Several tentative blade designs appear to meet the goal of acceptably low cost. Each utilizes some features that have already been included in earlier demonstrations, and each of these designs may also be considered a target design for some specific production quantity.

The cost of each of the proposed designs can be lowered when the presumed conservatism of the existing design inputs is minimized. The eventual "bottom-line" cost of any design will depend on the modified design inputs and also on emerging manufacturing techniques. This cost will be presented by those specialists of the manufacturing industries required by the selected designs. Although the selection of a particular design will depend heavily on the production quantity anticipated, it is believed that a technically viable and economically attractive design consistent with any given production quantity can be established.

DIFFUSER-AUGMENTED WIND TURBINES

The Diffuser-Augmented Wind Turbine (DAWT) is one of the advanced concepts being investigated to improve the economics of wind energy conversion. The project is aimed at increasing the output and reducing the cost, the off-duty time, and the technical risk of wind energy conversion systems (WECS). The DAWT appears to be best suited to large WECS for commercial power production because it permits a significant increase in the unit power output without extending the size of rotating machinery into the range where rotor dynamics causes excessive costs.

In the DAWT concept, the turbine rotor is enclosed in a diverging duct called a diffuser (see Figure 4.13). By recovering exhaust kinetic energy, the diffuser produces a greatly reduced pressure behind the turbine relative to that behind a free turbine. This causes more mass to flow through the turbine with at least as much pressure drop across the turbine. Because the wind power available to the turbine is the product of flow rate and pressure drop, much greater power output is possible than for a conventional WECS at the same turbine size and free wind speed. The DAWT also has additional operational and reliability advantages that arise from the diffuser's large nonrotating structure.

Figure 4.13: Conceptual Installation of a DAWT with 60-m Diameter Turbine

Source: COO/2616-2 (Pt. 1)

Conventional diffusers are limited to very small divergence angles because the boundary layer on their inside walls tends to separate if the flow decelerates too rapidly. Since conventional diffuser configurations must be very long to achieve significant area change, they are prohibitively costly to construct. Therefore, one research goal has been to apply modern techniques to reduce radically the size of the diffuser without sacrificing its performance, thereby producing a cost-competitive DAWT. The use of the interior supporting struts as guide vanes, or stators, ahead of the turbine blades, has been studied analytically and the economic tradeoff of the DAWT against ordinary WECS has been assessed.

A large number of models for diffuser designs have been tested in a small wind tunnel. Several configurations of each of two effective types of diffusers (ring wings and boundary layer controlled) have been investigated. Of these, the most cost effective configuration has been chosen as a baseline design. In tests, the aerodynamic effects of different turbines are simulated by interchangeable screens.

For boundary layer control (BLC), advantage has been taken of a situation which is unique to the application of diffusers in wind machines, the plentiful supply of high energy air just outside the diffuser's walls. In Figure 4.14 the boundary layer control technique is illustrated. Boundary layer separation is avoided by allowing external high energy air to create a jet along the internal wall heading downstream. The jet flow is strong because the pressure is subatmospheric inside the diffuser, and only a little flow is needed because the jet acts exactly where extra flow momentum is needed. In the wind turbine application, the externally supplied bleed air is only a minute fraction of the main diffuser flow volume so the additional equipment cost and performance penalty is very small, and the technique is particularly effective. In the baseline design, two annular slots are used in the diffuser wall. These slots enable the diffuser to produce almost a doubling of available wind power to the turbine. In addition to the pressure recovery inside the diffuser, the rapid flow divergence continues long after the end of the duct.

Figure 4.14: Flow Patterns and Velocity Profiles in a BLC DAWT

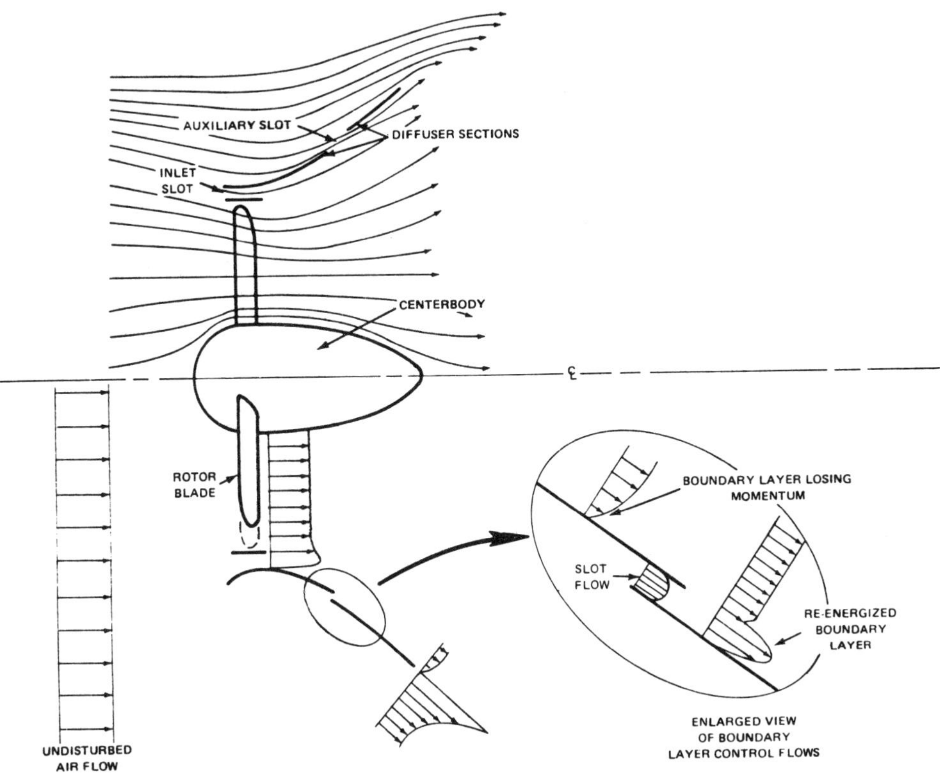

Source: COO/2616-2 (Pt. 1)

There is consensus that the major unresolved technical problem in the design of conventional Wind Energy Conversion System turbines (WECS) for very large power systems is in the blade dynamics of large diameter rotors. Because of this fact, integration of wind generators into a national or regional power grid is inhibited by the unacceptable reliability of very large units or the economic liability of many smaller units of comparable total power output. This technical factor interacts with the economic constraints associated with matching supply and demand schedules in variable wind, the low power density of wind, the high development risk of new system concepts, and the capital-intensive nature of wind power systems.

Many of these capital and performance restrictions of WECS can be reduced or eliminated by enclosing the wind turbine rotor in a suitably shaped duct. The nonrotating duct structure provides a compact diffuser section behind the rotor that produces a power augmentation of considerable magnitude (typically 1.5 to 2 times) for a given size rotor, as well as dampening gusts, lowering cut-in wind speed by raising the level of axial velocity significantly, and eliminating tower wakes.

Ducted wind turbines have been suggested periodically for over half a century. Early investigators tended to underrate the concept by various combinations of analysis and/or conceptual errors, failed to anticipate low pressures at the shroud exit, and assumed that diffusers would have to be very long and costly to avoid boundary layer separation. Early in 1960, an Israeli group (Kogan) repeated correctly much of the earlier analytical development and proceeded to some experiments concerning performance of two dimensional diffusers, duct inlet contour effects on axial misalignment of the incoming flow, and performance of short axisymmetric diffusers with exit plane ejectors.

Later studies by Igra showed that the exit plane of a Diffuser-Augmented Wind Turbine (DAWT) has, in fact, the greatly depressed pressure level that was hoped for in order to obtain large augmentation effect. Evidence is now clear that base pressure reductions on the order of one-half the wind dynamic pressure result from both viscous and inviscid interaction, and this kind of base pressure reduction raises the diffuser augmentation significantly. The diffusers in this application have the unique feature that a plentiful supply of high energy air exists just outside the diffuser wall.

However, the aerodynamic benefits of the DAWT must overcome the highly visible cost of the shroud structure. Therefore, the focus of research has been the technical challenge of extracting sufficient fluid mechanic performance from a very compact diffuser that will be cheaper to build than the incremental cost of larger diameter wind turbines to produce the same rated power. Such short diffusers are cost-competitive designs because they will result in reduced capital expense of electrical energy generation by wind energy conversion.

Diffuser Augmentation Theory

The results of one dimensional momentum theory applied to a DAWT show that the power available to a perfect ducted turbine can be made to increase significantly by a high exit/rotor area ratio, a high diffuser efficiency, an optimum ratio of inflow disk to free stream velocity, and a strongly negative base pressure coefficient. Optimistic values of diffuser efficiency and base pressure coefficient, operating with optimized disk loadings, lead to analytical predictions of augmentation ratio as high as 4 in the limit of infinite area ratio. Augmentation ratio is the ratio of power output of an ideal DAWT rotor to that of an unshrouded ideal rotor of the same size and in the same wind. Since structural cost increases rapidly with shroud area ratio, the most cost-effective DAWT will operate at modest area ratio, and a correspondingly smaller augmentation ratio. Present baseline configuration optimizes at an area ratio of 2.8, and has delivered augmentation ratios as high as 1.9.

Diffuser Development

Several different diffuser concepts have been suggested for a DAWT. The design criteria of a maximum subatmospheric pressure at the exit plane, a large pressure recovery within the diffuser, and the smallest possible structural cost imply the need for functional diffusers with equivalent half angles much greater than the conventional 3° to 6°, which are dictated by boundary layer separation. Nevertheless, compact diffusers for DAWT must maintain effective performance characteristics.

The two most promising design concepts have been chosen for further study. The first of these employs repeated injection of external air for boundary layer control. External high energy air from the wind is injected tangent to the wall, thereby adding axial momentum to the boundary layer. The additional momentum helps the boundary layer fluid to flow against the severe adverse pressure gradient and frictional losses that are present in the wall region of large angle diffusers. This can prevent the flow from separating from the wall, the primary cause of the failure of flow in large angle diffusers.

The second diffuser concept is the use of a diffuser constructed from short ring airfoils. Each ring airfoil produces a local aerodynamic pressure and velocity field as a result of the section contour. The low pressure along the internal ring surface induces more flow through the turbine. By the use of high lift wing contours for the rings, or by use of flaps, appreciable augmentation is obtainable.

Experiments have been conducted in a low speed, low turbulence level, free jet tunnel facility. The core region of a 29.2 cm diameter free jet flow is used as the test section in which uniform wind conditions are simulated. A velocity of 13.0 m/sec (43 fps) was used for the tests. Models were mounted slightly downstream of the nozzle exit plane. Boundary layer controlled diffuser models were constructed of stainless steel and aluminum sheet metal. Ring wing models were machined from aluminum bar stock. Over 150 configurations have been investigated in this program.

The instrumentation employed a variable reluctance differential pressure transducer. The measurements were derived from the combined use of a single static pressure probe and a single total pressure probe, each of which can pass right through the screens used. For the axial and radial pressure measurements, the probes were mounted on a motor-driven traversing mechanism. This device produces an electrical output proportional to its position so that pressure versus spatial position could be recorded.

Since a family of wind turbines is impractical to build for an exploratory investigation of small scale diffuser models, the turbine energy extraction by screens that dissipate the energy at the turbine station was simulated. The turbine performance is simulated by the screen loading. The power extracted per unit area is the product of the total pressure drop and local velocity. From measurements of disk loading and ratio of local to free stream velocity, the augmentation ratio, r, can be found. Repeated axial surveys at different radial positions indicate that r increases with radial distance from the axis. The weighted average over the entire cross section of the turbine simulator gives the total diffuser augmentation ratio, $\bar{r}$.

Model Test Results

The augmentation ratio was measured by performing hundreds of pressure surveys on a large number of model diffusers in a range of turbine disk loadings between 0.3 and 1.10. A peak $\bar{r}$ value of almost 1.9 can be achieved with a 30° diffuser half angle at an optimum disk loading coefficient of about 0.63, using boundary layer control (see Figure 4.15). The ring wing diffuser exhibits an increasing $\bar{r}$ value with disk loading. Although an optimum disk loading has not been encountered for the flapped NACA 4412 contour diffuser tested, an $\bar{r}$ = 1.6 is indicated for the maximum disk loading coefficient tested, 1.10.

Figure 4.15: Representative Model Test Results of Two Short Diffuser Designs

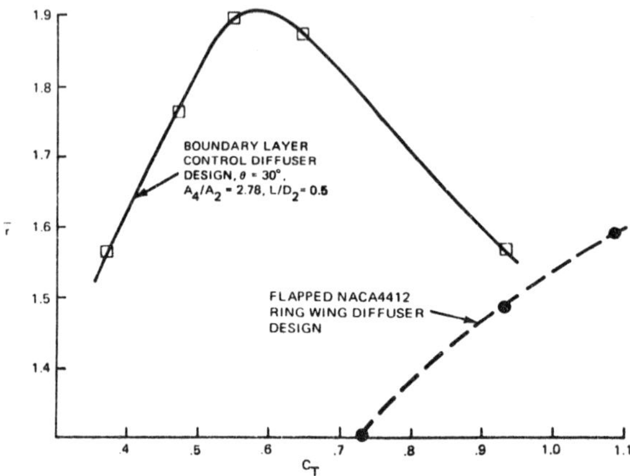

Source: COO/2616-2 (Pt. 1)

The ratio of local-to-free stream velocity for the boundary layer control (BLC) diffuser at peak $\bar{r}$ is 1.27. The value of this ratio is 0.9 for the ring wing design at $\bar{r}$ = 1.6. Both diffusers produce considerably greater inflow velocity to the turbine than the theoretically best velocity ratio value (0.67) for conventional unshrouded turbines, at an optimum disk loading coefficient of 2.0. For the BLC diffuser, where such a comparison is possible, the optimum performance from one dimensional theory should occur at a local disk loading of 0.65. This is computed from a measured reduction of exit plane pressure equal to 0.57 of the wind dynamic pressure and a measured induced velocity ratio of 1.27.

The low exit plane pressures that are always measured in the DAWT experiments were first noted by Igra. They are so large and so significant to DAWT performance, that their origin must be explained. Originally, the phenomenon was attributed to the strong shear layer peripheral to the wake, but recent analysis based on thin airfoil theory has indicated that the phenomenon is explainable completely by inviscid processes. This fact promotes confidence that low base pressures will remain when diffusers are scaled up to full size and high Reynolds numbers.

Economic Analysis

Economic assessment of the DAWT relative to a conventional WECS requires that the most competitive versions of each concept be identified and priced according to realistic and consistent costing rules. An analysis objective of comparing the capital cost per unit power of DAWT with that of the optimum conventional wind turbine was established. This involves the cost trend of production quantity (~100/year) wind turbines with size as well as the cost trend of a short diffuser of demonstrable augmentation capabilities. The rotor cost information has been generated by two different contractors under NASA-Lewis Research Center sponsored contracts.

The diffuser cost estimates were made by Grumman on the basis of a finite element structural analysis for a full scale BLC baseline diffuser design (2.78 area ratio, length/turbine diameter = 0.5, θ = 30°, $\bar{r}$ = 1.89). The total cost of only the diffuser and rotor elements has been normalized by the cross-sectional area of the turbine to facilitate comparison of the two systems. Two characteristics of these input data are clear:

- A 100% uncertainty exists in state-of-the-art turbine cost estimates, especially in the low to intermediate size range, depending on the estimating organization.
- The typical U-shaped average cost curve shows the initial economics of scale to about 25 meters diameter, and sharply rising costs as larger rotor diameters require more elaborate construction and advanced materials to overcome escalating aeroelastic stresses.

At this time, operationally reliable rotor diameters greater than about 65 meters are highly speculative. The most straightforward advantage of the DAWT is that it offers an alternative to replication of conventional turbines when desired unit output is greater than that of a larger-than-optimum WECS. With a DAWT, significantly greater power can be generated without the cost and time delay of additional land acquisition, legal hearings, grid terminals, maintenance and operating staff, or other costs of replication of size-limited wind power plant designs.

Other secondary benefits from using DAWT are tower cost reduction, elimination of tower wake and reduction of wind shear and direction problems, and greater annual on-line factor. The operational usefulness of the annual wind energy spectrum is broadened because the DAWT inlet acceleration lowers the minimum wind speed for cut-in of turbine power for fixed size and rpm rotor. Further, the duct can easily raise the high-speed cut-out, end of the wind spectrum because of the inherent capability of the diffuser to modulate flow by introducing spoilers or movable stators.

It is possible to compare the DAWT with conventional WECS in different ways, depending on the application. To illustrate the differences, two specific comparisons have been chosen: equal rotor diameters used in both systems and equal rated power output in both systems.

For the latter condition, the rotor diameter of the WECS is increased by the square root of the DAWT augmentation ratio, $\bar{r}$. Of course, it is not currently realistic if the WECS turbine must be increased much beyond 65 meters.

The two cost comparisons are shown in Figures 4.16 and 4.17 on the basis of the turbine and diffuser elements of the system only. That is, differences in the costs of the electrical generator, tower, foundation, control system, and mechanical shaft linkage or transmission system components have not been considered in these cost comparisons, nor have any other system economics of scale.

Figure 4.16: Cost Comparison of DAWT and Conventional WECS for Equal Rotor Size

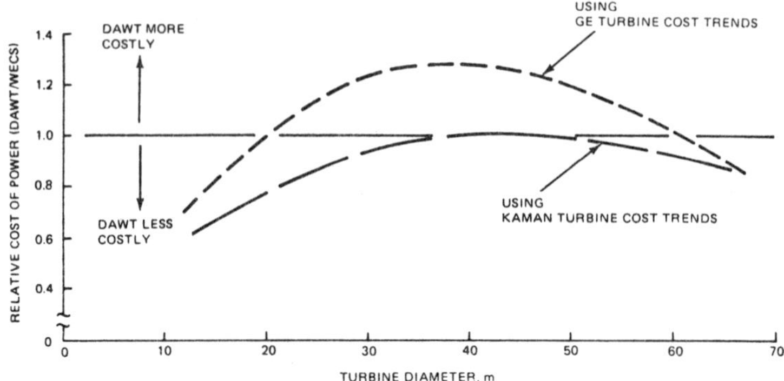

Figure 4.17: Cost Comparison of DAWT and Conventional WECS for Equal Rated Power

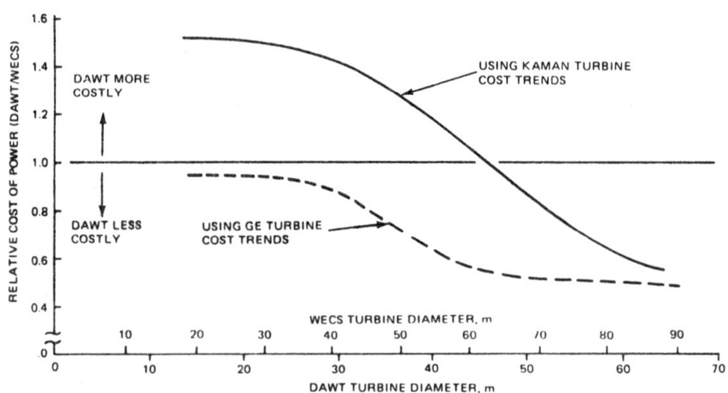

Source: COO/2616-2 (Pt. 1)

In the case of equal rotor size in both systems (see Figure 4.16) it is clear that DAWTs have an economic advantage for small rotors and very large turbine diameters. The economics in the intermediate size range are clouded by the diversity of authoritative cost estimates; the DAWT can be marginally less (i.e., 0 to 10%) or up to 25% more costly.

For equal rated power output, shown by Figure 4.17, the DAWT can be significantly cheaper (~50%) than a WECS for turbine diameters greater than about 35 meters. For smaller turbine sizes, the DAWT can be anywhere from marginally cheaper to significantly more expensive depending on which cost trend one actually experiences.

A usable annual wind pattern factor has not been included in the graphic results. Because of the inherent wind speed-up features of the diffuser inlet section, there can result a 5 to 50% greater annual power conversion for the DAWT compared to the WECS. Therefore, there is a probable real economic advantage for the DAWT, regardless of which turbine cost estimate is appropriate.

Stators

One of the goals of this investigation was to determine if the struts that will support the turbine within the shroud could be used to advantage as a variable stator. The objective was to provide varying preswirl to a constant speed, constant pitch, rotor so that it could operate efficiently throughout the wind range. A computer code was written to investigate this concept in the context of a DAWT, and several such designs were formulated and evaluated. They appear in most respects to be satisfactory in performance, although repeated attempts to broaden the region of high power coefficient suggested that the wind range over which the stator control could be effective was narrow.

This suggested a comparison of matched rotor installations with and without stator preswirl. Off-design performance increased substantially without the stator, both above and below the design wind speed. This surprising result is because the rotor airfoils gain less by remaining at an ideal angle of attack than the rotor, as a whole, loses. These losses are due to adverse change in relative velocity when the stator attempts to hold rotor angle of attack constant with fixed pitch and speed. The following conclusions have been reached:

(1) Controllable stators are useful only as a possible means of inexpensive protection against overload at high wind speeds.

(2) A constant speed, fixed pitch rotor is practical without a controllable stator if the rotor dynamic design can withstand the oscillations of rotor blades operating fully stalled at high wind speeds.

(3) Because struts are needed anyway, there may be some value in introducing a constant preswirl to produce better performance at low wind speed with less rotor blade area, sacrificing performance at high wind speeds when the power available is greater than that which can be used by a controlled generating system.

VORTEX AUGMENTOR CONCEPT

Under certain predictable conditions vortices appear in a flowing fluid. These vortices are real manifestations of an idealized whirling flow field in which the velocity follows concentric circular streamlines with a magnitude inversely proportional to radius. Vortices are efficient concentrators of kinetic energy: consider the tornado which is an example of an atmospheric vortex flow. The thrust of the work on vortex augmentors is to utilize the unusual aerodynamic characteristics of vortices to develop improved wind energy conversion systems.

The keystone element here is the generation and control of discrete vortices of high kinetic energy density by appropriate interaction of aerodynamic surfaces with natural winds of low kinetic energy density. Suitably designed turbines are utilized to extract energy from this compacted vortex field. This idea is termed the Vortex Augmentor Concept (VAC). The specific goals of the project are: (1) evaluation and assessment of the VAC in the context of its potential for wind energy conversion applications, and (2) development of a small-scale working model of a particular VAC system, identified early in the project as being the most suitable candidate for immediate application.

Two distinct types of devices were initially considered for the program: slender body vortex generators and bluff body vortex generators. A general discussion of these two types follows.

Slender Body Vortex Generators

A typical example of this type of vortex generator is the wing of an aircraft. Vortices emanating from the wing tips trail out behind the aircraft and, because of the concentrated swirling motions in these trailing vortices, they can pose a danger to trailing aircraft. This is illustrated in Figure 4.18 (18).

When jumbo jets are in routine operation, their vortex wakes will make air terminals even more hazardous for light aircraft. If a light plane flies through the wake transversely, it will be subjected to high normal force loading similar in effect to gust encounters. If it flies midway between the trailing vortices, it will experience a strong downwash, possibly stronger than its rate-of-climb performance; this kind of encounter is especially dangerous near the ground where the persistent settling tendency may cause a light plane to crash. The most improbable, and most hazardous, type of encounter is flight straight down the vortex core; this induces extremely high roll rates on the penetrating aircraft, which also may exceed its control capabilities. This was a factor in 1966's B-70 tragedy.

The potential dangers of interference between aircraft trailing vortices and following aircraft has prompted a great deal of research on vortex phenomena including an international conference in 1970 (19) and a two-day Professional Study Seminar in 1974 conducted by the American Institute of Aeronautics and Astronautics (20).

Similar vortex structures are produced by flow separation at sharp leading edges of low aspect ratio wings such as delta planforms. This type of flow field is illustrated in Figure 4.19 (18). The process of vortex formation is illustrated in this figure and the added contribution to lift due to the vortices is depicted. This increment in lift for such planforms provoked further research on vortex phenomena. Basic theoretical work in this field is presented in monographs such as that by Thwaites (21).

Figure 4.18: Vortex Generation by Aircraft Wings

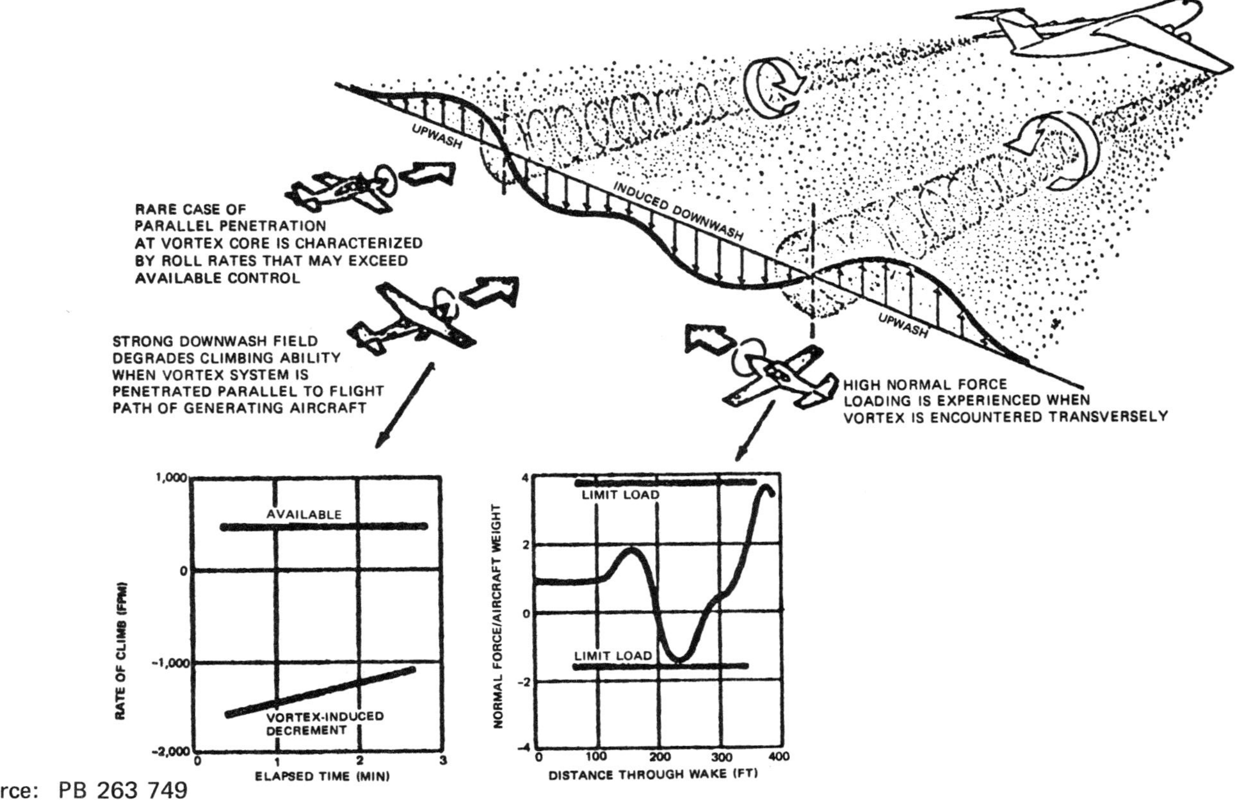

Source: PB 263 749

Figure 4.19: Vortex Generation by Delta Planforms

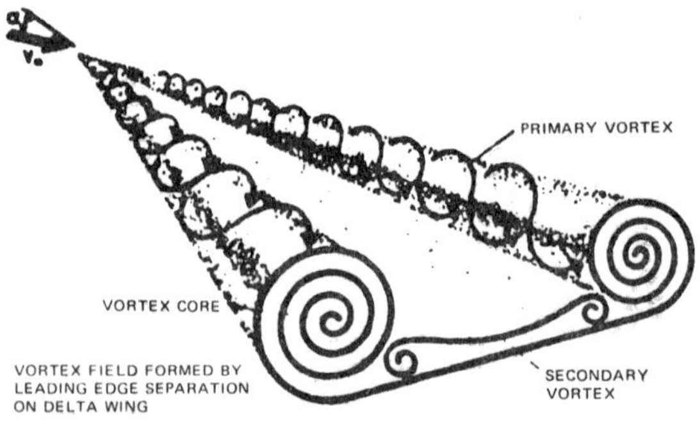

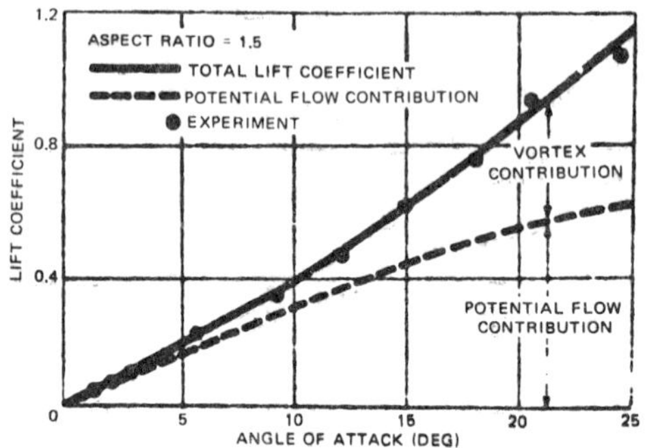

Source: PB 263 749

It is suggested that the energy in these amplified swirling flow regions, whose strength is dependent upon the attitude of the delta wing, can be efficiently tapped by suitably designed rotors. Both the axial and circumferential components of the velocity are greater than the undisturbed, oncoming flow velocity so that the absolute velocity in the vortex region, given by the square root of the sum of the squares of the two components, is substantially larger than the free stream velocity.

Thus, a small turbine wheel of a diameter on the order of a quarter of the semispan of the delta can be operating under higher velocity conditions than a turbine in the free stream. Such a system of velocity multiplication provides potential for smaller, highly loaded turbines with all their advantages, potential for altering incidence of the delta so as to maintain more nearly constant speed turbine operation in the face of varying wind speed, and potential for operation at lower wind speeds than in an unaugmented wind energy system.

It is evident then that there is a significant body of theoretical, experimental and practical information on vortices generated by flow past slender wings and bodies. The objective of this program includes the application of this data base to the development of the Vortex Augmentor Concept (VAC). Natural winds flowing past suitably designed bodies of the type discussed here can be forced to concentrate a significant portion of kinetic energy into these localized vortices for extraction by appropriate turbine design. In addition, proper attitude of the vortex generator can allow control of the position and strength of the vortices produced.

Bluff Body Vortex Generators

When wind flows by a bluff obstruction, for example, a billboard, its speed decreases near the center and accelerates to values much higher than the undisturbed wind speed at the edges. In the process vortices are generated all around the obstacle, sometimes in very complicated patterns. Such flow fields are generated on aircraft by such protuberances as fasteners, spoilers, flaps, antenna masts, etc., and on other obstacles such as islands, mountains, land structures, moving vehicles. A review of such flow fields in general is given by Sedney (22).

Initial studies of flow fields generated by bluff obstacles have been carried out in the Polytechnic laboratories (23). The experimental model is schematically illustrated in Figure 4.20. A particular configuration pertinent to the discussion is a thin square plate set perpendicular to the free stream. Velocity contours downstream of the obstacle are shown in Figure 4.21. In these plots the obstacle and velocities measured are on the undersurface of the plate. Note that near the edges of the obstacles large velocity overshoots occur; the gradients accompanying these overshoots are indicative of the vortex field alluded to previously.

Figure 4.20: Schematic Representation of the Flow Field Behind a Plate-Mounted Bluff Obstacle Placed Normal to the Flow (23)

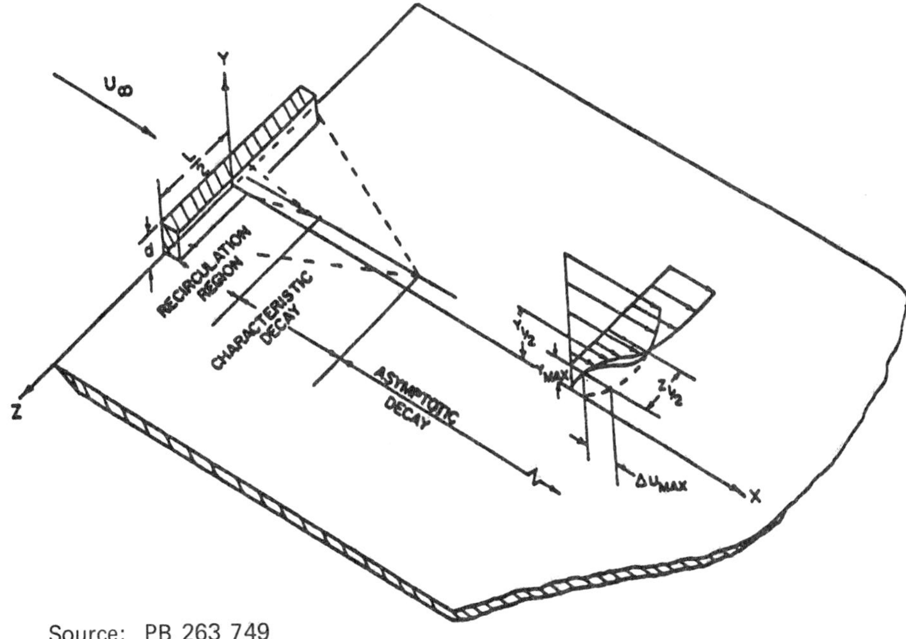

Source: PB 263 749

Figure 4.21: Velocity Field Contours at Various Stations Behind a Square Obstacle Placed Normal to the Flow

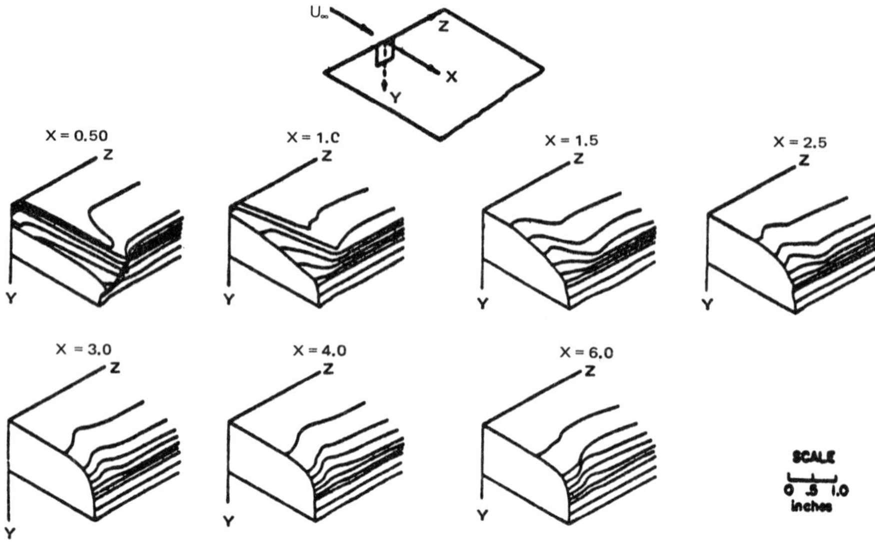

Source: PB 263 749

It is suggested that operation of a vertical axis rotor system, such as one of Savonius type, will exhibit improved performance if situated properly in this highly sheared flow field. That is, the advancing, or driven leg of the rotor, can be exposed to the high velocity overshoot while the lagging leg is exposed to the lower velocity in the lee of the obstacle. Variations in the magnitude of the thus-augmented velocity fields can be achieved by attitude control of the obstacle.

The vortices generated by bluff bodies can be utilized to drive suitably designed rotors; the strength and position may be controlled by appropriate changes in orientation and geometry of the obstacles. The state-of-the-art in analysis and experimentation on such systems is sufficiently far advanced to warrant evaluation of this second variant of the VAC in terms of wind energy conversion.

The Vortex Field Flow

Characteristics of the vortex field generated by delta-type aerodynamic surfaces are fairly well understood and have received considerable attention. Detailed studies of the velocity field in the vortex flow are not nearly so well documented. The graphs in Figure 4.22 show what are probably the only reliable velocity measurements readily available. The axial component of the velocity in the vortex, V_A, from the data presented by Hall (24), is seen to be substantially greater than the oncoming wind speed, V_∞. But in addition to this there is also a tangential component of velocity in the vortex V_T, which is on the order of magnitude of the oncoming wind speed, V_∞. The resulting absolute velocity in the vortex, i.e., the vector sum of V_A and V_T, will therefore also be substantially greater than V_∞.

Figure 4.22: Experimental Data on Velocity Field in Delta Wing Vortices

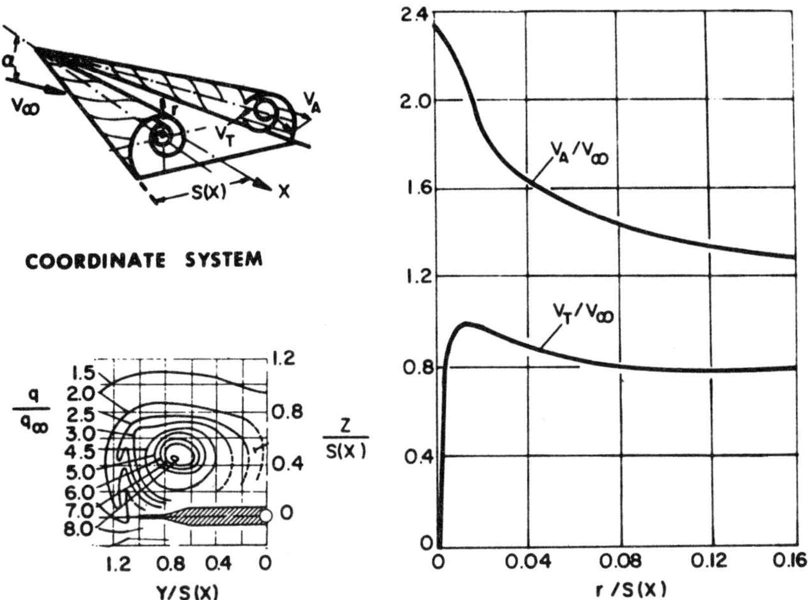

Sketch shows coordinate system. Graph on the right shows axial velocity V_A, and tangential velocity V_T from data presented by Hall (Ref. 24). The graph at the left shows contours of dynamic pressure $q=\frac{1}{2}\rho V^2$, where V is the total velocity and was taken from Hummel (Ref. 25).

Source: PB 263 749

This then is the augmenting effect of the vortex generator surface: it amplifies the wind speed in the vortex locality in that the swirling flow tends to concentrate the low kinetic energy density wind from a large area into a high kinetic energy density flow in a small (vortex) area. Another illustration of this effect can be obtained from the data given by Hummel (25). These data appear as the contour plot of Figure 4.22 and show lines of constant dynamic pressure, $q = \frac{1}{2}\rho V^2$. The numbers identified with each contour are therefore the square of the ratio of the magnitude of the local resultant velocity to the wind speed. It can be readily seen what a splendid job of wind focusing the vortex augmentor surface does.

The directed nature of the vortex flow field allows a choice in rotor operation as illustrated in Figure 4.23. If the rotor is to spin in the sense opposite to that of the vortex the resultant velocity is high. On the other hand, the rotor may be designed to spin in the same sense as the vortex so that a much lower resultant velocity is seen by the rotor for the same rotational speed. One may choose the mode best suited for a given application; in general, the operation of the rotor "in a tailwind" will be advantageous for high tip speed ratios since the drag force on the rotor, which is proportional to V_r^2, will be substantially reduced.

Figure 4.23: Velocity Diagrams for Rotor Blades in the Vortex Velocity Field

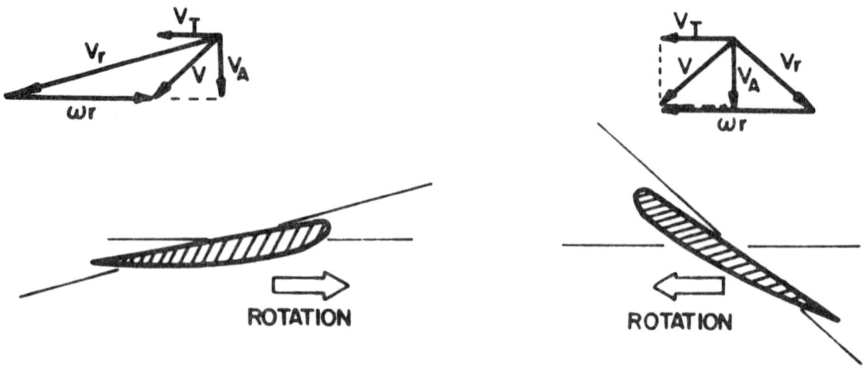

The diagram on the left is for rotor rotation against the vortex spin and that on the right is for rotor rotation with the vortex spin.

Source: PB 263 749

Power Possibilities: Using the velocity field information presented previously, the ideal performance of VAC systems may be calculated. A general computer program for rotor performance based upon blade element theory was developed and utilized for design purposes.

To give some indication of the possibilities inherent in the VAC, a calculation of the performance of a rotor operating in a wind of 7.5 m/sec (approximately 17 mph) as a function of rotor radius was performed for three conditions: (1) unaugmented operation, (2) operation in the VAC mode using the data of Hall, and (3) operation in the VAC mode using the data of Hummel.

The results of the calculation are shown in Figure 4.24 and they indicate that, for the same rotor radius, the VAC system will provide from twice (Hall's data, low angle of attack delta surface) to six times (Hummel's data, high angle of attack delta surface) the power output of a conventional system facing the natural wind. Or alternatively, for the same power output, a conventional system using a 24 m (80 ft) radius rotor could be replaced by a VAC system of 17 m (57 ft) radius (low angle of attack case) or as little as 9.8 m (33 ft) radius (high angle of attack case). Quite substantial improvements in power performance are therefore possible with VAC systems.

Power Testing of Various Configurations

At least ten basic vortex augmentor geometries have been studied in a program including over 100 tests resulting in close to 1,000 data points. Many detail changes including five different rotors and five different generators were tested. Velocity and power measurements were made, smoke flow and tuft grid flow visualization were used, and many results are in the form of still or motion pictures.

Figure 4.24: Comparison of Ideal Power Output of a Wind Turbine as a Function of Radius at a Free Wind Velocity of 7.5 m/sec (roughly 17 mph)

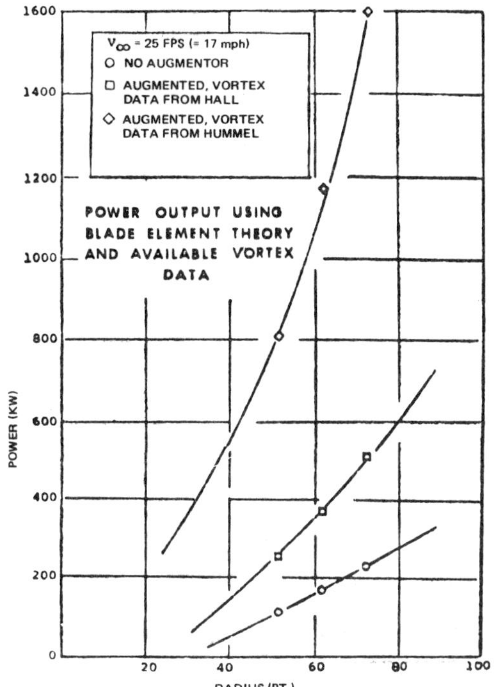

Note that 20% of the radius is excluded from the calculation to account for the hub effect.

Source: PB 263 749

Experiments on the performance of laboratory scale VAC system models were performed in the 1.2 x 1.5 m (4 x 5 ft) environmental wind tunnel of the Polytechnic Institute of New York Aerospace Laboratory (see Figure 4.25). Many different configurations were tested; only a limited number will be described here for brevity. A typical set of configurations includes:

(1) No augmentor: Rotor-generator assembly faced into the free stream;
(2) Clean delta: A flat half-delta (right triangle) mounted vertically, 75° sweepback of the chamfered leading edge, 2.18 m (87 in) long by 0.6 m (24 in) high;
(3) Ramp amplifier: Same as previous configuration with addition of ramp-like centerbody to further intensify vortex; and
(4) Flap amplifier: Same as configuration (2) except for vertical hinge line 1.9 m (75 in) back from tip. The surface aft of hinge line can then be used as flap, i.e., to effect a change in camber of the delta surface. The change in camber can be such as to intensify the vortex; in addition the flap surface can be used as a control element to aid in changing the angle of attack of the whole surface.

Figure 4.25: Schematic Diagram of the Environmental Wind Tunnel of the Polytechnic Institute of New York

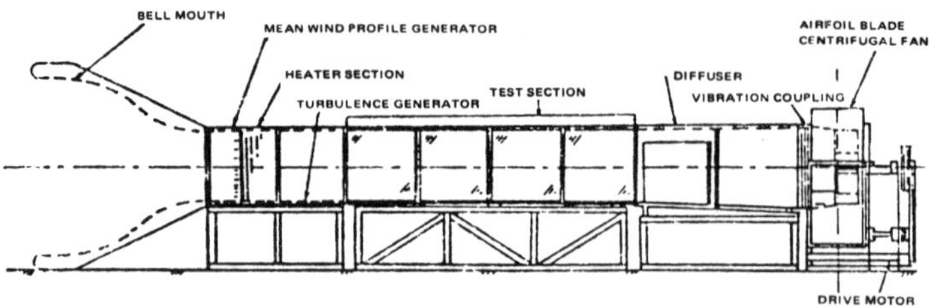

Source: PB 263 749

A 0.275 m (11 in) white pine, constant chord rotor (NACA 63618 airfoil section) was used in the test series presented in Figure 4.26. A commercially available bicycle generator was used in conjunction with load resistance box and oscilloscope for the power measurements shown in the figure. The VAC systems not only produced from two to over four times the power of the unaugmented system at 6 m/sec (20 ft/sec), but were able to run and carry a load at much lower velocities than the unaugmented system. The concentrating effect of the VAC promises to allow wind systems to operate at much lower wind speed, thereby opening up whole new markets for wind energy conversion systems.

Figure 4.26: Power Output vs the Cube of the Free Stream Velocity for Four Wind Energy Conversion Systems as Measured in the Polytechnic Environmental Wind Tunnel

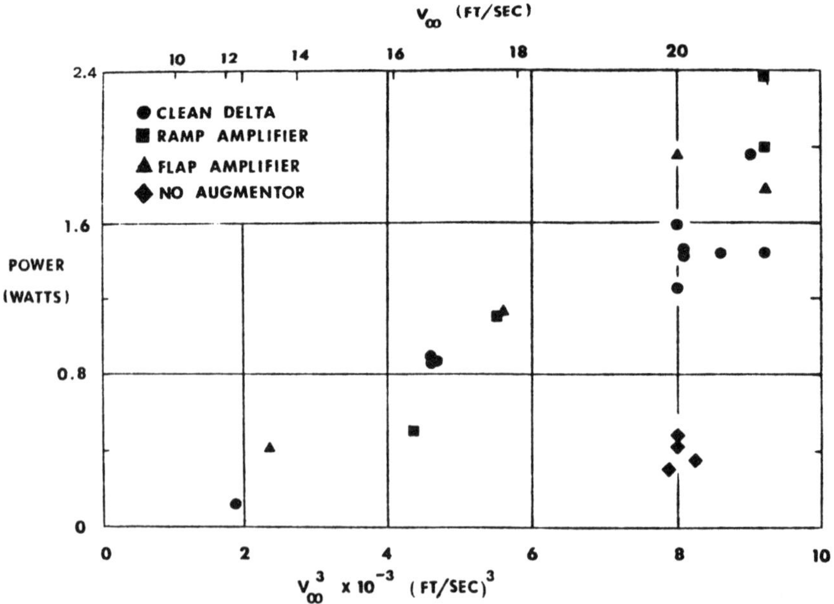

Source: PB 263 749

During this phase of the program it was concluded that bluff body vortex generators were inferior to the slender body type since they lead to augmentation values of around 1.5 as compared to 4 and more for the better slender body types. These types of augmentor systems were therefore excluded from further study in this program. With attention now solely on slender body vortex generators, an extensive wind tunnel investigation of vortex geometry and characteristics was performed.

ELECTROFLUID DYNAMIC GENERATOR

An electrofluid dynamic (EFD) generator is a direct energy conversion device which converts fluid dynamic energy into electrical energy. In an EFD generator, charges of one polarity are seeded, often by a corona discharge, into a flowing neutral gas. These charges are transported by viscous interactions (molecular collisions) with the neutral gas against an electrical potential gradient and produced dc power. EFD generators are typified by low current densities and high voltages (e.g., 100 kV to 1,000 kV), requiring on the order of 10^8 neutral molecules to transport one elementary charge against the high potential gradient.

Charge carriers for EFD generators can range from ions with high mobility to singly charged colloid particles with very low mobility. Efficiency is greatest with low mobility charge carriers. Analyses demonstrate that an EFD generator using charged colloid particles is well suited to electrical power generation from the wind.

Description

A schematic diagram of a simple EFD wind generator is shown in Figure 4.27. The generator consists of the following parts: (1) a mechanism for producing charged colloids; (2) an inlet electrode which also serves as an attractor electrode; (3) a collector electrode; (4) a high voltage power supply; and (5) a control system.

Figure 4.27: Schematic of an EFD Wind Generator

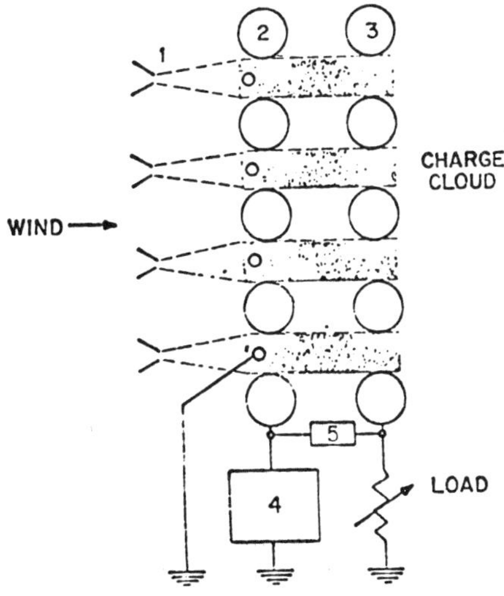

(1) COLLOID CHARGING SYSTEM
(2) INLET/ATTRACTOR ELECTRODE
(3) COLLECTOR ELECTRODE
(4) HIGH VOLTAGE POWER SUPPLY
(5) FEEDBACK CONTROL SYSTEM

Source: COO/4130-77/1

The high voltage power supply places the attractor electrode at a high voltage relative to the colloid producing mechanism. Charged colloid particles of one sign produced by the charging system are swept past the attractor electrode by the moving air towards the collector electrode. The collector voltage depends on the load and current, but the charged particles must be driven up a potential hill by the neutral particles in the air. Thus, electrical power is generated directly by the moving air with no moving parts required. The feedback control system senses the voltage on the collector and adjusts the attractor voltage, thereby controlling the output current in order to hold the output voltage constant.

Advantages and Disadvantages

Some of the advantages of EFD wind generators are the following:

(1) Power is generated at all wind speeds for which the generator is constructed to withstand.
(2) Power output increases with wind speed up to the highest speeds.
(3) There are no moving parts, with the possible exception of orientation devices, resulting in quiet operation, fewer foul weather problems, and the virtual elimination of dynamic loadings.
(4) Fatigue problems are minimized.
(5) Maintenance requirements are reduced, compared with those for conventional wind generators.
(6) Zero mechanical inertia allows gust energy to be utilized.
(7) Production of high voltage dc power is ideal for transmission from remote generation sites.
(8) Versatile cross section shapes are possible.
(9) Large sizes are reasonable.
(10) Direct conversion from kinetic power to electric power eliminates need for generator matching.
(11) All-electronic feedback controls allow rapid compensation for changes in wind velocity and loading conditions.
(12) Off-shore use is reasonable.
(13) Stationary designs that can produce full power for various wind directions are feasible.
(14) Two-dimensional diffusers with no circular sections can be used.
(15) Construction cost projections are low.

For central power production applications, low cost of fabrication would be the most significant advantage; for applications in remote regions, ease of maintenance would be very important.

There are certain disadvantages to an EFD wind generator, some of which are the following:

(1) A source of colloid particles (e.g., water droplets) is required.
(2) Corrosion and/or erosion problems resulting from the colloid particles are possible; such problems are minimized by using small size colloids.
(3) Safety problems may exist (in particular, there is no intrinsic sign of danger from such a device, since there are no moving parts).

(4) Colloid charging methods may produce ozone, NO_x, or nitric acid.
(5) High voltage dc is disadvantageous for some applications.

A radically simpler design for wind generators is apparently required for their widespread application in central power production. The EFD wind generator is probably the simplest wind energy conversion device that can be used to produce electrical energy. Although it is currently impossible to make detailed cost estimates of EFD wind generators, crude estimates (1976 dollars) suggest that a cost of $75/m^2 ($7/ft^2) or less would be reasonable.

Conventional windmills are currently limited in size both because of blade design problems and because the rotational speed of the rotor must be reduced as the blade diameter is increased in order to keep the tip speed at the value which maximizes efficiency. Additionally, low rotational speed causes difficulties in transmission design. For the EFD wind generator, there are no fundamental limitations on size; therefore, economics of scale can be realized.

EFD Wind Generator Theory

The theory of EFD devices is well developed and scaling laws are understood. Experimental evidence compares favorably with available theory and the feasibility of the EFD generator has been well established (26)(27). Because high power outputs and high power densities were desired, the main thrust of previous research was directed towards high velocities and high pressures in the conversion section. Because of electric breakdown considerations, the maximum pressure drop that can be achieved in a one-dimensional EFD generator at 1 atm is about 40 N/m^2; at 30 atm, the pressure drop could be 900 times as great, or 36,000 N/m^2. The pressure drop of 40 N/m^2 is quite suitable for wind generator applications. An ideal windmill extracts eight-ninths of the dynamic pressure of the wind; thus, 40 N/m^2 corresponds to a wind velocity of 8.58 m/sec (19.2 mph).

Theoretically, below 8.58 m/sec the performance of an EFD wind generator is not limited by electrical breakdown; above 8.58 m/sec, the generator performance is limited by electrical breakdown, so that the pressure drop in the conversion section would have to remain constant, with the power increasing linearly with the velocity through the conversion section. It is appropriate to consider 8.58 m/sec as the velocity at which a one-dimensional EFD wind generator reaches "rated" power; however, the EFD wind generator has an advantage over a conventional wind generator, since the output power can still increase linearly with wind velocity, whereas the output power of the conventional generator remains constant after rated power is obtained.

The ratio of the velocity through the EFD wind generator to the wind velocity could be kept close to the value for optimal performance by varying the current (and hence, the charge density) in the conversion section, for wind velocities below 8.58 m/sec. For wind velocities above 8.58 m/sec, the ratio of velocities approaches one asymptotically as the pressure drop in the conversion section becomes small compared with the dynamic pressure. Results of analyses are shown in Figure 4.28, where the curve labeled "ideal 1-D wind generator " is for colloids without slip.

Figure 4.28: The Performance of the EFD Wind Generator as a Function of Mobility

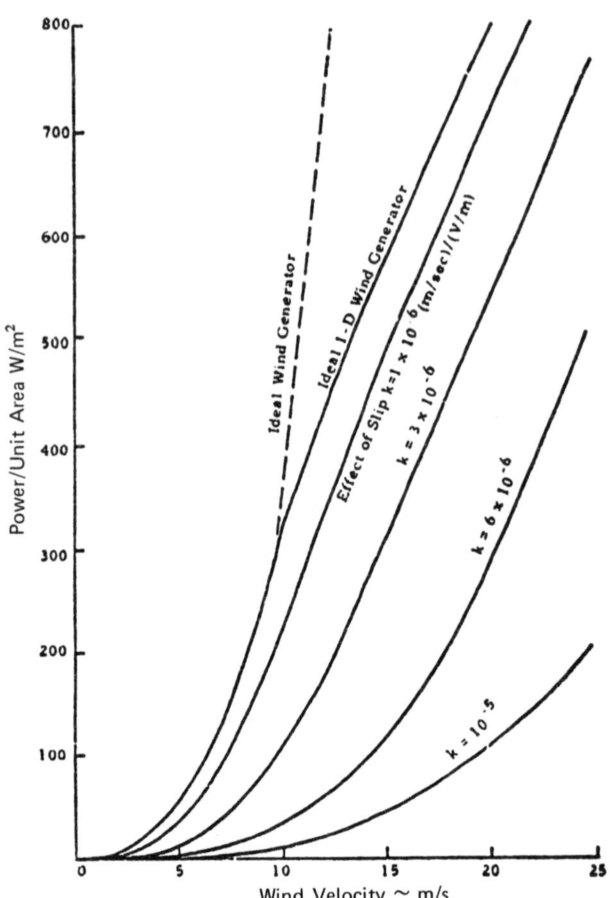

Source: COO/4130-77/1

If two-dimensional effects are included, then higher pressure drops can be obtained before breakdown limits performance; hence, higher values of velocity can be achieved before the EFD wind generator deviates from ideal performance.

Summary of Progress Achieved

Significant progress has been made toward obtaining practical EFD wind generation of electric power. Work accomplished has included the following:

(1) A substantial theory has been developed to predict the performance of EFD wind generators, including the effects of geometric parameters, electrical conditions, and drag.

(2) Computer studies of electric field effects have progressed, in which a realistic characterization of these effects is possible.
(3) Theoretical results have been used to aid in the design of an EFD wind generator test rig, and to predict its performance for comparison with experimental results.
(4) A small Eiffel wind tunnel has been designed and built.
(5) Instrumentation was designed and installed in the tunnel, including a drag balance.
(6) A preliminary EFD wind generator test rig was designed and installed in the tunnel; this test rig permits adjustment of various components in order to obtain data over a wide range of geometric parameters.
(7) Test data have been obtained with the original test rig and with modified versions of the rig.

Over 200 test runs have been made in the wind tunnel. Information from these tests, from computer models and other theoretical analyses, and from a comprehensive review of the literature is sufficient to allow the identification of specific objectives for continued research. The most important areas for research are identified as the following:

(1) Development of a system for the laboratory production of charged colloid particles with suitable mobility at high generation rates;
(2) Experimental determination of the performance of the EFD wind generator test rig with design-level charge densities and with charged colloids of proper mobility, and comparison of this performance with theoretical predictions; and
(3) Construction of a theory for energy-efficient charged colloid particle production systems, with experimental investigations of such systems.

It now seems clear, based on experience to date, that generation of charged colloid particles is the most critical problem in the development of EFD wind generators. Techniques of particle production and charging will be required to obtain particles with sufficiently low mobility, at sufficiently high rates, with sufficiently low power input.

CIRCULATION CONTROLLED VERTICAL AXIS PANEMONES

Studies have been undertaken at West Virginia University to determine the cost and performance benefits of a vertical axis device which features circulation control of the airflow across its reticulated rotor. Research in the theory and application of high-lift airfoils employing circulation control by blowing over a Coanda trailing edge has shown that this method is very efficient in the production of lift for low blowing coefficients. Coanda blowing applied to a vertical axis panemone might result in an efficient machine which could be competitive with horizontal types while having the known advantages of vertical axis machines. The relative independence of wind direction, the inherent aerodynamic inlet construction (Figure 4.29), and blade simplicity could result in favorable scaling to a large machine. Figure 4.30 illustrates the high-lift, variable angle of attack, circulation-controlled airfoils.

Figure 4.29: Circulation Controlled Wind Machine

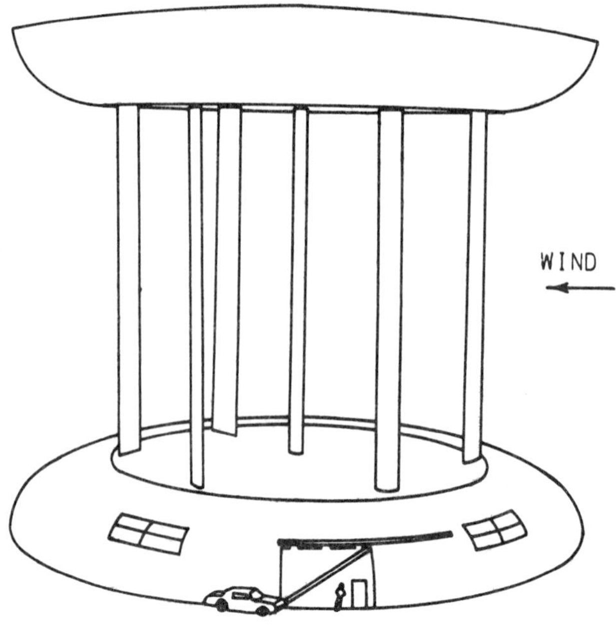

Figure 4.30: Circulation Controlled Airfoil

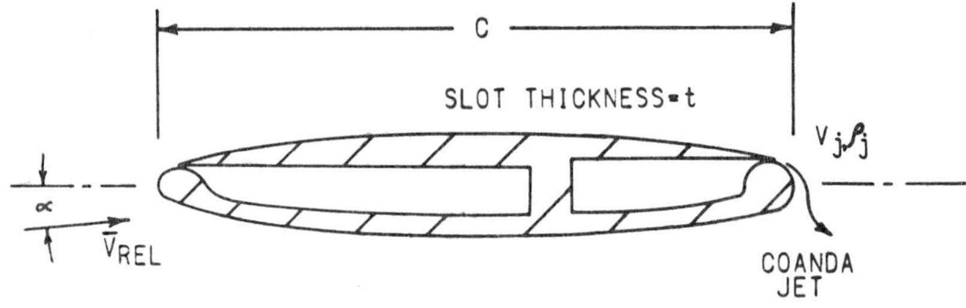

Source: NSF-RA-N-75-050

REFERENCES

(1) Weingarten, L.I. and Nickell, R.E., "Nonlinear Stress Analysis of Vertical-Axis Wind Turbine Blades," *Journal of Engineering for Industry, Transactions ASME,* Vol. 97, Series B, No. 4, p 1234-1237 (November 1975).

(2) Banas, J.F. and Sullivan, W.N. (editors), *Sandia Laboratories Vertical-Axis Wind Turbine Program Technical Quarterly Report Oct.-Dec. 1975,* Sandia Laboratories, SAND 76-0036 (April 1976).

(3) Strickland, J.H., "Aerodynamics of the Darrieus Turbine," *Proceedings of the Vertical-Axis Wind Turbine Technology Workshop,* Sandia Laboratories, SAND 76-5586, p II 29-58 (May 17-20, 1976).

(4) Thresher, R.W. and Meyer, R.E., "Darrieus Structural Studies at Oregon State University," *Proceedings of the Vertical-Axis Wind Turbine Technology Workshop,* Sandia Laboratories, SAND 76-5586, p III 10-27 (May 17-20, 1976).

(5) Templin, R.J., *Aerodynamic Performance Theory for the NRC Vertical-Axis Wind Turbine,* National Aeronautical Establishment of the National Research Council of Canada, LTR-LA-160 (June 1974).

(6) Wilson, R.E. and Lissaman, P.B.S., "Applied Aerodynamics of Wind Power Machines," Oregon State University (May 1974).

(7) Strickland, J.H., *The Darrieus Turbine: A Performance Prediction Model Using Multiple Streamtubes,* Sandia Laboratories, SAND 75-0431 (October 1975).

(8) Abbott, I.H. and Von Doenhoff, A.E., *Theory of Wing Sections,* McGraw Hill Co., Inc., New York (1949).

(9) Barzda, J.J. et al, *Structural Design of Blades for the Sandia 17-m Vertical-Axis Wind Turbine,* Kaman Aerospace Corporation, Phase I Final Report R-1473, Contract No. 02-7472 (August 1976).

(10) Weingarten, L.I. and Lobitz, D.W., "Blade Structural Analysis," *Proceedings of the Vertical-Axis Wind Turbine Technology Workshop,* Sandia Laboratories, SAND 76-5586, p II 137-150 (May 17-20, 1976).

(11) Biffle, J.H., "System Structural Response," *Proceedings of the Vertical-Axis Wind Turbine Technology Workshop,* Sandia Laboratories, SAND 76-5586, p II 168-179 (May 17-20, 1976).

(12) Blackwell, B.F. and Sheldahl, R.E., "Selected Wind Tunnel Test Results for the Darrieus Wind Turbine," *Proceedings of Vertical-Axis Wind Turbine Technology Workshop,* Sandia Laboratories, SAND 76-5586, p II 59-72 (May 17-20, 1976).

(13) Ham, N.H., "Aeroelastic Analysis of the Troposkein-Type Wind Turbine," *Proceedings of the Vertical-Axis Wind Turbine Technology Workshop,* Sandia Laboratories, SAND 76-5586, p II 185-204 (May 17-20, 1976).

(14) Fischer, J., "The Past and the Future of Wind Energy in Denmark," *Proceedings of the Second Workshop on Wind Energy Conversion Systems,* Mitre Corporation, MTR-6970, p 162-172 (September 1975) (Also NSF-RA-N-75-050).

(15) Rodeman, R., "Effects of System Imbalance," *Proceedings of the Vertical-Axis Wind Turbine Technology Workshop,* Sandia Laboratories, SAND 76-5586, p II 180-184 (May 17-20, 1976).

(16) Reuter, R.C. and Sheldahl, R.E. (editors), *Sandia Laboratories Vertical-Axis Wind Turbine Program Technical Quarterly Report April-June 1976,* Sandia Laboratories, SAND 76-0581 (November 1976).

(17) Reuter, R.C., Jr., "Tower Analysis," *Proceedings of the Vertical-Axis Wind Turbine Technology Workshop,* Sandia Laboratories, SAND 76-5586, p II 151-167 (May 17-20, 1976).

(18) Sforza, P.M., "Aircraft Vortices–Benign or Baleful," *Space/Aeronautics,* p 42-48 (April 1970).

(19) Olsen, J.H., Goldberg, A. and Rogers, M., *Aircraft Wake Turbulence and Its Detection,* Plenum Press, N.Y. (1971).

(20) "Vortex Wakes of Large Aircraft," AIAA Professional Study Series Short Course, (June 15-16, 1974).

(21) Thwaites, B. (Ed.), *Incompressible Aerodynamics,* Oxford at the Clarendon Press, Great Britain (1960).

(22) Sedney, R., "A Survey of the Effects of Small Protuberances on Boundary Layer Flows," *AIAA J.,* Vol. 11, No. 6, p 782-792 (1973).

(23) Sforza, P.M. and Mons, R.F., "Wall Wake: Flow Behind a Leading Edge Obstacle," *AIAA J.,* Vol. 8, No. 12, p 2162-2168 (December 1970).

(24) Hall, M.G., "A Theory for the Core of a Leading Edge Vortex," *J. of Fluid Mechanics,* 11, p 209 (1961).

(25) Hummel, D., "Untersuchungen uber das Aufplatzen der Wirbel an Schlanken Deltaflugeln," *Zeitschrift fur Flugwissenshaft,* 13, Heft 5 (1965).

(26) Minardi, J.E., "Computed Performance Characteristics of Electrofluid Dynamic Colloid Generators," *Trans. ASME Journal of Engineering for Power,* 93, p 183-91 (1971).

(27) Lawson, M.O., Fretter, E.F. and Griffith, R.W., "Report on Progress in Achieving Direct Conversion of a Major Fraction of Sonic Flow Kinetic Power into Electrical Power by Electrofluid Dynamic (EF) Processes," *9th IECEC,* San Francisco, CA (August 1974).

APPLICATION TO ELECTRIC UTILITIES

Material for this chapter has been based upon reports from the University of Wisconsin-Milwaukee and Oklahoma State University (NSF-RA-N-75-050); Brookhaven National Laboratory (BNL 50736); Bureau of Reclamation, Denver, CO (PB 273 582); General Electric Co. (ERDA/NASA/9403-76/1) and Kaman Aerospace Corp. (DOE/NASA/9404-76/1).

ELECTRICAL TECHNOLOGY

Basic Factors

The basic factors to be considered in discussing the electrical technology associated with converting wind energy to electrical energy are listed below.

(1) Type of output
- (a) dc
- (b) variable frequency ac
- (c) constant frequency ac

(2) Aeroturbine rotational speed
- (a) constant speed
- (b) nearly constant speed
- (c) variable speed

(3) Utilization of the electrical energy output
- (a) Battery storage
- (b) Other forms of storage
- (c) Interconnection with ac grid

Direct dc generation is practical only on a small scale at present, limited to about 10 or 20 kW. Aeroturbine speeds need not be constant and usually a battery storage system is employed. Small-scale power requirements at very remote places can be satisfied adequately using such systems.

Storing wind energy in thermal form for subsequent use in space heating can be effectively done using a variable frequency ac (or dc) system in conjunction with a heating coil-thermal storage arrangement. Obviously, aeroturbine speeds need not be constant. It is also possible to use a rectifier system to obtain dc, for use as dc or to be inverted to get constant frequency ac supply.

It has been recognized from the beginning (1) that large-scale generation of electrical energy from wind will be in constant frequency ac form, to be pumped into an existing utility grid. Initial attempts of significance employed constant speed aeroturbines driving conventional synchronous generators (e.g., 1.25 MW Smith Putnam Unit). Induction generators require essentially constant speed aeroturbines to generate constant frequency ac power suitable for pumping into utility mains.

Relative merits and demerits of synchronous and induction generators for wind energy conversion systems can be found in the literature (2). The technology of both synchronous and induction generators is highly advanced and very mature. NASA/Lewis Research Center has embarked on an intensive program to build and test experimental 100 kW constant speed-constant frequency units employing conventional synchronous generators (3).

Another approach that is getting considerable attention recently is to allow the aeroturbine speed to vary optimally with wind and employ variable speed constant frequency (VSCF) generating systems to obtain constant frequency power to be pumped into existing utility mains. Methods for obtaining constant frequency output from variable speed shaft fall into two broad categories (4).

A. Differential Methods
   - (1) Mechanical techniques to get constant speed and then employ synchronous generators.
     - (a) Variable ratio gears
     - (b) Planetary gear systems
     - (c) Hydraulic pump-motor arrangement
   - (2) Electrical techniques to get constant frequency (5)(6).
     - (a) Frequency makeup generator (7) (differential action by feeding slip frequency power to rotor).

B. Nondifferential Methods
   - (1) Static frequency changers
     - (a) ac-dc-ac link (8)
   - (2) Rotary Devices
     - (a) ac commutator generators (2)
     - (b) Cycloconverters and frequency changers (9)(10)
     - (c) Amplitude Modulated frequency changer (11)
     - (d) Field Modulation and demodulation techniques
       - (1) High frequency switching (12)
       - (2) Low frequency switching (13)-(16) (Field Modulated Generator System being developed at Oklahoma State University)

Figures 5.1 and 5.2 illustrate schemes to harness wind energy with and without the aid of a conventional utility grid. Because of the availability of economical high-powered solid-state switching devices such as thyristors and diodes, several research efforts are underway at present to develop VSCF generating systems.

Figure 5.1: Schemes to Harness Wind Energy Without the Aid of a Conventional Utility Grid

Figure 5.2: Schemes to Use Wind Energy as a Supplementary Energy Source

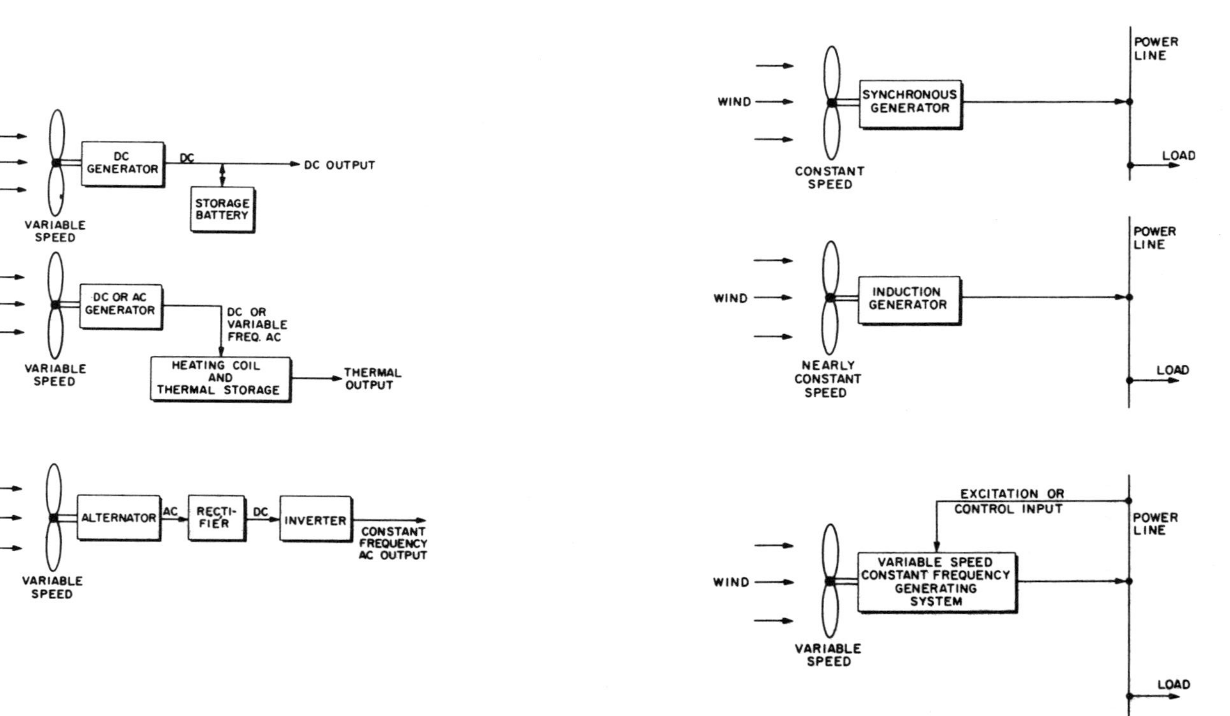

Source: NSF-RA-N-75-050

To mention only a few, researchers at University of Wisconsin, Madison, are working on judiciously synthesizing an ac/dc/ad link with a nonsynchronous generator driven by an aeroturbine; at the Milwaukee Campus of the University of Wisconsin, ac commutator generators and rotor fed induction generators are being considered as possible alternatives and at Oklahoma State University, Stillwater, field modulated generator systems are being actively developed for use in wind energy conversion systems.

Constant Frequency Wind Energy Conversion Systems

Various suggestions which have been made to convert the output of wind energy systems (WES) to constant frequency are discussed below.

Synchronous Generator: The most obvious suggestion is to run two synchronous generators of WES in parallel with the utility power system. Assuming that the utility system is large compared to WES, the machine will be in synchronism for varying energy inputs. The disadvantages are that the machine will draw power from the mains when motoring, and that during gusts, the tendency to run out of synchronism is very great. Bringing the machine back to synchronism and connecting to the system is tedious. This also may initiate larger system instabilities.

Induction Generator: If an induction machine is driven above synchronism, it acts as a generator. So long as the machine is stiff and running above synchronous speed, the induction generator supplies energy back to the system at the system frequency. However, it has the following three disadvantages: (1) it generates less energy from a given wind regime than the synchronous machine; (2) it operates at a low power factor due to heavy magnetization currents which are nearly in quadrature with voltage; (3) it is more expensive than the synchronous generator, but this is offset by less stringent speed control mechanisms.

AC-DC-AC (ADA) Conversion: The concept in ADA conversion techniques is to generate variable frequency ac from an alternator, convert to dc and then invert to constant frequency ac to tie with the utility system. Percy Thomas has suggested a variation of this approach by using a dc generator coupled to a rotary converter. Unfortunately, dc generators are limited in size and rotary converters are inefficient (17). ADA conversion can be elegantly achieved by solid state technology. Already ADA systems handling bulk power in the MW range are in use in dc high voltage links (18). The disadvantage is one of cost, except in the situation where WES is located at such a distance from the utility power lines that the economics of dc high voltage transmission can be exploited.

Special Excitation Systems:Field Modulated AC Generator – Of the different schemes that have been suggested, the only one of interest is the field modulated frequency down conversion system (19). In this system, an ac generator is excited by 60 cycle energy. The output, which is a modulated sine wave, is rectified, filtered and inverted to yield the necessary 60 cycle output. While it is an interesting concept, the system is complicated and costly. The need for ac excitation requires the machine to be completely laminated. The whole power output of the machine has to be demodulated by solid state techniques. Thus, the number of solid state devices used in a rectifier-inverter system (ADA) and in this system may be comparable, while this technique lacks the cost advantage of ADA which is also used in the HVDC transmission. Also, to keep a small modulating ratio which is important, the frequency and hence the speed of operation is high. Some of the techniques, particularly the first two, have been used in wind generating stations in the past. However, each of these had its own disadvantages. Two systems which could be more efficient and also more cost effective than the above were studied at the University of Wisconsin-Milwaukee. These are (1) ac commutator generators and (2) rotor fed induction generators.

AC Commutator Generator: The ac commutator generator has been mentioned by Golding (20) as a possibility, though it was not used in any practical machine in the past. In fact, it is an extremely good suggestion for small sizes, perhaps up to 1,000 kW for the following reasons (21)(22)(23):

(1) The output frequency of an ac commutator generator is that of excitation frequency, irrespective of the speed. This property is similar to a dc machine. The commutator simply establishes a magnetic field corresponding to field excitation, irrespective of the speed of rotation of the armature. Thus, if the machine is excited by 60 hertz, the machine will generate the same frequency ac (See Figures 5.3 and 5.4).

Figure 5.3: AC Commutator Generator System

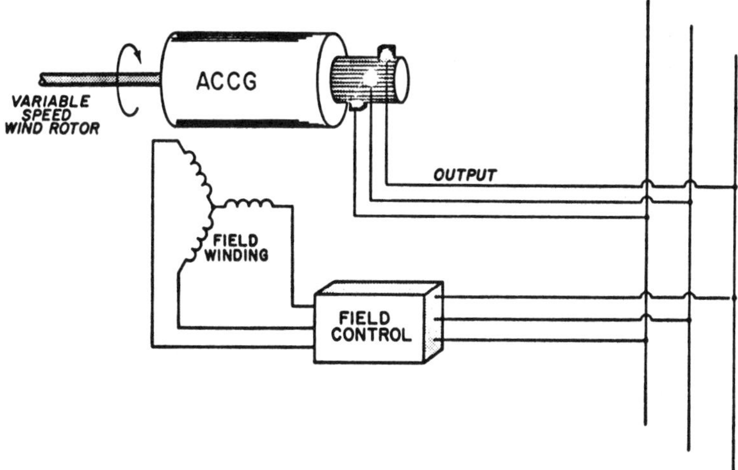

Source: NSF-RA-N-75-050

(2) The simplicity of the machine structure is obvious. The armature is just like a dc machine armature.

(3) Cost of the machine must be similar to the cost of dc machines and hence must be cost competitive.

(4) There is no peripheral equipment as in many other schemes.

(5) The machine could be designed and operated at a much better power factor than an induction generator which is well known to have problems of poor power factor.

Figure 5.4: Field Winding and Compensatory Winding for AC Commutator Generator System

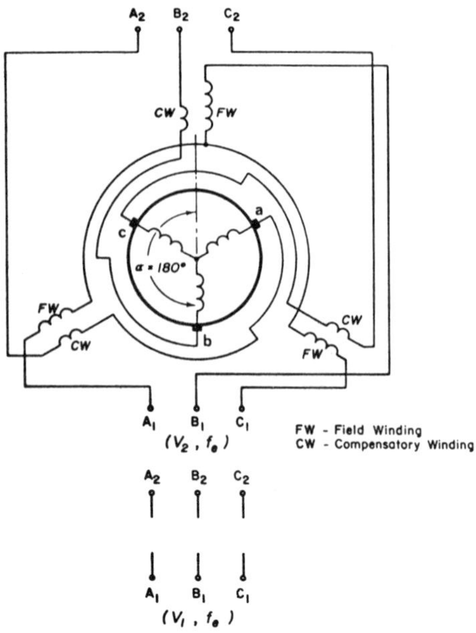

Source: NSF-RA-N-75-050

Rotor Fed Induction Generator: Induction machines operate as generators when operated at speeds greater than the synchronous speed. Such an induction generator has been used in large wind power plants in Russia and Europe (20). However, serious objections have been raised in the past as to the operation of these machines in parallel with power systems (24). Due to the low speed of rotation, the diameter must be large to get a reasonable watts value. Also, for satisfactory operation, it was thought that the airgap had to be large. This involves a large increase in excitation current which the network should supply resulting in a poor power factor.

The major advantage of the induction generator is that it generates constant frequency power irrespective of rotor speed and is inherently more stable than the synchronous machine. The latter point is of extreme importance as wind energy systems grow large in size. To overcome the long serious disadvantage to the induction generator (i.e., poor power factor), the rotor fed induction generator (RFIG) has been proposed to improve the power factor of the machine. The concepts are similar to the speed control of large induction motors.

Principle of Operation – The RFIG is similar in construction to the wound rotor induction generator. The rotor windings are brought out to sliprings. Power from the slip frequency generator can be fed to the rotor through these sliprings. A slip frequency generator generates voltages of slip frequency whose magnitude and phase can be varied. A schematic of RFIG is shown in Figure 5.5. Just as in an induction motor, by varying the phase of injected slip frequency voltages in the rotor circuit, power factor of the generator can be improved.

Figure 5.5: Rotor Fed Induction Generator System

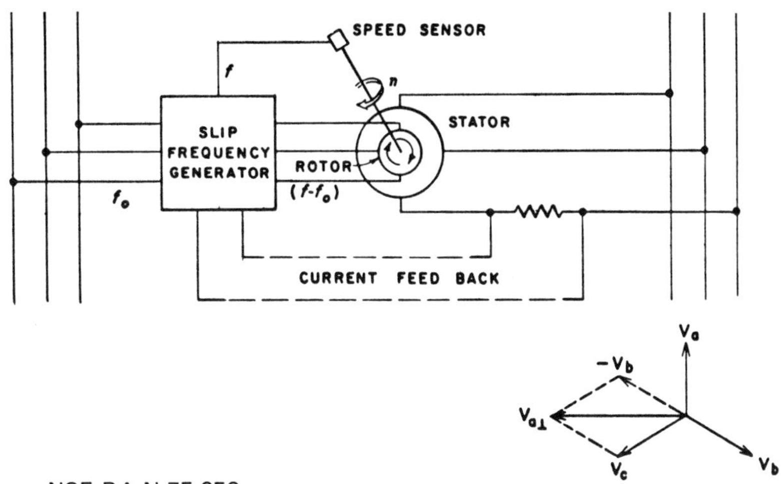

Source: NSF-RA-N-75-050

The power rating required for a slip frequency generator to make RFIG work with reasonable power factor rather than an objectionably low one is being studied. This will provide the cost comparison of this scheme with others which have been proposed.

Design of Slip Frequency Generator –In the past, slip frequency has been generated by Scherbius type machines. This adds another machine to the system making it costly and cumbersome. However, solid state devices are now available. The following power electronic techniques to generate slip frequency have been proposed.

(1) Frequency Modulator–A frequency signal f is generated by an instrument type generator on the main shaft. This is modulated by line frequency f_o and filtered to get a signal of frequency $(f - f_o)$.

(2) f_o-dc-sf_o Link–AC line voltage with frequency f_o is rectified and fed to the inverter. Thyristors are fired by a control signal so as to produce slip frequency voltages. If SCRs are used in both inverter and rectifier circuits, this system can feed power in both directions.

(3) Cycloconverter–Observing the fact that slip frequency is only a small fraction of line frequency f_o, design of a cycloconverter to generate slip frequency voltages using established techniques in power electronics will be straightforward. In all these systems, a power amplifier will be used when necessary between the slip frequency generator and the sliprings of the rotor.

The advantage of these schemes over others is the application of solid state devices to manipulate slip frequency power as compared to other schemes where solid state devices have to deal with the whole power output of the machine. Similar solid state techniques in induction motor speed control (25)(26) have resulted in power factor improvement to 0.95 with the solid state equipment rated for ⅓ power. As the slip frequency range in induction generators is not so large as in motors, the cost advantages may be more significant.

ROLE OF WIND POWER IN ELECTRIC UTILITIES

The vast majority of wind machines built historically had power ratings in the kilowatt and subkilowatt range. From a contemporary utility perspective, considerably larger machines are necessary to reduce the overall costs of site selection, site preparation, installation, maintenance, access roads, transmission lines, protective switchgear and machine components. Although there are a multitude of machine design possibilities (e.g., Savonious, Darrieus, vertical axis articulated blade, American farm windmill, diffuser augmented turbine) only the familiar propeller type, horizontal axis machine has proven its potential for operation in the megawatt power range (1).

Wind generators deployed by utilities in the near future, and in particular, the machines to be installed by the Department of Energy (DOE) on various utility systems, will, therefore, employ this design concept and are likely to be characterized by a two-bladed, horizontal axis rotor with a diameter of 150-300 feet, blade pitch and yaw controls, a rotation frequency of 30-60 rpm, and a 0.2-2.0 MW rated capacity, constant speed, synchronous generator. The key factor constraining the size of these machines is expected to be blade structure technology.

Even the largest wind generators now envisioned, 1.5-3.0 MW in rated capacity, are still very small by contemporary utility standards. New coal or nuclear-fueled generators have power ratings of about 1,000 MW. The simplicity of the wind generator in comparison to these huge plants is counterbalanced by the need for large numbers of machines: about 1,500 1.5 MW units, operating at a capacity factor of roughly 0.35 over a year, would be needed to match the annual energy production in one such coal or nuclear plant.

The protective switchgear, monitoring equipment and transmission interties required for reliable utility operation, plus land leasing or acquisition for so many machines add substantially to the basic machine cost. (In one recent study (27) it was estimated that costs for transmission lines and protective switchgear range from about $100 to $300 per kW, depending on the number of units in an array, the spacing between units, and the distance to the grid intertie.) In addition, power outages due to a lack of wind are more frequent than forced outages due to conventional plant breakdowns so that greater reserve requirements are likely to be needed with wind power.

As discussed in the previous section, a variety of mechanical to electrical power conversion schemes have been proposed for use on wind machines. These include (28)(29): dc generator with inverter, variable shaft speed-constant electrical frequency ac generator, induction generator and synchronous generator. The primary advantage of the first two is that they are, in principle, somewhat more efficient since they permit wide variations in the speed of the rotor and thus have a higher aerodynamic to mechanical power factor. However, they are heavier and more costly than the induction and synchronous generators which require nearly constant and exactly constant speed operation respectively; the gain in efficiency may be outweighed by these and other factors. The constant speed, synchronous generator is currently favored due to a combination of beneficial features:

(1) the potential for resonant structural vibrations coupled to the low rotor frequency is minimized by having a constant rotor frequency.

(2) the generator places no reactive power demands upon the grid.

(3) it is a well-known technology.

(4) weight is less than most alternative schemes.

Several machine design studies, (30)(31)(32) have focussed on developing specifications for machines which minimize the cost of electricity generated. (See later sections in this chapter for discussion of the General Electric and Kaman Aerospace designs). The design problem is complicated by the difficulty of estimating cost-performance characteristics of the machine components. For example, production costs vary tremendously as a function of production quantity so than machine component costs are highly dependent on whether the total production run is 10 units or 1,000 units. Design studies to date indicate that the optimized design should have a performance curve as shown in Figure 5.6.

Below the cut-in velocity, no net power can be delivered to the grid and the machine does not operate. Above the rated velocity (corresponding to the rated capacity of the generator) blade pitch is adjusted to maintain a constant output.

Figure 5.6: Characteristic Machine Response

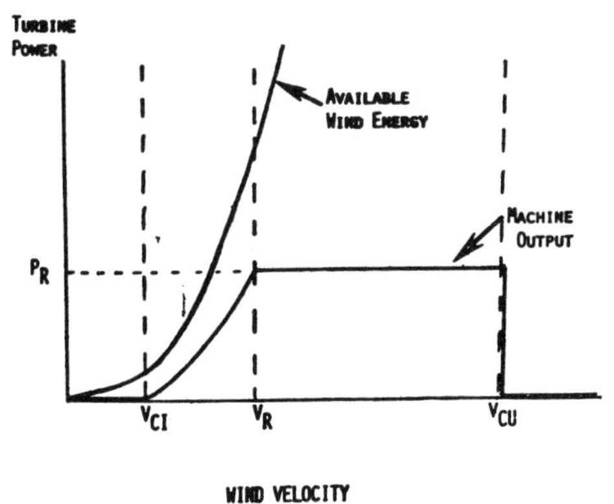

Source: BNL 50736

Typically, the cut-in velocity is roughly half the rated velocity and the rated velocity is slightly higher than the mean wind speed of the wind regime for which the machine is being designed. Although it may appear from Figure 5.6 that vast quantities of energy are being lost at the higher wind velocities due to the plateau in the performance curve, the wind velocity frequency distribution is such that winds seldom blow at the higher velocities and a relatively small fraction of the total energy is contained in such winds. Above some velocity, the cut-out velocity, the machine is shut down to avoid damage from high winds.

As an example of the results of design studies, it is estimated that a 1.5 MW machine optimized for an 18 mph mean wind speed would have a capacity factor of roughly 0.35 and would produce about 7×10^6 kWh per year.

Variation in the output of wind turbines is caused by atmospheric phenomena and is manifested on several time scales of interest to utilities. Short-term variations, caused by gusting, can affect a single turbine's output; these are essentially averaged out over the spatial dimensions of even relatively small arrays.

Longer-term variations are caused by meteorological factors, the diurnal cycle, and seasonal changes and will affect the output of a large array as well as single units. All fluctuations have the potential for creating problems for utility operations: subsecond time scale variations can produce network stability (synchronization) problems; second-to-minute scale variations can result in grid security (loss-of-load) problems; and longer scale variations create various dispatching (scheduling of generating units) problems with consequences regarding the mix of power generating units required by the utility. The degree to which variations in power output pose difficulties for utilities is dependent on the fraction of total load being met by wind power.

These various problem areas are under study and their impact on the technical and economic feasibility of wind power use is not yet clear. (Preliminary results from one of these studies conducted by the General Electric Space Division presented at the Third Biennial Wind Energy Conference in Washington, D.C. September 1977, indicate that these problems can be dealt with easily given conventional utility control and protection systems.) If it is determined that the potential problems associated with stability and security are serious enough to require attention, possible solutions might, for example, involve:

(1) (Stability) Ensuring that wind generators are tied into grids that are sufficiently large that they always have considerably larger power inputs from conventional generators than from wind machines.

(2) (Security) Changing spinning reserve requirements.

Characteristic average diurnal wind power cycles are shown in Figure 5.7 for low, moderate, and high wind velocity regimes. Note that in the low and moderate regimes, which contain the bulk of total available wind power, the averaged cycles show peaks that correspond roughly to daily utility load peaks.

This correspondence tends to increase the value of wind energy to utilities since the highest operating cost conventional units are generally employed during the peak load periods.

Figure 5.7: Diurnal Cycle of Averaged Power Available in Various Wind Regimes

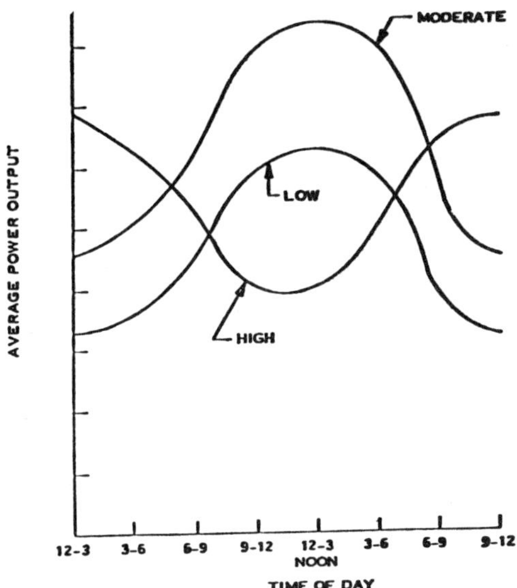

Source: BNL 50736

Analyzing the Economics of Wind Power

An economic evaluation of wind power in a utility context involves a comparison of the total cost of wind turbines to a utility with the "value" of wind power (the total monetary savings that results from the displacement of conventional power). The determination of value requires that, besides computing the monetary savings resulting from decreased fuel consumption, additional savings that might be achieved due to changes that can be made in the purchases of conventional generating equipment must be accounted for.

In general, a utility purchasing wind machines will be able to alter both the net capacity of and the type of conventional generating units that it would otherwise have purchased and experience monetary savings on both counts. These capacity related savings increase the value of wind power to utilities above the levels computed from fuel savings alone.

An analysis of the economics of the utility use of wind power requires the consideration of a variety of factors. Besides siting and machine design, which are factors common to the economics of all wind power applications, other factors specific to utilities must be considered such as reliability constraints, diurnally and seasonally varying loads, and the purchase and scheduling of a variety of alternative generating systems with different combinations of fixed and operating charges. Seemingly obvious conclusions regarding how wind generators might be employed by utilities (e.g., that a storage system would be used in conjunction with wind generators, or that wind energy would displace the most expensive conventional fuels) may, upon deeper analysis, prove to be erroneous. The complexity of the subject is compounded by the inappropriateness of conventional utility analytical tools for studying wind power. Conventional generation planning (i.e., determining future purchases of new capital equipment) and dispatch analysis techniques assume regularities in the daily, weekly, and seasonal loads, and also assume utility control over the output of generating units except for infrequent forced outages.

The major exception to these assumptions is the treatment of run-of-the-river hydropower plants, the output of which can vary with rainfall. However, the output of such hydro plants does not vary as rapidly or as frequently as would the output of a wind turbine and the total capacity of such plants is generally a small fraction of a utility's total capacity. Wind generators are a sharp departure from conventional generators in that they exhibit significant and frequent variations beyond the control of the utility.

Traditional utility analytical methods exploit the load regularities and the utility's control over conventional generating units by adopting simplifying techniques such as the use of monthly and annual load duration curves. In doing so, they do not deal with the chronological behavior of the system. Such methods cannot be used directly to analyze the potential role of wind generators in a utility; either they must be augmented by aditional analyses which explicitly account for wind variation or, alternatively, techniques must be employed which analyze the load, conventional generation, and wind power output simultaneously and in a chronological fashion. These latter techniques have the virtue of being conceptually more straightforward, but they are cumbersome and difficult to use in practice.

Capacity Credit

Although the term "capacity credit" does not have a well-defined meaning in a power systems context, it has been used in reference to wind and solar generated electricity to refer to magnitude of the conventional equipment capacity that can be replaced by a unit of capacity from these variable sources. Often it is assumed that the credited capacity of wind power is zero; i.e., wind power is so unreliable that, for all practical purposes, no decrease in the conventional capacity otherwise needed is possible.

The wind generated energy is then said to contribute as a "fuel saver" such that its contribution to the utility is measured only by the fuel not consumed when wind power is available(30)(33). The economic value of wind machines is then computed as the cost of this "saved fuel."

The "capacity credit" and "fuel saver" concepts are based on an overly simplistic view of utility operations. Their use ignores the complexities of utility systems that result from the variety of conventional generators employed, each of which has a different combination of reliability, thermal inertia, efficiency vs. load characteristic, and fixed vs. operating costs. Most importantly, the "no capacity credit" assumption ignores the fact that it is possible for a utility to save considerable monies in capital investments by purchasing peaking type generators (low fixed charge, high operating cost units) instead of expensive base load generators (high fixed charge, low operating cost units) while retaining the same total conventional capacity. Such a shift would make sense when combined with the purchase of wind generators if it were determined that the extra peaking type units could be used primarily to provide the needed level of system reliability while burning relatively little fuel. In this case, the total cost of wind generators would be offset by some combination of fuel and capital expenditure savings even though it was assumed that no net "capacity credit" was assigned to the wind machines. There is no direct relationship between "capacity credit" and utility economics and it is utility economics which are ultimately of interest.

The "capacity credit" concept is troublesome because its use is an attempt to define with one number for any and all utilities what is in fact a multifaceted characteristic that is unique to a specific utility.

Because utilities usually own and plan purchases of a variety of generating unit types, the substitution of wind power for conventional generation generally involves changes in the purchase or retirement of several different types of conventional units.

Due to differences in the reliability of different types of generating units, they cannot be substituted for one another on a kilowatt for kilowatt basis while retaining a constant level of system reliability. Thus the hypothetical "capacity credit"—the net reduction in all conventional capacity due to the use of windpower—will differ from one utility to another depending on previously existing equipment and the new equipment purchase schedule that would otherwise have been followed.

A computation of the economic value of wind power, based upon the wind characteristics, cost structure for conventional power generation, and reliability constraints associated with a specific utility, provides a direct, meaningful measure of the competitiveness of wind power. "Capacity credit" is a confusing detour around this central question.

Some Initial Results

The results available to date regarding the economics of wind power use by utilities are general and preliminary and require elaboration by more detailed studies. Nevertheless, a few tentative conclusions have emerged that provide insights into the role wind power might play in electric power generation.

The least expensive method of incorporating wind generators into a utility network is to feed the power directly into transmission grids whenever wind power is available without the use of separate storage or back-up capacity dedicated solely for the wind array.

Other generating units in the system would be constantly adjusted to meet the difference between the total load on the system and the wind array output. Thus, a greater number of generating units capable of rapid changes in output (on time scales of minutes to hours), such as gas turbines, hydropower units, or storage devices such as batteries, are likely to be needed if wind machines are employed than otherwise; large steam units, with outputs constrained to vary slowly, cannot easily provide back-up for wind generators. In addition, more peaking units are required by the utility in order to provide reserve power for periods when wind outages coincide with peak loads.

The availability of storage can alleviate dispatching problems arising from wind variability. However, storage and wind power are not as closely coupled as is sometimes believed. From a utility perspective, storage units are simply a method of providing peaking capacity and, as such, must compete with alternative peaking devices in any application. Storage units at each windmill, or even at each windmill array, are especially unlikely as they are prohibitively expensive; a utility investing in storage would want to use it for a variety of functions and would site the storage devices with the entire system's operation in mind.

An attractive combination of storage and wind power occurs in situations where dammed hydropower is currently being used for baseload power but, with the addition of more hydroturbine units, could be converted to peaking duty at a modest cost. In particular, it has been suggested that hydropower authorities in the western United States, seeking new power sources to supply growing demand, might find wind power an attractive possibility (30)(34).

An additional factor favoring this application is the advantageous tax and financing regulations governing capital investments by public power authorities which tend to make capital intensive wind power systems relatively less expensive for the authorities than for private utilities.

Another favorable application for wind power is to provide power for islands or remote communities in high wind regimes. Often power in such areas is supplied by diesel or turbine plants burning expensive fuels at low efficiency. Not only is the cost of power from these plants expensive, thus rendering wind power more competetive than elsewhere, but these rapid response units provide ideal back-up power for wind systems.

As an example of some specific results, a General Electric Company study using a utility generation planning model indicates that the value of wind power varies from $200 to 400/kW depending upon the region of the country and the wind regime within that region (35). These results assumed 1975 fuel and plant costs. About half of the value could be attributed to fuel savings and half to capital expenditure savings for conventional capacity with the assumed 1980s mix of generating units.

By the year 2000, with a significant fraction of the mix assumed to be composed of capital intensive nuclear units, capacity related savings comprise about two-thirds of the value. A 20% increase in plant costs, and fuel escalation rates of 2%/year for coal and uranium fuel, 5%/year for oil, and 15%/year for gas results in a value of about $700/kW in the northeast United States. These latter assumptions provide a more realistic view of the value of wind power in the future.

Estimates of the production and installation costs of wind machines are very sensitive to assumptions regarding the cost of the first machine, the effects of production experience, and mass production. Most estimates range from $500 to $700/kW for the 100th unit produced when a 90% learning curve (cost drops by 10% for each doubling in cumulative production quantity) is assumed. The General Electric Company submitted a bid of $1,586/kW (in 1975 dollars) to the Energy Research and Development Administration for the second of two 1.5 MW wind machines. Based on this cost, the 90% learning curve results in just under $700/kW for the 100th unit. A comprehensive analysis of the environmental costs, positive and negative, of wind power compared to conventional generating alternatives has not been made. Clearly, air and water pollution are likely to be reduced if wind power is used instead of additional conventional fuel burning power plants. At this time, the only significant negative "environmental" impacts of wind power that are known are TV and microwave interference and "pollution" of the visual environment. If sited carefully, the latter problems can be minimized. Thus, a comprehensive comparative analysis is likely to reveal net noneconomic benefits of wind energy.

HYDROSTORAGE

Technologically, the crucial element limiting widespread application of wind and solar environmental energy sources is the fluctuation of the power output compounded by fluctuations in power demand. The power grid with diverse energy sources can reduce the first to some extent, but the problem can be eliminated only by employing some form of energy storage. While many forms of storage suggest themselves, hydrostorage provides a viable option using existing technology.

Power fluctuations can be compensated by pumping water into a storage reservoir above a hydroelectric power plant during periods of energy surplus, and releasing it through the power plant to supplement the energy on the power grid during an energy deficit. As with any energy storage system, there is a loss, leaving an input/output ratio on the order of 3:2 or 4:3.

Conventional and pumped-hydrostorage facilities have existed for many years. Installed capacity in the Western United States at the end of 1976 was 46 GW. Studies by the Federal Power Commission (36) and the Corps of Engineers (37) found nearly 700 sites (primarily new sites) in Arizona, California, Nevada, Utah, Idaho, Oregon, Washington, and western Montana suitable for pumped-storage facilities, with a potential plant capacity of over 2,000 GW. With few exceptions, site selection was based on office reconnaissance-type studies considering such criteria as topography, reservoir size, availability of water, source of pumping energy, and operating pattern (daily/weekly and seasonal storage). Figure 5.8 shows the potential pumped-storage sites in the Pacific Northwest.

Although the reserve capacity of existing hydropower facilities seems sufficient for the first 5 years of a system development, past experience indicates that, after authorization and appropriation of funds, at least 5 years are required to bring hydroelectric facilities online.

The wind-hydrostorage energy system displays obvious environmental advantages over conventional and nuclear power plants. No fuel is burned, no thermal pollution is caused, no hazardous wastes remain, and the only nonrenewable resources consumed are those necessary to build and operate the devices.

Some environmental changes would occur, principally as use of land for wind farms and hydrostorage facilities. WECS spaced about 10 rotor-diameters apart would occupy less than 1% of the wind farm area exclusively, allowing other uses of the rest of the land, such as grazing or farming. Hydrostorage facilities might affect certain species of fauna and flora. Visual and noise impact would be considerable only if the wind farm were in a populated area, an unlikely situation.

Figure 5.8: Potential Pumped-Storage Sites in the Pacific Northwest (Daily/Weekly Cycle)

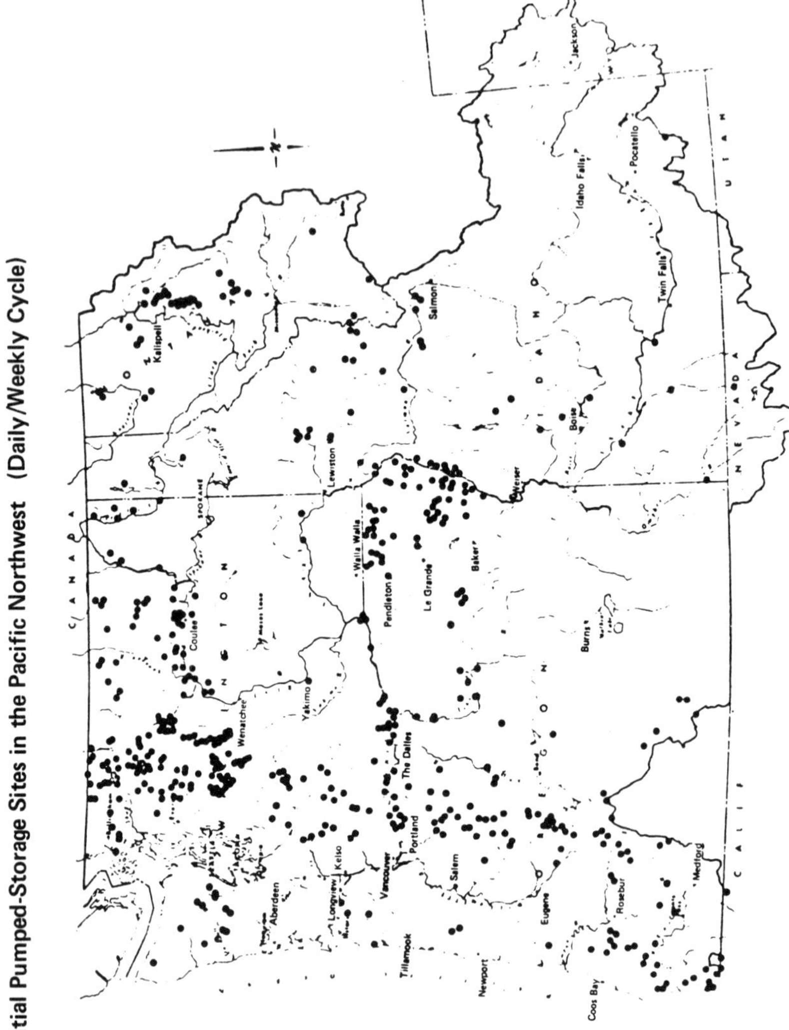

Source: PB 273 582

Possible television interference could be mitigated by providing cable TV from a remote antenna for the few households affected. It is not unlikely that a wind farm would become a sight-seeing and recreational attraction of positive rather than negative environmental value. Before a wind energy system is implemented, a thorough assessment of environmental and socioeconomic impacts must be carried out.

GENERAL ELECTRIC DESIGN STUDY OF WIND TURBINES FOR ELECTRIC UTILITIES

This study was performed by General Electric's Advanced Energy Systems. The program was directed by the NASA Lewis Research Center's Wind Power Office, Power Systems Division, for the Energy Research and Development Administration and is an integral part of the Federal Wind Energy Program. This effort was one of two parallel studies directed by NASA to evaluate the use of wind turbines to generate electrical power in a cost effective manner. General Electric was supported in this study by the Hamilton Standard Division of the United Technologies Corporation.

Two wind turbine generator (WTG) systems were studied in detail: a 500 kW unit, assumed to operate at a 12 mph median wind site, and a 1,500 kW unit, assumed to operate at an 18 mph median wind site. The capital costs of these units (based on 1975 dollars) were estimated to be $935/kW and $430/kW for the 500 kW and 1,500 kW systems, respectively. These capital costs assume that 100 units are being procured and that all development has been completed such that the WTG is a "mature commercial product". The capital costs associated with the initial developmental units are substantially higher.

The capital costs include: the program management of a 3½ year, 100 unit, fabrication and installation program, cost of interfacing with the utility, equipment and subsystem assembly costs, site preparation costs, system contractor overhead and profit. The capital costs plus anticipated maintenance costs were used in conjunction with an economic model to estimate a typical utility's energy generation cost using WTGs. The cost of generating energy was calculated to be 4.04 ¢/kWh and 1.57 ¢/kWh for the 500 kW unit in the 12 mph site and the 1,500 kW unit in the 18 mph site, respectively. While these energy generation costs include the utility's cost of financing the purchase and maintaining the WTGs, it does not include power distribution or utility overhead costs.

The overall objective of this study was to establish the preliminary designs of cost effective wind turbines that are compatible with utility company requirements. Specifically, the program's activities were directed at: establishing the design and operation requirements for a WTG, which is linked to an electrical utility network, defining the WTG design and operational concept most suitable for this application, determining the most cost effective operating conditions as a function of sites having median winds between 9 and 21 mph and generating two preliminary designs and cost data, one system at 12 mph median wind site and one at an 18 mph mean wind site.

Assuming that the results support continued investigation of WTGs for this application, it is intended to use the product of the study (and the parallel effort being directed by NASA) as the basis for the final design and construction of prototype systems.

System Requirements

This section describes the requirements to which the WTGs were designed. These requirements were either provided by NASA-Lewis at the beginning of the study or resulted from the knowledge acquired as the study progressed. The subsections below present these requirements relating to: electrical utility interfaces, operational requirements, cost requirements, reliability/availability requirements and development risk.

Electrical Utility Interfaces: The wind turbine generators (WTGs) are designed to connect with typical electrical utility network lines. In general, it is expected that these lines will be of the "distribution" or lower "subtransmission" voltage class. The specific technical considerations involved in connecting the WTG to distribution or subtransmission circuits concern: voltage fluctuations, power factor, circuit relaying and line reclosing, telephone interference and stability.

Voltage Fluctuations – Voltage dips associated with the wind turbine generators may occur during starting and stopping of the units and from the output variation resulting from wind velocity changes. For WTGs using synchronous generators, voltage dips resulting from starting could be completely eliminated by automatic synchronizing which is common practice in utility power generation. With regard to induction generators, switching of associated capacitors (used for power factor correction), will also cause voltage dips.

Frequent and/or excessive voltage fluctuations would be annoying to utility customers and must be avoided. Infrequent voltage dips of a larger magnitude may be tolerated; larger voltage fluctuations (due to the WTG operation) may be tolerated if they occur infrequently, however, more frequent fluctuations will become an irritation to users of electrical equipment unless their amplitude is reduced as indicated.

Power Factor – A power factor approaching 1.0 is normally preferred by electrical utilities since it maximizes net energy transfer efficiency. Power factors of less than 1.0 result in reactive power which increases the utility equipment requirements. Reactive power is measured in volt-amperes reactive (1,000 var = 1 reactive kilovolt-ampere, abbreviated kvar).

Since kilovars cannot be economically transmitted over long distances due to excessive I^2R losses, utilities will use substantial amounts of fixed and switched shunt capacities to supply the kvar near the points where they are consumed. The result is that distribution system power factors approach unity, typically 0.98. In considering induction generators, which have power factors of about 0.8, the kvar requirements should be supplied at the WTG location so as to raise the power factor as close to unity as possible.

Circuit Relaying and Line Reclosing – A typical distribution circuit will generally have automatic reclosing of the feeder circuit breaker to quickly restore service in the event of temporary faults such as those caused by lightning flashovers. In addition, particularly at the lower voltage levels, such as 24.9 kV and below, there will typically be one or more automatic reclosers spaced along the circuit performing the same task for faults beyond their location. The line reclosers and feeder circuit breaker have carefully coordinated overcurrent trips based on the assumption of radial feed from the substation. Reclosing may be done automatically, but only in the proper sequence, i.e., after the generator is tripped the circuit may be reclosed and, if successful, the wind generators put back on the line.

Telephone Interference – Since all generators produce a small amount of harmonics in the voltage wave, there is the potential for harmonic current flow which could cause interference in closely coupled telephone circuits. This has rarely been a problem and has been solved in most cases where it occurred by ungrounding some capacitor banks which acted as a "sink" for certain harmonics, or in some cases relocating the capacitor bank.

Stability – Electrical instability in a WTG is most likely to arise from wind gusting conditions. Loss of stability, or synchronism, in a synchronous generator results in high pulsating current flows in the connected circuits and attendant voltage fluctuations. Pull-out of an induction generator, i.e., operation beyond the peak of its slip-torque curve, results in abnormally high reactive current flow (approaching locked rotor values) with similar detrimental results. These conditions generally will require tripping and "restarting" of the unit with some accompanying outage time to the unit.

Operational Requirements: The principal operational requirement is for the WTG to produce electrical energy at a cost which is competitive with conventionally produced energy. The specific operational requirements addressed in this study are presented below.

Design Life – All static components including the tower were designed for a minimum service life of 50 years. Dynamic components were designed for a minimum service life of 30 years, but, may include periodic maintenance and replacement.

During normal operation the units will be designed for unattended, fail-safe automatic operation as well as manual control. The units will be designed for a minimum availability of 90% over the service life with special consideration given to servicing and maintenance of critical areas.

Environmental Requirements – The units will be designed to withstand the range of atmospheric environments experienced from New England to Alaska or the Caribbean area to hot desert climates. The unit must therefore be capable of operation in snow, rain, lightning, hail, icing conditions, salt water vapors, windblown sand and dust and in temperatures extremes of –51° to +49°C. If cost effective, designs adaptable to local severe conditions with minimum change will be acceptable.

Wind Speed and Gust Model – In order to calculate annual WTG energy production, velocity duration curves were supplied to GE by NASA-Lewis. The wind profiles used are shown in Figure 5.9 for sites having median winds between 8 and 24 mph. The WTG is required to operate during normal wind gusting conditions and survive gusts which exceed cut-out velocity. In addition, the WTG must be designed to withstand a maximum steady wind speed of 53.6 m/s (120 mph) at 9.1 meters (30 feet) above ground.

Velocity versus time relationships for the assumed wind gusts are shown in Figure 5.10. As shown in the figure the maximum amplitude of an individual gust is nearly twice the steady state level of the wind; the maximum value of the gust occurs at one-half the gust period.

Figure 5.9: Velocity Duration Curves

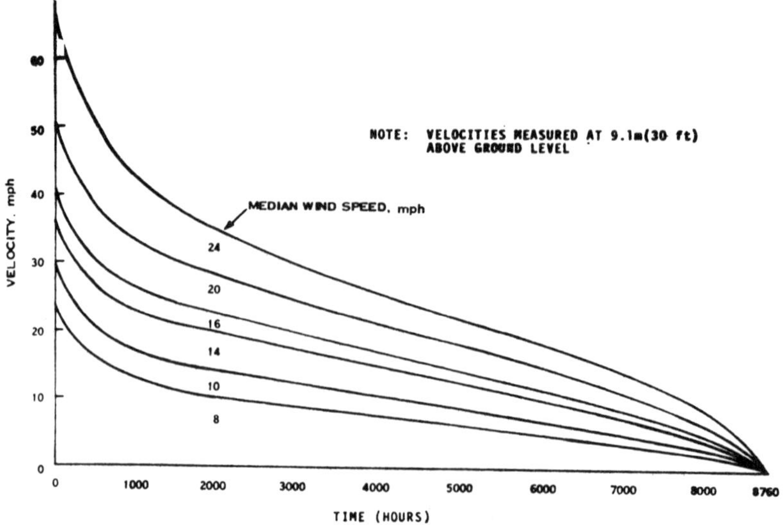

Source: ERDA/NASA/9403-76/1

Figure 5.10: Wind Gust Model

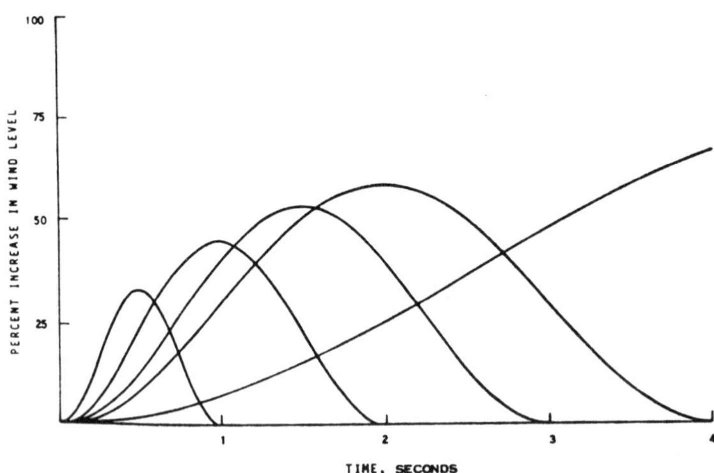

Source: ERDA/NASA/9403-76/1

Also of importance is the fact that gusts of shorter duration reach lower amplitudes, but the rate of increase in the wind speed is higher.

Cost Requirements: An objective of the study was to develop two preliminary designs which represent minimum cost systems in the 12 and 18 mph median wind sites. Minimum cost was to be based upon ¢/kWh and include capital, operational and maintenance costs. Within the framework of minimum cost, consideration was given to architectural aesthetics and public acceptance.

Reliability/Availability Requirements: The WTG was designed for a minimum availability of 90% over the service life with special consideration given to servicing and maintenance of critical areas.

Development Risk: It was the intention of this study that the WTG preliminary designs could lead to final design and construction within 2 years. Therefore, the designs must utilize the latest proven design, material and fabrication technology insofar as its use minimizes electric power generation costs. Whenever possible, the technology will have a base of proven experience.

Preliminary Design

This section describes the two preliminary designs which were generated in the study: a 500 kW unit and a 1,500 kW unit which operated in 12 mph and 18 mph median wind sites, respectively.

System Design: Approach – The preliminary designs had the following features incorporated:

Constant rotor speed
Variable pitch control
Single rotor per tower, placed downwind
Propeller type blades
2 blades per rotor
Fixed ratio gearbox transmission
Synchronous generator
Truss or concrete tower
Microcomputer based control system

Based on the results of the parametric analyses it was recommended to NASA that the preliminary design focus on a 500 kW system for the 12 mph median wind site and a 1,000 to 2,000 kW system for the 18 mph median wind site. NASA elected to proceed with a 500 kW unit in a 12 mph site and a 1,500 kW unit in an 18 mph site. In addition, the blade clearance height from the ground was directed to be 15.2 m (50 feet) instead of the previously used 6.1 m (20 feet); this change was prompted by data from the parametric analyses which suggested a cost advantage in raising the clearance height. In addition, a larger ground clearance results in a flatter wind velocity profile across the rotor disc, thereby offering the potential for lower cyclic stresses.

The first step in the preliminary design was to use an updated version of the design optimization code to obtain the operating conditions for the minimum cost systems at the aforementioned power levels and wind sites. Results from the optimization code are shown in Table 5.1; these conditions formed the basis for the ensuing design work.

Table 5.1: System Characteristics

Power rating, kWe	500	1,500
Annual energy, kWh	1.88×10^6	6.62×10^6
Capacity factor	0.42	0.51
Rotor diameter, m (ft)	55.8 (183)	57.9 (190)
Rotor speed, rpm	29	40
Rated velocity, m/s (mph)	7.27 (16.3)	10.1 (22.5)
λ Rated (λ design = 10)	9.0	9.5
Tip speed, m/s (ft/sec)	84.7 (278)	121 (398)
Cut-in velocity, m/s (mph)	3.54 (7.92)	5.11 (11.4)
λ Cut-in	18.5	18.4
Cut-out velocity, m/s (mph)	17.9 (40)	22.3 (50)
λ Cut-out	3.67	4.18
Hours above V_{ci}	6,257	6,568
Hours at rated	2,067	2,718

Source: ERDA/NASA/9403-76/1

The operating conditions identified by the computer code were then used in the aerodynamic and structural design of the blade and hub. This work was also supported by a dynamic analysis of the rotor/tower interactions. Following the design of the rotor subsystem, the main shaft, pintle and tower designs were initiated. Design of the generator and gearbox transmission proceeded directly from the results of the computer code. The design of the control system was a continuing process involving design trade-offs with each of the major subsystems. Primary emphasis was placed upon minimizing the WTG energy generation cost. Reduction of these costs involves two aspects: minimization of capital expenditures and maintenance costs and maximization of the annual energy produced.

Initial design efforts had shown that the most effective way to minimize capital costs was to reduce the structure and overall weight in the system. The relationship between weight and cost is evident in many industrial products which do not contain high technology components requiring exotic manufacturing techniques or materials. In order to reduce weight it is necessary to lower the structural loads on the system such that the system reliability is not compromised.

Therefore, cost effective design requires an approach which avoids high wind loading conditions which can result from rapid changes in wind direction or high wind velocities. This can be accomplished by means of a properly designed control system which maintains the WTG in the least vulnerable condition under adverse environments such as rapidly shifting winds or wind gusting situations exceeding the WTG cut-out velocity.

Maximization of the energy production is achieved by rating the system at the most favorable power level and optimizing the design operating conditions, velocity ratio (λ) and rpm. In addition, it is necessary to ensure that the system operates under the broadest range of wind conditions possible. This requires a control system which: provides for synchronization with the network at the lowest wind speed possible; allows continous operation under varied wind gusting conditions; provides system control up to the maximum practical cut-out velocity and allows for synchronization under intermediate and high wind conditions to allow restart from a maintenance or high wind cut-out condition.

Key Design Features – An exterior view of the upper portion of the 1,500 kW preliminary design is shown in Figure 5.11.

Figure 5.11: Exterior View of 1,500 kW WTG Nacelle Area

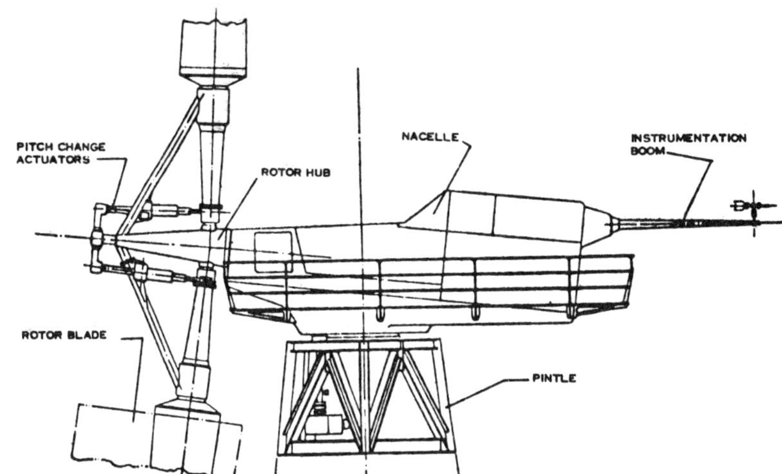

Source: ERDA/NASA/9403-76/1

In the illustration the rotor subsystem can be seen at the left. The hub portion is about 7.6 m in diameter while the blade chord and blade root thickness are 3.6 m and 1.5 m, respectively. The mechanical power transmission and generator are housed within the nacelle; at the right of the nacelle is an instrumentation boom which points into the wind. The entire nacelle is mounted atop the pintle which contains the azimuth drive mechanism. As seen in the illustration, the main shaft is inclined 6 degrees to the horizontal in order to reduce the rotor overhang distance required for tower clearance.

Also important, is the fact that the blades are coned at 3° which lowers the steady state shank bending stresses and assists in maintaining the desired blade/tower clearance. The angle of inclination and the cone angle have been selected so as to achieve the objectives stated without seriously compromising the effective blade diameter.

A view of the 1,500 kW WTG with the nacelle cover removed is provided in Figure 5.12. In this illustration the main shaft, supported by two bearings, is shown. Separating the main shaft from the gearbox transmission is a flexible coupling. The coupling will compensate for small misalignments between the main shaft and the gearbox which are present at assembly or developed over the life of the unit. The transmission is oriented such that the high speed shaft is on top where it is accessible for maintenance. Attached to the high speed shaft, at the left of the transmission, is a hydraulic brake which is capable of bringing the system to an emergency stop in 20 seconds.

Figure 5.12: Preliminary Design Layout

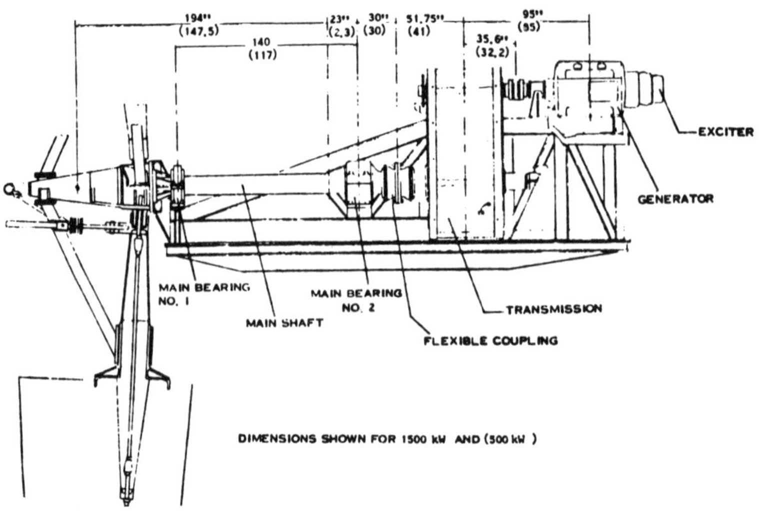

Source: ERDA/NASA/9403-76/1

Another flexible coupling joins the transmission high speed shaft to the generator shaft; located on the generator shaft is a torque sensor. The role of the torque sensor is to measure the torque input to the generator; therefore its effectiveness is not compromised by the presence of flexible couplings elsewhere in the drive train.

Vertical orientation of the transmission results in a favorable position for the generator. In the design shown, the generator and associated electrical leads and auxiliaries are easily accessible.

The pitch change design is a mechanical mechanism utilizing proven commercially available hardware (jack screw actuators, angle gearboxes, universal joints, a T gearbox and a pitch change input shaft). The design relies on electric power for start-up only. During normal operation this design utilized the kinetic energy of the main shaft to effect changes in pitch, thereby, eliminating any dependency on electrical or hydraulic systems during critical situations.

The control system is comprised of a microcomputer, sensors and actuators; the microcomputer and the isolated sensors are powered by an uninterrupted power supply (UPS). The UPS system, which consists of a battery coupled to a trickle charger (to maintain a full charge at all times), and the use of a redundant microcomputer provide the necessary reliability to the control system. The principal functions of the control system are to provide for system start-up, shut-down, nacelle turning, power regulation, mutual protection of the network and the WTG, data computation and control of the system during maintenance and emergency modes.

System Cost Summary – This section describes the anticipated cost of commercial WTGs based upon the designs. In making this projection, it was assumed that the development of WTGs had been completed and that considerable operational experience had been gained with prototype WTGs operating in the electrical utility application. With this assumption the role of the system contractor becomes:

Identifying the specific requirements of the utility;

Performing site planning;

Purchasing and/or manufacturing all subsystems;

Assembly and installation of the units and

Operational check-out of all units

Specifically, it was assumed that 100 production units would be fabricated and installed within a 3½ year period and that the facilities for producing these units were available. A summary of the capital costs (1975 dollars) on a per unit basis is shown in Table 5.2. Due to the fact that these costs are being estimated from a limited experience base with this technology, some of the figures are subjective. Therefore, an explanation of these costs is provided below.

Table 5.2: System Cost Summary (1975 Dollars)

System Task	500 kW WTG 12 mph Median Wind	1,500 kW WTG 18 mph Median Wind
Program management	2,000	2,000
Final design	1,500	1,500
Systems integration	2,500	2,500
Equipment costs		
Rotor	154,000	185,000
Transmission	100,000	162,000
Electrical	37,400	60,700
Controls	13,600	13,600
Tower (concrete)	39,300	55,900
Bedplate/pintle	19,100	36,400
Subtotal	363,400	513,600
Assembly costs		
Nacelle	8,600	8,600
Site delivery	5,600	9,300
Erection	14,700	14,700
System check-out	2,500	2,500
Subtotal	31,400	35,100
Site costs		
Land	1,500	1,500
Control building	3,500	3,500
Total direct labor and material	405,800	559,700
Gross operating margin	60,870	83,955
Total cost	466,670	643,655

Source: ERDA/NASA/9403-76/1

Program Management: The total system contractor program management cost was estimated to be $200,000 or $2,000 per unit, through labor and overhead. Program management tasks would include program scheduling, purchasing, financial control and any technical direction required. Over the 3½ years it was assumed that the program management manpower level would scale down from three to one-half a man.

Final Design: A total of $150,000, or $1,500 per unit, was included in the first 6 months of the program to account for a limited amount of redesign from previous units. This is approximately a 6 to 8 man level, and includes labor and overhead.

Systems Integration: During the initial stages of the program, system requirements and interfaces must be identified, a specifice site erection plan must be generated, and a configuration control procedure established. A total of $250,000, or $2,500 per unit, was allocated for these functions which represents a manpower level of four (4) scaling down to one-half (0.5) from program start to finish.

Equipment Costs: Equipment costs represent vendor quotations for established commercial products except for the rotor. Quantity order discounts of 15% were included on the transmission and electrical equipment; system contractor's overhead expenses are included, estimated to be 15%. The rotor (blades, hub and pitch change unit) cost is the least certain of the equipment costs.

An initial cost figure, based on present day knowledge and experience was developed by United Technologies Hamilton Standard Division and resulted, in part, from their discussions with Hercules, a fabricator of filament wound products. Subsequently, downward revisions were made to this estimate to reflect the anticipated experience afforded by the development cycle prior to obtaining a 100 unit order. The final cost of $185,000 for the 1,500 kW rotor reduced to a figure of $5.3/lb, where approximately ½ of the weight is in the hub, a structural steel weldment.

Assembly Costs: The assembly costs of $35,100 and $31,400 for the 1,500 kW and 500 kW units, respectively, include the nacelle assembly, site delivery, and system erection and check-out tasks. These functions are assumed to be conducted by a trained crew who would develop considerable expertise and efficiency over the course of the program. The tower erection cost is not included here, but is covered under the tower equipment cost.

The nacelle assembly cost assumes assembly of the transmission, generating equipment and pintle to the machine bedplate at the system contractor's plant. Approximately 12 man weeks of technician effort, through labor and overhead was assumed.

Site delivery costs were calculated from standard industry quotations based on weight for a 1,000 mile delivery radius. The cost of erecting the system, $14,000, includes the cost of crane rental as well as labor required to perform mechanical and electrical connections. The erection procedure is described on p 187. A separate cost of $2,500 per unit was included for an operational system check-out.

Site Costs: The site costs assume that $1,500 per unit is required to perform soil bearing tests and to accomplish any clearing of brush and trees. Foundation costs are included in the tower equipment cost. A cost of $3,500 per unit was allocated for a control building located under the tower base.

The control building, which housed much of the electrical equipment and controls, is assumed to be a prefabricated enclosure. A system contractor fee of 15% was added to all costs and represents the gross operating profit margin of the business. Total capital costs of $466,670 and $643,655 were calculated for the 500 kW and 1,500 kW units, respectively.

First Unit Costs: The costs of Table 5.2 reflect estimates for a "mature commercial product" environment. Costs of early units will be substantially higher due to many factors. Recently, General Electric contracted with NASA-Lewis to construct and install two 1,500 kW WTG power systems (Mod-1). The engineering time and purchased parts quotation of the second unit, devoid of recurring costs, was employed to generate Table 5.3 which is a representative first unit cost summary for the 1,500 kW WTG system.

Table 5.3: Typical Prototype Production Unit Cost ($K) Breakdown, 1,500 kW WTG

	. . .Prime Contractor . . .				. . . Subcontractors. . . .				
Systems Task	**(A)**	**(B)**	**(C)**	**(D)**	**(A)**	**(B)**	**(C)**	**(D)**	**Total**
Program management	22	–	2	9	–	–	–	–	33
Final design/engineering support	22	–	4	10	–	–	–	–	36
Systems integration	67	–	18	32	–	–	–	–	117
Equipment costs									
Rotor/blade	–	–	–	249	237	223	23	181	913
Transmission	–	170	–	64	–	–	–	–	234
Electrical/controls	–	217	–	81	–	–	–	–	298
Tower (steel)	–	99	–	37	–	–	–	–	136
Pintle	–	15	–	6	–	–	–	–	21
Assembly costs									
Nacelle	–	53	–	32	–	24	–	9	118
Site delivery	–	13	–	5	–	–	–	–	18
Erection	–	–	–	83	–	140	21	61	305
System check-out	49	–	–	22	–	–	8	3	82
Site costs									
Land preparation	45	–	–	86	101	33	–	50	315
Control building	–	5	–	2	–	–	–	–	7

Note:

(A) Labor: Based on average aircraft industry labor rate of $10.80 per hour, OH rate of 107% used for all prime contractor labor costs.

(B) Parts: Includes catalog/off-the-shelf type items/raw materials. Suppliers OH, G&A, margin burdens included but not definable.

(C) Other: Includes travel and living, publications, computer, etc. – direct cost.

(D) Burden: Based on average aircraft industry G&A/material handling rate of 25% and assumed margin of 10%.

Source: ERDA/NASA/9403-76/1

Note that aircraft industry average hourly, overhead, general and administrative and fee rates were used in Table 5.3 rather than those particular to the specific contractors.

The total cost of Table 5.2 (for the 1,500 kW unit) can be related to the $2.633 million first unit cost by means of the learning curve to establish the cumulative quantity of units necessary to achieve the lower cost. A 90% learning curve is quite conservative when compared with historical data on items with similar technological complexity. This means that for each doubling of cumulative production the average unit cost (for all units) will be reduced to 90% of the previous value. Thus if the first unit costs $100, the first two will average $90, the first four will average $81 etc. Solving for the average cost of 100 unit production runs for a first unit cost of $2.633 million and 90% learning yields:

Units	1,500 kW WTG Average Cost
1-100	$1308 K
101-200	1046
201-300	966
↓	↓
3401-3500	650
3501-3600	647
3601-3700	644

Thus, if a 90% learning rate is achieved the total cost of Table 5.3 will be reached when cumulative 1,500 kW unit production is approximately 3,600 to 3,700 units. This is consistent with the mature commercial product assumption and well within potential WTG implementation possibilities.

A detailed economic model was used to calculate the energy generation cost for these 2 systems. The following assumptions were made in this model:

Straight line depreciation over the system life (50 years);

Capitalization method–50% bonds, 50% equity;

Cost of capital–9% interest, 11.5% return on equity;

48% tax rate on profits;

Use of a sinking fund to repay bondholders at end of 50 years and

Additional capital costs for rotating parts and maintenance costs included over the system life

Based on these ground rules the energy generation costs are 4.04 and 1.57¢/kWh for the 12 and 18 mph systems, respectively. A breakdown of the energy generation cost is provided in Table 5.4. As shown in the table maintenance (labor and materials) costs were calculated to be 16% of the total energy cost or $12,400 for the 500 kW/12 mph system and $16,100 for 1,500 kW/18 mph system annually.

Table 5.4: Breakdown of Energy Generation Costs

System	Start-Up Costs	Capital Fund	Equity Return	Debt Interest	Federal Taxes	Maintenance	Total
	(¢/kWh)						
500 kW, 12 mph median wind site	0.0921	0.1071	1.419	1.110	0.6492	0.6626	4.04
1,500 kW, 18 mph median wind site	0.0535	0.0414	0.5498	0.4303	0.2517	0.2432	1.57

Source: ERDA/NASA/9403-76/1

System Weight Summary – The weight of the system is important since, as for most industrial products, it will relate directly to cost. For a system such as the WTG, the weight of individual subsystems is particularly important since undesirable dynamic interactions (related to weight) can occur between subsystems. Specifically, the rotor/tower dynamic analysis performed showed that it was advantageous to reduce the nacelle weight since higher nacelle weights caused a reduction in the tower natural frequency. Also, higher blade weights result in a necessarily stronger, heavier and more costly mechanical power transmission to accommodate the increased load, as well as decreasing the tower natural frequency.

Table 5.5 provides a weight summary for both the 500 kW and 1,500 kW WTG systems.

Table 5.5: Weight Summary

	Power Level			
	500 kW.		1,500 kW	
	lb	kg	lb	kg
Rotor	27,000	12,147	35,000	15,875
Gearbox	24,000	10,886	46,000	20,875
Generator	4,500	2,041	11,200	5,080
Main shaft	4,000	1,814	5,000	2,268
MS bearings	2,300	1,043	3,500	1,588
Couplings	2,700	1,225	5,200	2,359
Bedplate/pintle	15,000	6,803	24,000	10,886
Subtotal	79,500	36,060	129,900	58,920
Tower				
Concrete	450,000	204,000	650,000	295,000
Steel truss	86,000	39,000	118,000	53,500
Total				
with concrete tower	530,000	240,000	780,000	354,000
with steel truss tower	165,000	75,000	248,000	112,000

Source: ERDA/NASA/9403-76/1

Total weight aloft for the 500 kW and 1,500 kW systems is approximately 36,060 kg and 58,920 kg, respectively. The rotor/nacelle configuration was designed such that the static loading provided a zero moment about the azimuth bearing.

The rotor subsystem weights include the hub, blades and pitch change mechanism. The higher weight of the 1,500 kW rotor subsystem is directly attributable to the larger loads upon the blades of this unit. Further refinements of the blade design could result in some weight reduction. Some conservatism may also be present in the weight of the bedplate/pintle subassembly, since weights for these structures were calculated on a preliminary basis. During a final design effort, a detailed computer analysis might allow some relaxation in the safety factors assumed, which were generally on the order of 1.5.

System Fabrication and Assembly: Various approaches were evaluated to determine the minimum cost fabrication and assembly procedure for the WTG. The preferred approach is to maximize the amount of assembly performed at the plant and to minimize the number of site operations.

The pacing items in the WTG fabrication cycle under production conditions are the rotor, transmission and generator. Normal lead times for these subsystems are expected to be 26 to 30 weeks. The WTG fabrication process can be divided into 5 main areas as follows:

Assembly of the nacelle at the plant;

Fabrication of the rotor at the plant;

Site preparation and erection of the tower;

Mounting of the nacelle and rotor with the tower and

Completion of system electrical connections and system check-out

Based upon a preliminary assessment it is estimated that 16 man-weeks would be required to perform the nacelle assembly. Basically, this process would begin with attachment of the pintle and the main shaft subassemblies to the bedplate, attachment of the transmission gearbox, insertion of the variable pitch input actuators, attachment of the generator, installation and connection of instrumentation and electrical leads and attachment of the nacelle housing. The nacelle could be transported to the site by means of a flatbed truck. The rotor would be shipped directly from the factory to the site in a minimum of 3 subassemblies, two blades and the hub. An alternative, which may result in lower transportation as well as tooling costs, is to have a bolted joint in the blades. In either approach, complete assembly of the rotor subsystem would be performed on the ground at the site.

Prior to shipment to the site, the entire rotor subsystem would be checked out. However, the extent of the required inspection for a mass production effort has not yet been determined. It is anticipated that blade deflection, hub deflection and variable pitch change functional tests would be made on a sampling basis as a minimum.

Erection times averaging 6 to 8 weeks are reasonable expectations for truss towers. Since the weight of these towers is only about 40 to 60 tons, the foundation requirements are not as extensive as for a concrete tower. In addition, a full cure of 28 days would not be required for the foundation before the steel sections could be erected. The steel beams used in this type of construction can be easily transported by truck.

In constructing a concrete tower, the precast sections can be trucked directly to the site from a central location to the raw materials supplier/tower subcontractor, or, the raw materials can be trucked directly to the site where the precast sections could be made.

The weight of the concrete towers is estimated to be 225 and 325 tons for the 500 kW and 1,500 kW systems, respectively. Therefore, larger foundations will be required for these structures. The heavier weight of the concrete tower may also require that a full cure period of 28 days be observed for the foundation before the tower is erected. Consequently, total erection time for the concrete tower is expected to be 8 to 12 weeks.

The equipment can be hoisted atop the tower in 1 or 2 steps. Preferably, the rotor subsystem will be attached to the nacelle on the ground and the entire unit hoisted atop the tower. Alternatively, the nacelle would be put in place first and the rotor hoisted to be mated with the nacelle. The deciding factor as to which approach is used will be the weight of the nacelle and rotor as well as the available crane weight and height capacity.

Following installation of the WTG major subsystems, final electrical connections will be made within the nacelle. In addition, a control building will be erected at the tower base which will house the microcomputer, various electrical equipment and electrical connections to the utility.

The final step in the assembly process will be to check out the system. This procedure will consist of establishing the continuity of all control and electrical circuitry, allowing the system to break away under low wind/no load conditions and then operation under load. The time required for the check-out of production units will be dependent upon the experience gained at the time.

System Operation: The WTG must be designed to be controlled in a predictable manner under various conditions ranging from start-up to emergency shutdown situations. Operation of the system can best be explained from the power ratio versus velocity ratio curve shown in Figure 5.13 for the 230XX airfoil series.

Figure 5.13: System Operation

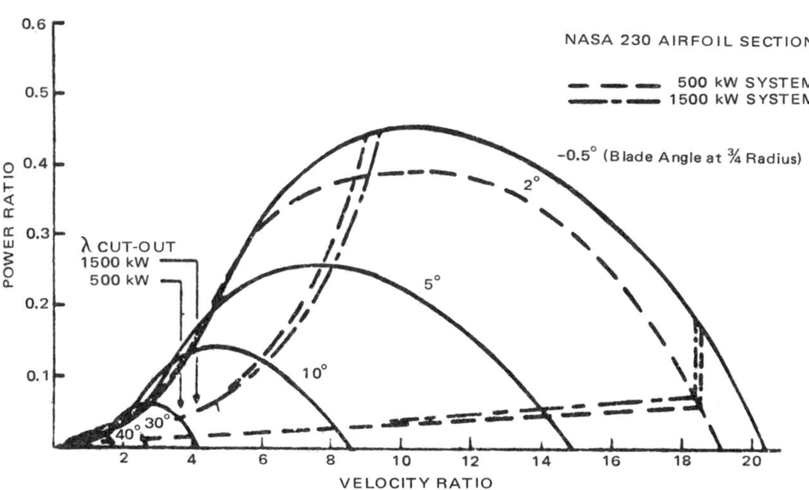

Source: ERDA/NASA/9403-76/1

Power ratio is defined as the power derived from the wind to that which is available (0.593 is theoretical maximum for horizontal axis machines), and velocity ratio (λ) is the blade tip speed divided by wind speed. Also shown in Figure 5.13 are lines of constant pitch angle (B); as shown, the optimum performance is achieved at a λ of 10 and a pitch angle of -0.5°. At rated wind conditions the 500 kW and 1,500 kW designs operate at λ 9.0 and 9.5, respectively. Lower wind speed conditions result in a higher λ condition; therefore, the tendency is to move along the crown of the B = -0.5° curve for most of the operating time. At wind conditions greater than rated, power regulation is achieved by pitching to higher blade angles until cut-out wind velocities are achieved.

System Start-Up – Start-up can occur from low, intermediate or high wind conditions; it is expected that start-up from low wind conditions which is discussed here, will be the most common start-up occurrence. When the wind sensor indicated that the wind speed is equal to or greater than the cut-in velocity (V_{CI}) for approximately 5 minutes the control system will sense, via the azimuth position indicator, any misalignment between the rotor axis and the prevailing wind direction and command the azimuth drive motor to align the rotor axis with the wind. Simultaneously, the Bendix type pitch change drive motor will be engaged to cause a change in the blade pitch from B ¾R = 90° to the maximum rotor breakaway torque position of B ¾R = 72°.

The blade pitch angle is determined from the output of a linear voltage differential transformer (LVDT) attached to the pitch change actuation shaft located in the pitch change power supply. Having the proper blade pitch and cut-in velocity, the rotor will break away autonomously as the rotor axis is aligned with the wind. When the shaft speed sensor indicates a main rotor shaft speed of 0.5 rpm, the control system will command the pitch change mechanism to change B ¾R from 72° to 2° and the rotor will then accelerate to near synchronous rpm in approximately 7 minutes provided the wind is maintained.

At 10% under synchronous speed the control system will take over speed regulation and by modulating the blade pitch and monitoring the utility voltage, phase, and frequency, utilizing the potential transformers, synchronize the generator to the utility network. Power in the rotor at this time will be sufficient to provide for generator friction and windage core losses, and power train running torque.

System Cut-In – When the WTG approaches synchronous speed conditions the system will be brought to synchronous speed by changing the blade pitch angle. Figure 5.14 illustrates the blade pitch angle movement required and associated speed error as a function of time, before mechanical stability is achieved under steady state wind conditions. Synchronization requirements include: matching phase (rotation and polarity), voltage amplitude and achieving constant rph at 1,800 (frequency of 60 Hz).

Having satisfied the utility requirements for synchronization, the control system will simultaneously command the main circuit breaker to close, thus connecting the WTG to the network. Next, the blade pitch will be changed from B ¾R = 2° to B ¾R = -0.50° at which time the electrical power output will be approximately 5 kW as determined by the current transformer and potential transformer data.

The control of torque transients due to wind gusts subsequent to cut-in will be controlled by increases in the generator field strength in response to increases in torque as indicated by the torque sensor on the generator input shaft. The control system will house this field-torque feedback loop to maintain on-line stability.

Cut-In to Rated Power – Under these wind conditions the system will operate along the B = -0.5° line.

Figure 5.14: Generator Speed Response Under Constant Wind Conditions

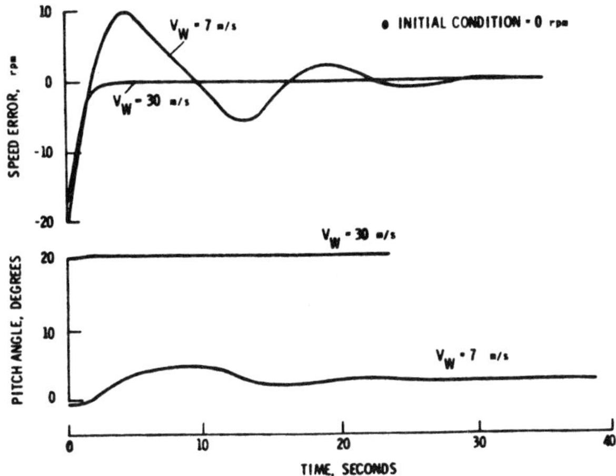

Source: ERDA/NASA/9403-76/1

For wind velocities greater than cut-in and less than or equal to rated velocity, the regulation of power is controlled solely by changes in generator field strength in response to changes in the torque sensed by the generator shaft torque sensor. The control system will therefore issue no blade pitch change commands for wind speed less than rated for the purpose of maintaining power. Primary power control is based on torque instead of wind speed.

For wind direction changes ≤15° for wind velocities less than rated there will be no change in rotor axis azimuth unless this azimuth variation is sensed for a period greater than 6 minutes. If a wind direction change occurs greater than ±15° a slip clutch on the azimuth drive will allow the rotor axis to slip in the direction of the rapid change. The slip clutch is located on the output shaft of the azimuth drive speed reducer. The value of the slip torque would be preset and controlled during slip.

To control gusts, blade pitch change and field modulation will be used in response to the torque sensed on the generator input shaft. The control system will initiate command signals to either or both pitch change and field strength depending upon the magnitude and response rate of the gust.

Rated to Cut-Out Conditions – For wind velocities above rated velocity, the control system will maintain constant rated power output based on data from the current and potential transformers located on the output of the synchronous generator. As power increases above rated, the control system will sense the current increase and increase pitch to maintain rated conditions. Should the power decrease, the control system will automatically try to reset blade pitch to -0.50°, thereby increasing the power output. The equilibrium blade pitch angle will always be that angle which results in rated power output for wind speed between rated and cut-out.

Gust control will be effected by proper application of either generator field modulation or blade pitch change in response to variations in sensed torque.

High Wind Cut-Out – Wind speed cut-out velocities of 40 and 50 mph were selected for the 500 and 1,500 kW systems, respectively. Above these wind velocities, only a small fraction of the site annual wind energy exists.

When measured wind velocities are greater than cut-out speed for periods greater than 6 minutes the control system will increase blade pitch to bring the generator output current to 25% of the rated value in approximately 10 seconds. When this value is achieved the tie-in circuit breaker will be opened and blade pitch increased to prevent excessive overspeed and maintain turbine rpm to some value below synchronous value as long as the wind velocity is greater than cut-out.

Low Wind Cut-Out – When the measured wind velocity falls below cut-in the generator will motor since the wind turbine will not have sufficient power output to supply the parasitic electrical and mechanical losses. This motoring mode will be sensed by the current and potential transformers and after a period of 6 minutes of monitoring, the control system will command the circuit breaker to open and thereby disconnect the WTG from the network. The control system will set the blade pitch angle to 2° to start the resynchronizing process. If, however, the wind does not return to cut-in speed after 6 minutes the blade pitch will be set to 90° or full feather, minimum drag configuration. The wind turbine will lose rpm slowly and the primary brake (located to the high speed through shaft on the speed increaser) will be proportionally applied to stop the turbine blades in the horizontal position.

Emergency Shutdown – Emergency shutdown may be required in various circumstances, therefore, the WTG has been designed to achieve shutdown as expeditiously as possible. In general, when a condition requiring shutdown is recognized the primary brake will be applied, the blade pitched to full feather and the circuit breaker opened.

Subsystem Design: The major WTG subsystems are the rotor, tower, transmission, generator and controls. Each of these is discussed briefly to illustrate the level of technology identified in this study for cost effective WTGs.

The transmission and generator are essentially commercially available items. Therefore, neither technology or fabrication development is involved with these subsystems.

Concrete or steel truss type towers on the order of 150 feet high are also state-of-the art. The unique design aspect for this application is the ability to define the structural requirements resulting from the dynamic and static loads on a minimal cost basis.

The use of a microcomputer in the control system represents a technology which is now being used in many commercial applications. The use of a microcomputer will not require any technology development and represents an important asset to the widespread use of WTGs. The rotor design is based upon state-of-the art propeller technology. Consequently, the tools to analyze the rotor subsystem from a structural and aerodynamic standpoint are sophisticated and their accuracy has been verified from the large hardware data base available. Filament winding fabrication is also well developed. Therefore, the development aspect of the rotor subsystem is in scaling previously developed units to the diameters required for economical WTGs. While the challenges should not be minimized, only a modest development risk appears to be present due to the well developed supporting technologies involved in the design.

Rotor – The effective aerodynamic blade diameter is 55.8 m and 57.9 m for the 500 kW and 1,500 kW units, respectively. In order to compensate for blade deflection the mechanical lengths are 56.4 m and 58.7 m. The aspect ratio (blade length per average chord width) is identical for both blade sets.

The rotor blades are made of a filament wound material having a spar/shell construction. This approach was found to provide the necessary stiffness for the least cost. The blade design has a bolted section at the blade midplane region. Use of a bolted section is believed to offer low transportation and tooling costs for the same reliability.

A critical feature of the rotor subsystem design is the hub/blade attachment. Obviously a rigid and reliable hub design is needed to achieve the predicted aerodynamic performance.

The third major component of the rotor subsystem is the variable pitch change mechanism which is critical to the overall operation and reliability of the WTG. Under normal operating conditions pitch change capability is required for start-up, synchronization and cut-out situations. Failure to change pitch under operating conditions above rated wind speed could result in blade overload and/or overspeed conditions. The principal features of the pitch change mechanism are:

No power is required to maintain a fixed blade angle

Excess load capability is available to clean jams such as icing in the blade bearing areas

A spring loaded brake to increase blade pitch will automatically feather the blades in the event of loss of the control system input signal.

A feedback signal to identify blade angle position is incorporated.

The actuators located at the hub assembly end of the pitch change are industrial catalog items. The power train which runs through the main shaft also used industrial components and an automotive type disc brake. Therefore, the unit cost and availability are consistent with the overall system objectives.

Tower – Both truss and concrete tower designs were generated for the preliminary design WTG systems. While the concrete design offers lower cost and aesthetic appeal, the truss tower approach has the advantages of being adaptable, at reasonable cost, to sites having poor soil conditions and presents less of a tower shadow to the rotor.

The load criteria used for the towers assumes: continuous cyclic loading at rated wind conditions, intermittent loads due to an instantaneous doubling of the rated wind velocity and a 120 mph storm condition. One aspect of the tower design which was investigated in considerable detail was the effect of dynamic interactions occurring between the rotor and tower upon the tower design. The analytical investigations pursued in this regard showed that definite economic advantages could be realized by "tuning" the rotor/tower subsystems so as to minimize rotor/tower dynamic load factors. The effect of tower shadowing was also included in dynamic interactions investigations. In general, tower shadowing exerts an appreciable impact on tower designs, especially from a dynamic viewpoint, because it is the shape of the towers which determines the nature of the forcing functions.

The concrete towers presented in the preliminary design were based upon a reinforced concrete, annulus geometry approach which offers a low cost design. Schematic diagrams of the concrete tower designs are shown in Figure 5.15; due to the small difference in blade diameters, both towers were assumed to be 42.7 m in height.

Figure 5.15: Tower Design Definition

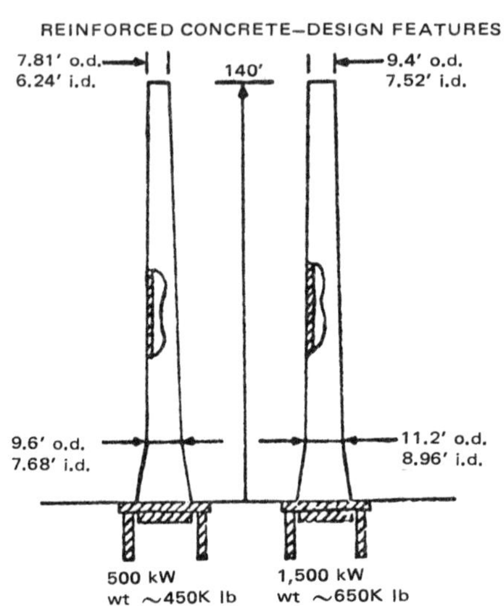

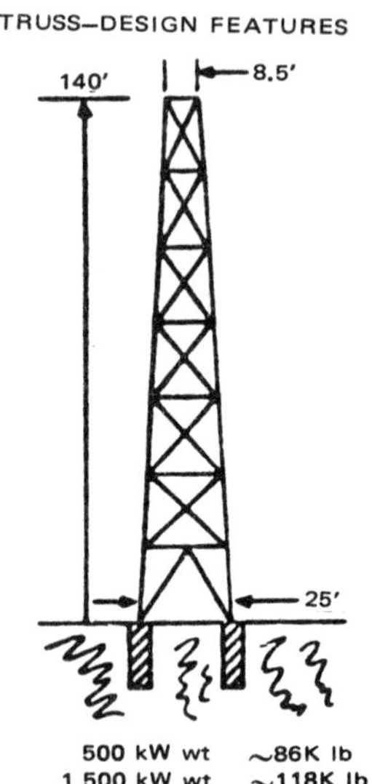

Source: ERDA/NASA/9403-76/1

The thin wall design approach used results in an economic use of materials and good dimensional control during the fabrication process. The concrete towers were designed in ten 4.27 m sections in order to remain within the limits of conventional truck transportation capability. A precasting fabrication technique was selected over a slip forming approach on the basis of preliminary cost estimates. Fabrication of the sections could be done at the factory, at the site or some intermediate location depending on the site location, accessibility, materials availability and local labor rates. Also shown in Figure 5.15 is a schematic of the truss tower designs. All members specified for the 500 kW and 1,500 kW truss towers are commercially available structural sections. Main supports and lower bays are comprised primarily of higher strength steels while the upper bays and most cross members are lower strength steels. In utilizing bolted construction and minimizing welds the intent has been to duplicate electric transmission line tower technology which is in a mature state of development. Since WTG towers experience greater loadings, however, it has not always been possible to choose convenient member shapes; I-beams were required for their load capability for some of the members. Since those type sections and a symmetrical tower are not convenient for easily bolted joints, some welding will be required. It should also be noted that galvanized steel is proposed for the towers, with a three to five thousand dollar repaint being required every seven to ten years.

As in transmission line tower technology, the most efficient construction approach appears to be to prepare the members at a central location, transport, and then erect bays at ground level and hoist the bays aloft with a crane. The advantage, as with precast concrete, is to minimize aloft construction. Secondary design refinements could also perturb the 6.10 m bay height presently defined. The truss towers both have the same envelope sizes, a distinct feature that should be maintained. The idea is to maintain as much commonality as possible with the advantage being that for a large number of towers, the same point on a learning curve can be achieved more quickly. As in power towers, deep rooted foundations will be utilized.

Mechanical Power Transmission – A schematic diagram of the mechanical power transmission is shown in Figure 5.16.

Figure 5.16: Mechanical Power Transmission Subsystem

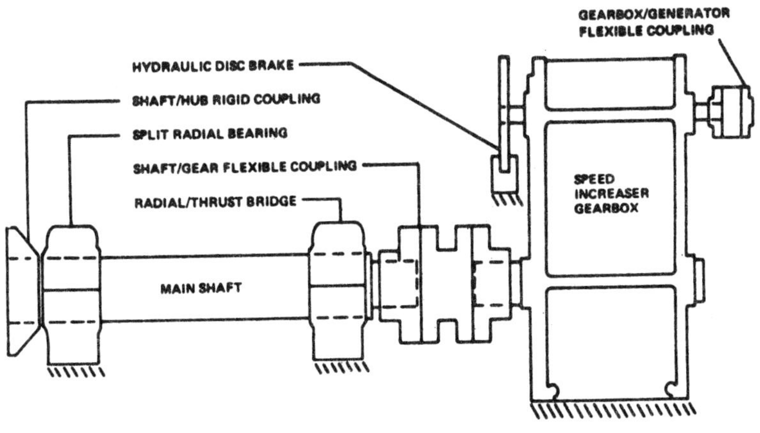

Source: ERDA/NASA/9403-76/1

The principal features of the transmission design are: all of the components are commercially available; a shock factor of 1.4 has been incorporated in all of the transmission components; ease of maintenance; allows for small shaft misalignments and incorporates system braking capacity. As shown in the figure the torque is transmitted from the hub to the gearbox by means of the main shaft.

The 1,500 kW system main shaft has an i.d. of 22.9 cm and an o.d. of 38.1 cm and is made of 4340 steel. Respective dimensions for the 500 kW system are 20.3 cm and 34.3 cm. The main shaft is supported by a radial bearing at the rotor end and a radial/thrust bearing at the transmission end of the system. A flexible coupling connects the main shaft to the low speed transmission shaft. This arrangement, plus the fact that the rotor can be alternately supported by the bedplate, allows for complete teardown and maintenance of the shaft, bearings and coupling if necessary, without disturbing the rotor assembly or other subsystems.

The gearbox is segmented in 3 sections and is oriented with the high speed pinion at the top. Since this is the area in which the most severe wear will occur, it has been made most accessible. The high speed shaft is connected to the generator shaft by means of another flexible coupling in order to compensate for potential misalignments.

A hydraulic brake, located on the rotor side of the high speed shaft provides emergency stopping capability in the event of a network outage or pitch change subsystem failure.

Electrical Equipment – The electrical subsystem consists of the electrical generating equipment and the equipment necessary to control and protect the generator. For the WTG this includes: auxiliary power (drawn from the utility network) to supply the controls with electricity when the WTG is not producing power, and emergency power for the controls when all connection to the utility network has been interrupted for any reason. In addition, the electrical subsystem includes equipment necessary to control the temperature of components in the nacelle and equipment necessary to maintain the control building in a benign environment. The control building protects the control equipment from severe weather conditions and other natural environments which are likely to be encountered.

All electric subsystems equipment except the generator, exciter, and lightning protection units are housed in the control building at the base of the WTG tower. All electrical equipment which is aloft (including drive motors for blade pitch change, azimuth gears and hydraulics) are enclosed in contiguous metal such that lightning striking any part of the WTG aloft will be directed to ground through the tower.

The major elements of the electrical subsystem are: the generator, main circuit breaker, an engine-alternator emergency power source, an automatic transfer switch, a pair of control power transformers (auxiliary and emergency power), a grounding resistor, heating and ventilating air-conditioning equipment, lightning protection equipment, voltage and current instrument transformers, an assortment of small contactors and relays, and cabling for the power and control signals.

The approach is to provide electrical power for pitch and azimuth control and protective functions even when the wind is insufficient for generating purposes or too strong for stable power generation. In the event that electrical power for control is interrupted, the emergency alternator-generator source is automatically started and the control power lead is transferred to the emergency set. As long as control power is available the system operation for all modes is autonomous; for abnormal condition the system will protect itself. Should service be required, the WTG will automatically alert a remote location and respond to queries of its condition. The master logic for this scheme is contained in a microcomputer. The data acquisition and data transmission scheme associated with this microcomputer are described in the section on the control subsystem. An uninterrupted power supply (UPS) is provided to insure that the logic process is possible when control power has been lost. A separate battery has been provided for starting the emergency set without control power.

All major electrical loads of the WTG are protected by circuit breakers and all circuit breakers have an operating time which is inverse to the overcurrent condition. The characteristics of the breakers are selected such that the circuit breakers closest to the load operate first under overload condition. A neutral grounding resistor is provided so that generator fault conditions will not be propagated beyond the WTG main contactor and the outage will be confined to the WTG.

For small generators (between 0.5 and 1.0 MW) the standard protective devices are not cost effective. Usually overcurrent, anti-motoring, thermal overload, and loss of potential transformer protection is provided. By utilizing the microcomputer and instrumentation transformer, it is possible to develop algorithms to perform these functions without adding protective relaying. In addition, synchronizing, ground sensing, and phase imbalance/rotor heating can also be provided with little increase in the instrumentation required for the WTG.

A single circuit breaker capable of interrupting 350 MVA is utilized to connect the WTG to the utility network. Control power is obtained for the auxiliary transformer from this circuit breaker by tapping in to the utility before the outbreak contactor. This auxiliary transformer is fused for 350 MVA interruption of a nominal 10 kVA load. When the auxiliary transformer is not supplying the control power, only the magnetizing current of the auxiliary transformer is being supplied by the utility network.

Control Subsystem – The control subsystem consists of the equipment necessary to control the pitch of the blade, the azimuth of the wind-rotor axis, and the generator operation. It also consists of the instrumentation necessary to measure wind, torque, temperature, pressure, electrical power and other parameters which are significant to the successful operation of the WTG and to record the transmitted information of significance to locations remote from the WTG. The control subsystem also consists of the equipment necessary for operation of the WTG in the maintenance mode.

The control subsystem includes all the components necessary to the above functions except for those components which are included as an integral part of major WTG elements. For example: the pitch control mechanism is a part of the wind rotor; the azimuth drive motor is a part of the pintle section of the tower; and the wind rotor brake and transmission oil heater are part of the transmission gearbox.

All instrumentation and electronics that are not required to be aloft are protected in a control house at the base of the WTG tower. The master control for the logic of operation is provided by a microcomputer in conjuction with data acquisition electronics. For on-off control functions, output signals from the microcomputer are buffered and amplified to drive relays and contactors which apply power to drive at full power as long as the signal is present. For proportional control codes, outputs are decoded to time function for time-ratio-control of silicon-controlled-rectifier circuitry that meters the ac power for the proportional control drives.

The computed data and status data output from the microcomputer is suitable for transmission over a leased phone line. The computer also constructs a coded signal as an input to a telephone call unit which will automatically dial a predetermined telephone number when conditions warrant outside attention to the WTG. Data may also be obtained by dialing the WTG site modem and acoustically coupling the phone set to a teletypewriter. The agent may then, from his remote headquarters site, command and interrogate the WTG to determine the status of the WTG and the nature of maintenance, if any, required. A maintenance crew arriving at the WTG site may acoustically couple a portable teletypewriter to the microcomputer and in this manner exercise the WTG and determine additional information on the condition of the WTG.

The various inputs and outputs of the microcomputer are isolated by optical coupling, high impedances and other techniques common to data acquisition technology so as to protect the electronics from damage which might otherwise be caused by large-power switching transients and lightning, In addition, a measure of redundancy is provided by using two microcomputers, either of which is programmed to control the critical functions of the WTG. Noncritical functions are divided between the microcomputers. In the unlikely event that one microcomputer fails to function properly only half of the noncritical functions will be temporarily lost.

An uninterrupted power supply insures that power sufficient to supply the data acquisition electronics and microcomputer is available even when the auxiliary power from the utility network is temporarily interrupted. In this circumstance, the microcomputer logic will initiate the operation of a standby engine generator set as a temporary measure, so that WTG control is maintained and the condition can be transmitted to the remote headquarters.

Study Conclusions

The principal objective of this study was to evaluate the economic incentive for utilizing WTGs in broad scale electrical power generation applications. During the program, conclusions were drawn concerning this question as well as more specific design, operating approach and fabrication alternatives. Both the overall study conclusions and the more specific results pertaining to the design are presented below.

Overall Results: The principal results of this study included the following:

Site Selection – The median wind speed of the site had an overwhelming effect upon the WTG economics.

Power Level – Dramatic cost reductions were available by designing systems at power levels at or greater than 500 kW; significant cost reductions continued to accrue up to the 1,500 to 2,000 kW range.

Cost – The cost of generating energy with a 1,500 kW unit at an 18 mph site and a 500 kW unit at a 12 mph site are estimated to be 1.57 and 4.04¢/kWh respectively, based on 100 production units. Respective capital costs were estimated to be $430/kW and $935/kW. The rotor assembly was determined to be the largest single contributor to the overall system cost.

Electrical Utility Integration – No major technical difficulties were uncovered in terms of linking the WTG in an electrical utility network. The WTG can provide electrical stability under steady state as well as gusting wind conditions and provide operational plant factors in the 40 to 50% range. (Plant factor is the ratio of the annual energy produced to that which would have been produced if the plant operated continuously at rated power. Also called capacity factor).

The importance of site selection is clearly demonstrated by the data shown in Figure 5.17 where the cost of generating energy is plotted against median wind speed.

Figure 5.17: Effect of Median Wind Velocity on Electrical Energy Costs*

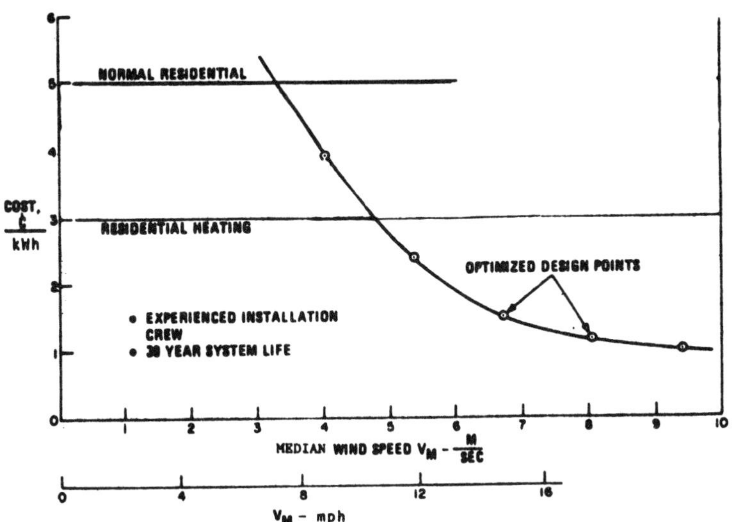

*Energy distribution costs not included

Source: ERDA/NASA/9403-76/1

The data shown in this figure were generated by optimizing the WTG for each wind site on the basis of cost: then the locus of minimum cost system for each wind site was selected to create Figure 5.17. Note that the cost data shown in this illustration is parametric in nature and is presented to indicate cost trends rather than absolute cost levels. The intent is to clearly illustrate the large decrease in costs at the higher median wind speed sites. The second conclusion presented above is supported by the results shown in Figure 5.18; these data show the cost trend as a function of power level for various median wind speeds.

Figure 5.18: Effect of Median Wind Speed and Rated Power on Cost (¢/kWh)

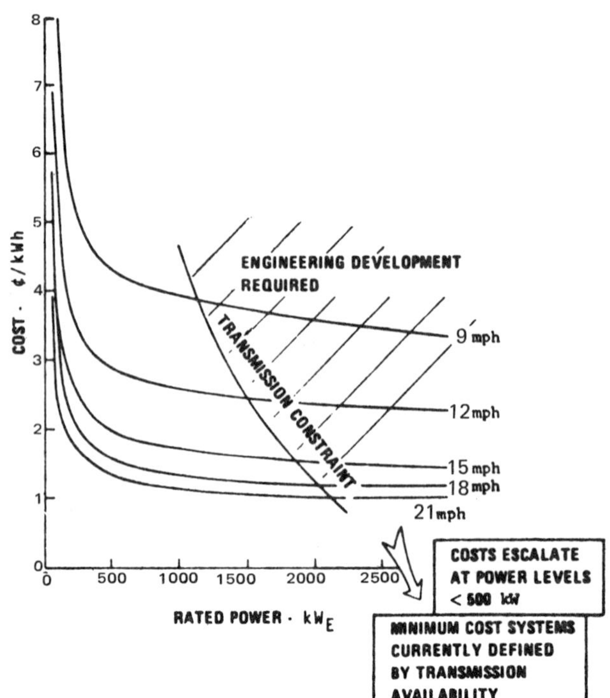

Source: ERDA/NASA/9403-76/1

At power levels up to 500 kW a sharp reduction in cost occurs due mainly to the economies of size. Above 500 kW further decreases in energy costs are apparent, however, at a slower rate. This is true due to the increasing expense of the transmission as a percentage of the total system cost; in fact, current limitations in going to higher power levels are dictated by the availability (in commercial, off-the-shelf, form) of larger transmissions. This constraint is also shown in Figure 5.18. Again the reader is urged to note that the data shown in Figure 5.18 is parametric only and should be used as a means of gauging trends.

The ability to design economically for power levels higher than that permitted by the transmission constraint line is also hampered by unknowns concerning the rotor subsystem design. Figure 5.19 illustrates the optimum blade diameters consistent with the cost/power level data shown in Figure 5.18.

This data, which is again parametric, was done for a blade clearance height from the ground of 6.1 meters. The major point to be made from Figure 5.19 is the fact that large blade diameters are required to achieve high power level/low cost systems. This study determined that large blade diameters could best be achieved using propeller as opposed to helicopter technology. This design approach results in a simpler fabrication procedure, a lighter and more reliable blade, and has the least impact on the overall system design. Also apparent in Figure 5.19 is the desirability of sites having high median wind speeds.

Figure 5.19: Cost Sensitivity to Rotor Diameter

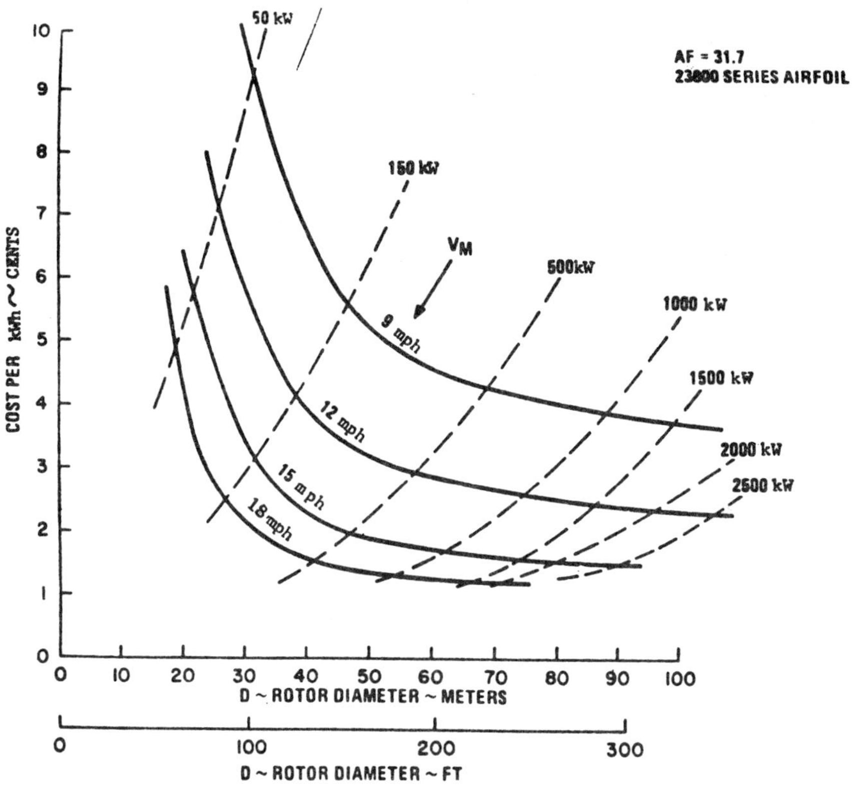

Source: ERDA/NASA/9403-76/1

The rotor subsystem was found to constitute about 40% of the total system cost. Therefore, the system design philosophy must be consistent with reducing the cost of this element. A major reason for the high cost of the rotor subsystem is that it must be designed for any unusual wind conditions that may occur over the lifetime of the unit. During the study, it was determined that the use of an intelligent control system which could avoid situations of destructive winds upon the blades would result in appreciable rotor subsystem cost reductions. Further efforts in this area may reduce the rotor subsystem to a smaller fraction of the system cost.

The fourth major conclusion resulting from this study is the fact that WTGs can be successfully integrated within an electric utility network. The WTG would be linked to the utility by means of a 4,160 V line to a transformer which would provide the interface with the network. Voltage fluctuations of less than 3 to 5%

can be maintained by the WTG under both steady state and severe gusting wind conditions. Synchronization of the WTG System with the network under a variety of environmental conditions has been investigated in detail and can be performed routinely by modulation of the blade pitch angle. Although both synchronous and induction generators can be utilized in the WTG, the use of a synchronous generator is favored due to the higher power factor offered by this approach. In addition, the transient performance of the synchronous generator is better characterized making the entire system more amenable to a detailed analysis.

Conceptual Design Results: The principal conclusions found are listed below.

WTG Configuration – The optimum configuration consisted of placing the power transmission/generator equipment atop of the tower. Efforts to place this equipment on the ground were precluded by the unavailability of commercial hydraulic transmissions or of right angle drives in the required size range. The potential cost benefits associated with equipment on the ground are lower maintenance, assembly and tower costs.

Placing the rotor downwind from the tower is favored due to the inherent stability of this configuration to changes in wind direction. Also of importance is that this configuration minimizes the rotor overhang required to accomodate rotor blade deflections. On the basis of the work performed, cyclic stresses on the rotor due to wind shear were judged to be significantly larger than those due to tower shadow.

The use of multiple rotors per tower was also considered during the conceptual design phase. In comparing a 100 kW system having one rotor to a 100 kW system having 3 rotors atop a single tower it was estimated that the cost of the 3 rotor system would be 1.5 times that of the single rotor system. This conclusion can be extrapolated to higher power levels since the rotor cost as a fraction of total system cost remains nearly constant over the 100 to 1,500 kW power range; however, the mechanical power transmission cost constitutes a larger fraction of total system cost at higher power levels.

Operating Mode – Systems operating at constant rpm and at constant velocity ratio were considered. The constant rpm system was preferred because it resulted in lower tower and blade loads as well as a lower cost transmission and generator. The single advantage of a constant velocity ratio system is the potential for higher energy capture (3%); however, in the investigations performed the increase in energy capture did not offset the higher system capital costs.

Both variable pitch and fixed pitch systems were considered. Variable pitch system using blade rotation was selected due to its excellent response characteristics under changing wind conditions.

Rotor Subsystem – A two bladed system was selected, as opposed to three, on the basis of lower cost. The technical acceptability of this decision was verified later in the program when studies of rotor/tower dynamic interactions and load magnifications were made. A rigid as opposed to a teetered hub was also selected on a tentative basis at this point in the study. Later, a dynamic analysis of rotor/tower interactions confirmed the technical acceptability of a rigid hub in conjunction with 2 blades; this analysis was performed during the preliminary design portion of the study. Filament wound blades were selected as the lowest cost approach. This concept is amenable to an efficient fabrication technique, provides a lightweight structure and requires minimum maintenance when formed in a propeller type structure. The use of a helicopter type blade, balanced at the quarter chord point, was unattractive due primarily to the following reasons: higher fabrication costs, heavier weight and higher maintenance costs. In addition, the propeller concept had been successfully demonstrated in Europe by Professor Dr. Hutter and was judged to have less development risk.

Tower – Many tower concepts including truss, concrete, tube shell and pole designs were evaluated. The truss and concrete approaches appeared to cost the least with the truss having the advantage of better cost predictability while the concrete offers a more aesthetic design.

Transmission Subsystem – Mechanical, hydraulic and electrical type transmissions were considered. The mechanical concepts such as: gearbox, belts, chains and combinations of these were judged to be the most cost effective. Of these the gearbox was preferred when maintenance aspects were addressed.

Generator – Either synchronous or induction generators can be accommodated in the WTG; however, the synchronous generator is preferred due to its higher power factor and the fact that it is better characterized.

Control Subsystem – A microcomputer was selected as the main control element in the system. The microcomputer offers an inexpensive approach to handle the many operating situations which can occur. Design emphasis on the control system permits a relaxation of the design requirements on other system components.

Preliminary Design Results: Using the component concepts previously identified and the quantitative results of the parametric analyses, 500 kW and 1,500 kW units were designed to operate in 12 and 18 mph median wind sites, respectively. The specific results of the preliminary design work are outlined below.

Cost – Costs to generate power were estimated on the basis of 100 production units and include the system contractors overhead and fee. The costs for the 500 and 1,500 units are $466,670 and $643,655 respectively, or $935/kW and $430/kW. The corresponding energy generation costs are 4.04 and 1.57¢/kWh. If the preliminary design costs quoted above are coupled with the parametric trends it is possible to estimate costs for power levels and mean wind sites other than the two preliminary design conditions.

Fabrication Approach – The approach used in erecting and checking out the WTG could have a significant impact on overall system cost. The logistics outlined during the preliminary design for these procedures included the following:

Delivery of the rotor subsystem directly to the site;

Erection of the tower by a local contractor;

Assembly and checkout of the nacelle at the factory and

Delivery and installation of the nacelle and installation of the rotor.

Variations in these procedures are possible depending upon the size of the WTG and the site location and conditions.

Rotor/Tower Dynamic Interactions – Tower and rotor costs can be minimized by analyzing these two major subsystems simultaneously. Proper selection of operating speed and natural frequencies of the rotor and tower can result in less cost yet higher reliability by reducing system weight and strength requirements. Therefore, the rotor/tower system should be dynamically "tuned" so as to provide the most cost effective design. Failure to assess the dynamic load factors could result in excessive cyclic stresses and fatigue of critical subsystems.

Rotor Subsystem Design – Two rotor blades mounted on a rigid hub were found to offer the least cost rotor subsystem. Both 2 and 3 bladed systems and teetered as well as rigid hubs were evaluated. The increased cost and sophistication of the teetered hub was found to be unnecessary if large overturning moments due to wind directional changes can be avoided by the control system.

Tower – Analysis of both truss and concrete towers for the preliminary design indicates that for sites with good soil conditions, the concrete tower is lower in cost. Also of importance is the fact that concrete towers appear to be more aesthetically pleasing to the majority of people. A comparison between concrete and truss towers is shown in Table 5.6 for 2 power levels. These costs include the system contractors purchasing cost and profit.

Table 5.6: Tower Cost Dollars

	500 kW	1,500 kW
Reinforced concrete		
Small quantity	45,200	64,000
Production attainable	39,300	55,900
Steel truss		
Small quantity	90,700	122,400
Production attainable	73,800	98,100

Source: ERDA/NASA/9403-76/1

System Stability – Both the electrical and mechanical stability of WTGs were investigated under a variety of conditions. Electrical stability can be maintained by generator field control under most wind gusting conditions. In situations where there is a high impedance line to the network, blade pitch control may be required in addition to field control for low frequency wind gusts.

Mechanical stability must be provided in order to synchronize the WTG. Gain factor and speed loop stability analysis indicate that synchronization can be achieved under steady state and gusting situations by varying the blade pitch angle.

KAMAN AEROSPACE DESIGN STUDY OF WIND TURBINES FOR ELECTRIC UTILITIES

This section describes work done by Kaman Aerospace Corporation which was a parallel study to the General Electric study described above.

Conceptual Design

Many basic concepts within the broad ground rules were studied at the outset of the conceptual design phase in arriving at a basic configuration. From these studies, the subsystems were identified and an arrangement of components emerged as follows:

(1) A rotor subsystem to extract energy from the wind and convert it to useful torque to drive the electric generator.

(2) A controls subsystem to position the wind generator system (WGS) relative to the wind direction, start and stop the WGS, control the rpm of the rotor, regulate power output, monitor the system for faults and initiate emergency shutdown, and record and transmit certain data to a control station.

(3) A structure subsystem to support and protect the generating components. The structure includes foundation, tower, turntable upon which the generating components rest and which interfaces the stationary structure with the rotating structure, and an enclosure to protect the generating components from the elements.

(4) A drive subsystem to transfer and convert the high-torque, low rpm of the rotor to the low-torque, high rpm input to the generator. It includes the rotor drive shaft, speed-increaser gearbox, and shafting from the gearbox to the generator. Ancillary items also considered part of the drive subsystem include blade pitch control mechanism, turntable drive mechanism, parking brake and a hydraulic power system for the orientation and pitch controls.

(5) The electrical subsystem to convert the energy of the rotor to electric current and feed the electricity to the grid. It includes the generator, cables, protective relaying equipment and controls, a transformer, and main circuit breakers.

A considerable number of concepts for arrangement and placement of subsystems were eliminated during the course of the study because of technical complexities and cost disadvantages. As one example, intuitively it might seem desirable to house heavy generating equipment on the ground, thus saving weight and space atop the tower. However, such an arrangement requires a high-torque, high-angle gearbox and considerable shafting, with bearing supports, to route the output torque of the rotor to the generating equipment. No such gearboxes were presently available commercially, and even if they were, they would be very heavy, thus largely offsetting the weight of the generating equipment. Therefore, to keep cost and risk low, the decision was made to house the generator in the enclosure atop the tower.

This analysis is typical of many which were necessary to pin down a concept within the guidelines of the program. Certain considerations were amenable to specific analysis while others were based heavily on engineering judgment. As the analysis expanded, the general outline of the WGS and its elements took clearer form, and is shown schematically in Figure 5.20.

Candidate Systems: From consideration of feasible alternatives, eight candidate systems evolved which met the study guidelines and from which all major component options, except the structure, could be evaluated consistently. The candidates included various combinations of rotor, gearbox, and generator and are listed in Table 5.7. Choices for the structure, which includes the tower itself, the turntable which carries and orients the dynamic components, and the foundation, were not considered a part of the candidate configurations in the study, since it was determined very early that the type of tower and foundation could be selected largely independently of the dynamic components, then sized to accommodate these components and the imposed loads. Alternatives for structural configurations are discussed separately.

Figure 5.20: Typical Wind Generator System Elements

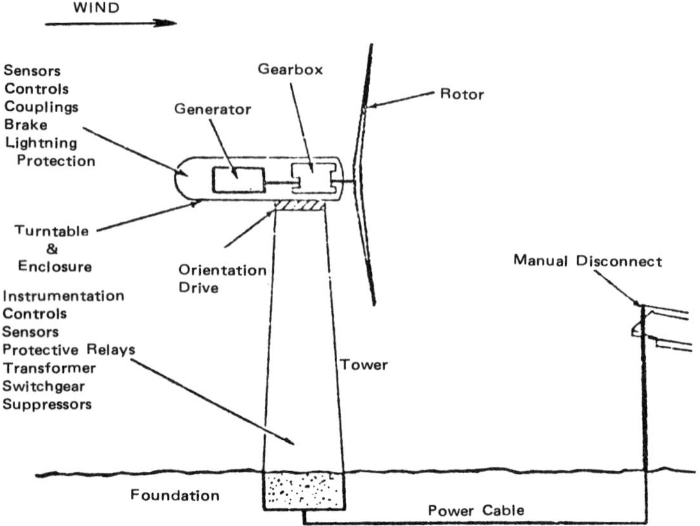

Source: DOE/NASA/9404-76/1

Table 5.7: Candidate Configurations

Concept	Rotor	Gearbox	Generators
1	variable pitch constant speed	fixed ratio gear	AC synchronous
2	variable pitch two-speed	two-speed ratio gear	AC synchronous
3	fixed pitch constant speed	fixed ratio gear	AC synchronous
4	fixed pitch variable speed	fixed ratio gear	DC generator DC motor/AC generator
5	fixed pitch variable speed	fixed ratio gear	AC synchronous transformer and rectifier DC motor/generator
6	fixed pitch variable speed	fixed ratio gear	AC synchronous transformer and rectifier solid state inverter
7	fixed pitch variable speed	variable ratio (hydrostatic)	AC synchronous
8	fixed pitch constant speed variable diameter	fixed ratio gear	AC synchronous

Source: DOE/NASA/9404-76/1

As shown in Table 5.7, candidate concepts consisted of various combinations of the three principal components: rotor, gearbox and generator. The rotor, because of its inherent mechanical and aerodynamic complexity and its sheer size, was the critical element influencing the overall design. Therefore, the rotor became the pacing component and the other parts of the dynamic system were selected to be compatible with it.

Rotor – Rotor concepts considered, apart from obvious consideration of diameter, number of blades, number of rotors per tower and the like, were characterized by their operating mode and method of torque regulation. The operating mode may be either constant or variable rotational speed. Torque regulation is necessary to limit rotor output torque at wind speeds higher than the rated wind speed for which the system is designed, to prevent mechanical and electrical limits from being exceeded.

A rotor with fixed pitch blades is very attractive in terms of simple, low-cost blade and hub design. Such a rotor would normally operate at a variable rpm which changes with the wind speed, resulting in operation at peak aerodynamic efficiency. However, it requires additional complexity in the gearbox and generating subsystems, so that the electric power generated is at the constant frequency required for the power grid.

Constant rpm rotors, on the other hand, usually have more complicated variable pitch blades and a more complicated hub. The operating rotor rpm is then chosen to maximize aerodynamic efficiency at a particular wind speed. Aerodynamic efficiency, therefore, suffers a small penalty at all other wind speeds. As noted below, aerodynamic efficiency is not an important factor when operating in wind conditions higher than rated wind speed.

For constant rpm rotors, the drive and electrical subsystems are simplified and are more efficient, which compensates for the reduced aerodynamic efficiency. This type of rotor is controlled by a synchronizing torque through the ac generator, which causes the rotor to maintain the selected rpm.

As mentioned previously, rotor torque regulation is necessary to prevent drive train and generator limits from being exceeded when the wind speed is higher than rated. This may be done several ways. Blade pitch variation can be used to reduce aerodynamic efficiency, either by direct mechanical control of the blade root or indirectly by varying the pitch of a servo flap near the blade tip. Either method is applicable to the constant rpm, variable-pitch rotor, but not to the fixed-pitch variable rpm. Torque regulation for the latter could take the form of drag flaps, aerodynamic spoilers, or a mechanical variation in blade length. Of course, any of these techniques adds complexity which tends to negate the simplicity ostensibly enjoyed by the fixed-pitch rotor.

Gearbox – A speed-increasing gearbox is required to convert the relatively low rotor shaft rpm (less than 50), to the high rpm (greater than 1,000) required by the generator. Methods considered include belt and chain drives, variable-ratio hydrostatic devices, two-speed gearboxes, and fixed ratio gearboxes. The belt and chain drives available simply do not have enough torque capacity and were dropped from consideration. The simplest, least costly gearbox has a fixed ratio. It is readily available commercially in a variety of sizes. The variable-ratio and two-speed gearboxes, also available commercially, were initially considered feasible alternatives, and were retained in the candidate concepts.

Generator – For a constant speed gearbox output, the AC synchronous generator and the induction generator are attractive candidates since they are designed to operate with a constant rpm input. The AC synchronous generator is most widely used and accepted in the industry, although the induction generator also has applications. A variable-speed input to the synchronous generator, such as would result from a variable-speed rotor driving a fixed-speed or two-speed gearbox, required additional devices to synchronize the frequency of the generator output to that of the electric grid.

For those rotors, the conversion of rotor power to the desired constant frequency electric output can be accomplished by using the variable-speed rotor to generate DC power, either by a DC generator or a combination of AC synchronous generator with a transformer and rectifier, then converting the DC power to 60 Hz AC to feed the grid. This latter step can be done either by the DC motor driving an AC synchronous generator or a solid state inverter. Each of these variable rpm rotor approaches suffers from efficiency, weight, and cost penalties compared with a constant rpm rotor driving a synchronous AC generator. From the considerations just discussed, the eight feasible candidate concepts of Table 5.7 were retained for further evaluation.

Evaluation of Candidate Systems: Candidate Systems – Systems were evaluated by sizing them with a consistent set of ground rules. They were sized for the two prescribed median wind speeds. A 100 kW system was used for the 5.4 m/sec median wind speed, and 1,000 kW system for the 8 m/sec median wind speed. Rated wind was defined as the lowest wind speed at which the WGS develops rated power.

The yearly energy output, weight, and costs were determined with emphasis placed on distinguishing differences between concepts. In the evaluation, the main criterion was energy cost, although capital cost was also an important consideration.

Using concept 1 as a base for comparison, both the 100 kW and 1,000 kW systems showed the same relative cost characteristics. For example, candidate system 4 (fixed pitch, variable rpm) shows a relative direct capital cost ratio of about 1.26 and a relative energy cost ratio of about 1.04. The relative costs apply to both the 100 kW and 1,000 kW systems. The results of this analysis, therefore, are largely independent of rated power and hence, median wind speed.

Those systems which drive the AC synchronous generator directly (concepts 1, 2, 3 and 8) have the lowest energy and capital cost. Of these, the variable diameter rotor (concept 8), though feasible, has high technical risk. The two-speed rotor (concept 2) also has greater technical risk due to the need to design the rotor and tower to avoid sustained operation at rotor and tower resonant frequencies. This leaves concepts 1 and 3 as the most attractive. However, it seemed prudent also to carry one of the variable speed candidates through conceptual design for more detailed evaluation prior to final concept selection. Of those, concept 5 was chosen from among the 4 variable-speed candidates primarily on the basis of lowest costs.

It should be pointed out that in addition to the cost considerations, good technical reasons exist for eliminating most of the candidates. Any variable speed electrical system requires converting variable shaft rpm into the constant frequency required to feed the utility grid, as discussed above. Although there are several combinations of equipment that can be used to perform the conversion, cost and weight favor a variable-rpm AC generator driving a transformer-rectifier, driving a DC motor, driving an AC generator or a solid-state inverter. Obviously, this system, with a number of elements in series, suffers several major disadvantages.

DC machinery is substantially heavier and more expensive than AC equipment of equal capacity. It is also less efficient, especially at partial power, and requires more maintenance. The solid state inverter offers the potential of better reliability and lower maintenance, but still suffers substantial cost penalties. The cost of controls and protective equipment also goes up. Finally, cascaded components reduce the net efficiency of the system considerably.

Tower: The tower supports the rotor and generating subsystems and serves as the fixed platform upon which they rotate and are oriented into the wind. It must absorb the static and vibratory forces imposed by the rotor and resist the wind forces acting on the tower itself. The fatigue strength of the tower must be great enough to withstand the rotor-induced vibratory loads, including effects of startup and shutdown cycles, gust variations, tower shadow, and gravity for a 50-year service life. The stiffness of the tower must be selected so that coupled rotor-tower natural frequencies avoid integral multiples of the rotor operating frequency.

As stated previously, it was determined that the candidate concepts for the dynamic components could be formulated and evaluated independently of the structure. However, several tower configurations were considered and evaluated. To minimize cost, existing commercial materials and standard construction techniques were used for all tower concepts. This limited the practical choices to structural steel and concrete, so the focus was on configuration rather than materials. Candidate concepts were:

Steel Truss – This concept uses commercial structural steel beams. Its advantages are low cost, ease of modification if necessary during the development program, and wide availability. The main disadvantage is appearance, but another is the fact that the stairs to the nacelle are exposed. The truss was the most promising concept at this point.

Steel Shell – This is a truncated circular cone, fabricated from rolled conical segments, field-welded in position at the site. The principal advantage is its appearance, while disadvantages are high cost and limited availability. Although the steel shell was less competitive, it was retained for further study.

Guyed Pole – This design consists of a circular steel cylindrical shell of relatively small diameter supported by pre-tensioned steel cables at the top and intermediate levels. To achieve enough stiffness to carry the bending loads imposed by the rotor, the support cables must have a large preload. This increases the foundation loads, particularly the foundation for the pole. Consequently, the foundation costs drive the cost of the structure subsystem up so that it becomes more expensive than either the steel shell or steel truss. Further, it was found that this design was not practical for a high power system because it required many large diameter cables at a high angle relative to the pole to produce an acceptable natural bending frequency, and such an arrangement makes it difficult to avoid interference between rotor and cables. Consequently the guyed pole was eliminated early in the study.

Reinforced Concrete – This is also a truncated circular cone shell cast at the site. It requires heavy steel reinforcing for adequate strength and stiffness. Its advantages include ready availability, good appearance, and a sheltered stairway. However, the walls have to be thick, and the required reinforcing drives up weight and cost. This concept was also retained for further study.

Precast, Posttensioned Concrete – This consists of a truncated circular cone, constructed of factory-cast, matched segments, minimally reinforced for handling. Provisions are made during casting for inserting steel tensioning rods at the site during erection. Factory casting permits use of high strength concrete. Posttensioning assures that the concrete remains in compression under all conditions of load, but this does not require as much steel as the reinforced concrete concept. Advantages are good appearance, sheltered stairs and low cost in mass production when the cost of the forms is amortized over many units. The prime disadvantage is the high cost of small production quantities. This tower was also considered a promising alternative.

Table 5.8 shows the results of an early conceptual design study. Later, during concept optimization and preliminary design, additional studies of the various tower and foundation types were made.

Table 5.8: Summary of Total Tower Costs, Dollars

	..... Truss		..... Shell		Guyed Pole
	100 kW	1,000 kW	100 kW	1,000 kW	100 kW
Total structural cost	7,700	34,200	23,900	69,800	17,300
Foundation cost	17,300	31,200	16,600	35,600	22,900
Basic cost total	25,000	65,400	40,500	105,400	51,200
Auxiliary provisions cost	13,799	23,100	13,700	23,100	14,800
Total tower cost	38,700	88,500	54,200	128,500	66,000

Source: DOE/NASA-9404-76/1

Selection of System for Optimization: Evaluation of the eight concepts left three feasible alternatives from which one was to be chosen for optimization. This required the conceptual design of those three candidates. The three systems were sized and evaluated on the basis of energy and capital costs. Results are shown in Tables 5.9 and 5.10. Concept 5, the fixed pitch, variable-speed candidate that was carried along in spite of early indications that it was not a favorable choice, is clearly shown to be inferior. Besides being heavier and more costly, it has significantly higher development risk.

Concepts 1 and 3 differ principally in the method of regulating torque. Concept 1 does this with a pitch control actuated by a servo flap, while concept 3 uses a fixed-pitch drag flap system.

Table 5.9: Summary of Weight Above Foundation, kg

	Concept 1 Variable Pitch . . Constant Speed. . .		Concept 3 Fixed Pitch . .Constant Speed . .		Concept 5 Fixed Pitch . .Variable Speed . .	
	100 kW	1,000 kW	100 kW	1,000 kW	100 kW	1,000 kW
Rotor	3,858	14,630	3,343	12,689	4,414	16,155
Drive system	2,451	14,605	2,451	14,605	3,036	17,429
Electrical system	1,334	3,459	1,334	3,459	1,479	3,808
Control system	347	372	347	372	347	372
Total dynamic system	7,990	33,065	7,475	31,125	9,278	37,764
Pintle and drive	5,645	19,465	5,645	19,465	5,645	19,465
Weight on tower	13,635	52,531	13,120	50,590	14,923	57,229
Tower*	6,841	29,166	6,841	29,167	7,847	31,026
Total system	20,476	81,697	19,961	79,747	22,770	88,255

*Excluding foundation.

Source: DOE/NASA/9404-76/1

Table 5.10: Conceptual Design WGS Costs, Dollars

	Concept 1 Variable Pitch . . Constant Speed. . .		Concept 3 Fixed Pitch . .Constant Speed . .		Concept 5 Fixed Pitch . .Variable Speed . .	
	100 kW	1,000 kW	100 kW	1,000 kW	100 kW	1,000 kW
Rotor						
Blades and flaps	30,105	68,945	29,920	66,440	38,125	83,335
Hub	20,025	88,865	19,645	86,165	24,900	109,560
Total	50,130	157,810	49,565	152,605	64,025	192,895
Tower						
Structure	7,950	35,440	7,950	35,440	9,745	38,530
Equipment	13,615	23,040	13,615	23,040	14,110	23,540
Foundation	25,300	41,200	24,300	41,200	29,000	43,700
Total	46,865	99,680	46,865	99,680	52,855	105,770
Pintle and drive	12,960	52,090	12,960	52,090	12,960	52,090
Drive system	19,475	110,150	19,470	110,115	23,980	130,675
Electrical system	28,695	79,740	28,695	79,740	53,960	219,685
Controls system	11,160	11,630	11,160	11,630	11,160	11,630
Total WGS	169,285	511,100	168,715	505,860	218,940	712,745
Erection and installation	16,160	24,625	16,160	24,625	17,525	33,725
Total installed cost	185,445	535,725	184,875	530,485	236,465	746,470

Source: DOE/NASA/9404-76/1

Concept 3 holds an insignificant weight and cost advantage. This is, however, offset by the technical considerations. Even without pitch change bearings, the mechanism to actuate and control the drag flap is as complex as the mechanism to actuate and control a servo flap. The fixed pitch rotor also presents a difficult problem when starting in low winds and stopping in high winds because of the inability to feather the blades. Pitch feathering mechanisms or augmented starting would likely be required. This rotor also operates at substantially higher thrust, thus increasing structural requirements.

Finally, a fixed pitch rotor must often operate in the vortex ring state, an aerodynamically unstable regime with possible wide fluctuations in thrust. This characteristic is extremely difficult to predict and could cause serious control problems.

These technical difficulties cause the risk of the variable rpm rotor to be considerably higher than the constant rpm candidate, without offsetting saving. The choice, therefore, was clearly for concept 1, the variable pitch, constant rpm rotor driving an AC synchronous generator through a fixed ratio gearbox. The steel truss tower was selected for the optimization phase. Its low cost, ready availability, ease of modification, and low risk, made it the obvious choice.

Computer Program: A mathematical model of the wind generator system and its environment was developed concurrently with the conceptual design task. In its entirety, the model describes the interrelationships of environmental variables (median wind speed, altitude and wind shear gradient, for example), rotor parameters (diameter, tip speed, chord, etc.). drive system (number of gear meshes), generating equipment (speed, efficiency), and structure (height, type), to produce a particular rated power. It yields weights and costs for the components, and calculates key economic and operational parameters required to evaluate the system, such as unit energy cost, direct capital cost and plant factor (plant factor, sometimes called capacity factor is defined as the fraction of energy actually produced by the system compared to the amount that could be produced if the system operated at rated power all year).

Concept Optimization

Basic Considerations: The mathematical model discussed in the preceding section was the vehicle for the optimization task. WGS major components and subsystems were optimized iteratively, and tradeoff and sensitivity analyses were made. The output variable that was optimized (minimized) was energy cost (¢/kWh). The concept optimization task included several cycles conducted to bring the major system parameters successively closer to their final values. The values selected for the preliminary design phase which followed were either the result of optimization or were determined from other design considerations.

It should be kept in mind that the optimization used a parametric model which had limited configuration detail. Inputs to the model were shaped by results of numerous separate detailed studies and tradeoffs. In order to proceed to the preliminary design phase, many more decisions were made which used the results from the model and detail studies, as well as Kaman's and Northeast Utilities' background knowledge and experience.

Major Parameters Affecting Energy and Direct Capital Costs: During the optimization phase, three parameters were found to be dominant. Most important is median wind speed ($\overline{V}$), since it determines the amount of energy available to generate power. The other two are system rated power (P_R) and the wind speed at which the WGS produces rated power, the rated wind speed (V_R). Rated power sizes the drive and electrical subsystem. Rated wind speed dictates rotor diameter and tower height. Thus, these three factors determine WGS size and yearly energy output.

The median wind speeds investigated ranged between 3.6 m/sec (8 mph) and 10.7 m/sec (24 mph), but the emphasis was on examining 5.4 m/sec (12 mph) and 8 m/sec (18 mph). Rated power ranged from 50 kW to 3,000 kW.

Figure 5.21 relates energy and capital costs to the median wind speeds of 3.6 m/sec (8 mph) and 10.7 m/sec (18 mph) for the selected rated wind speeds of 9.3 m/sec (21 mph) and 11.5 m/sec (26 mph), respectively. The curves shown were drawn using data developed during preliminary design and, therefore, represent the best information available.

Two important conclusions can be drawn from the plots of Figure 5.21. First, site median wind speed is the most important parameter affecting energy and capital costs. Second, and of equal significance, for given median wind speed and rated wind speed, energy cost and capital cost do not change rapidly with changes in rated power when the cost is near its minimum. To illustrate, consider the $\overline{V}$ = 8 m/sec (18 mph) curve on the energy cost plot. Energy cost is a minimum for a rated power of approximately 2,300 kW. The curve is quite flat in the neighborhood of this point so that rated power can be varied several hundred kW in either direction without causing a significant change in energy cost.

This is fortunate because it allows considerable latitude in selecting other system parameters, such as direct capital cost, rotor diameter, and cut-in wind speed with minimal impact on energy cost. It allows the designer flexibility in selecting variables to meet particular user requirements, technical risk limits, unit cost goals and operational requirements.

Systems Selected for Preliminary Design: The optimization program, and other qualitative and quantitative considerations, led to selection of the characteristics of the two wind generator systems. A low power system rated at 500 kW and a high power system of 1,500 kW were selected for the preliminary design phase. The 500 kW system has a 45.7 m (150 ft) diameter rotor and is optimized for the site with 5.4 m/sec (12 mph) median wind speed. The 1,500 kW system has a 54.9 m rotor and is designed for an 8 m/sec (18 mph) median wind speed (see Table 5.11).

Table 5.11: Systems Selected for Preliminary Design

	Low Power	High Power
Median wind speed, m/s (mph)	5.4 (12)	8 (18)
Rated power, kW	500	1,500
Rated wind speed, m/s (mph)	9.3 (21)	11.5 (26)
Rotor diameter, m (ft)	45.7 (150)	54.9 (180)*

(continued)

Table 5.11: (continued)

	Low Power	High Power
Rotor solidity	0.03	0.03
Rotor speed, rpm	32	34
Energy cost, ¢/kWh	7.0	2.8
Capital cost, $/kW	846	499
Unit cost, $	423,100	749,000
Plant factor, %	29	43

*Selection of 180-ft diameter rotor is supported by minimum energy cost calculations made subsequently, which were based on preliminary design data. See Figure 5.21.

Source: DOE/NASA-9404-76/1

Figure 5.21: Energy and Capital Cost Variations with Rated Power

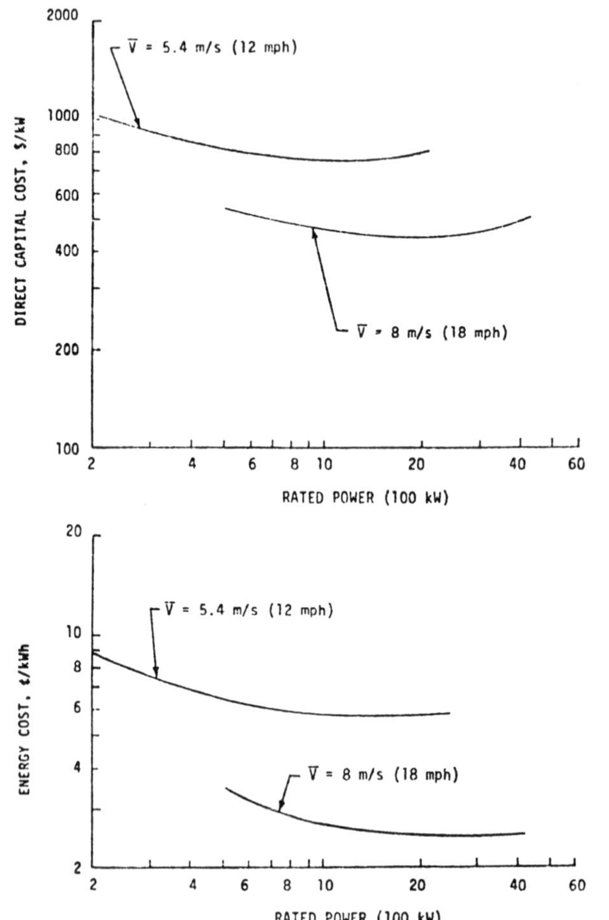

Source: DOE/NASA-9404-76/1

Site Adaptability: The two systems selected are adaptable to sites whose median wind speeds cover a range from about 3.6 m/sec to 10.7 m/sec. This is illustrated by Figure 5.22.

Figure 5.22: Energy Cost Variation with Median Wind Speed

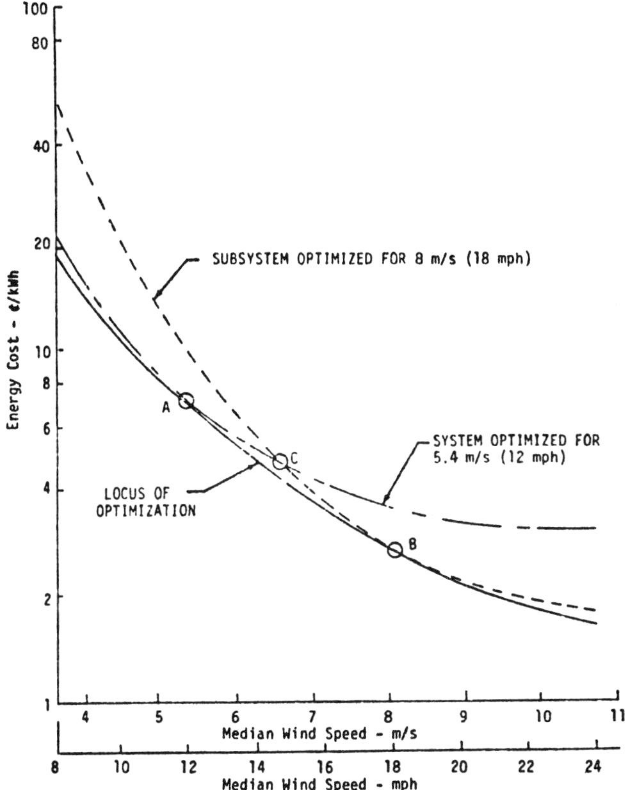

Source: DOE/NASA-9404-76/1

The locus of optimized systems line is drawn through points plotted for systems optimized (systems minimizing energy cost) at sites with a range of median wind speeds. The 500 kW system was optimized at 5.4 m/sec and hence, intercepts the locus of optimized systems at point **A**. The 1,500 kW system intercepts the locus at point **B** at its 8 m/sec design median wind speed. The significant point about this figure is that together, the selected systems cover a range of median wind speeds from 4 m/sec to 10 m/sec with less than one-half cent per kWh penalty attributable to their not being optimum at a particular median wind speed. The 500 kW system could be used up to about 6.3 m/sec, where the 500 kW line intersects the 1,500 kW line (point **C**), and the 1,500 kW system from that point on.

The conclusion that can be drawn from this discussion is that only two WGS designs should provide adequate coverage for sites with a wide range of median wind speeds, spanning most of the attractive locations in the United States. This means only two standardized WGS are needed, with the resulting economies of large-volume production of standard components.

Costs of Major Subsystems: The costs of major subsystems are presented simply as functions of rotor diameter or rated power, whichever is more appropriate for the particular subsystem under consideration. For example, rotor subsystem cost, Figure 5.23, is primarily a function of rotor diameter, whereas drive subsystem cost, Figure 5.24 is primarily a function of system rated power.

Figure 5.23: Rotor Subsystem Cost

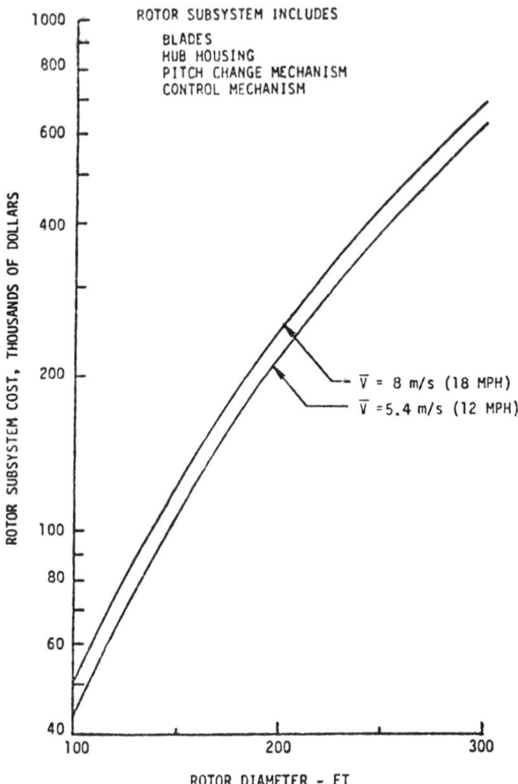

Source: DOE/NASA-9404-76/1

These costs are shown in Figures 5.23 through 5.27, based on the results of parametric cost analysis of the preliminary design, performed after the main body of work on the study was completed. Figure 5.28 shows energy cost as a function of rotor diameter.

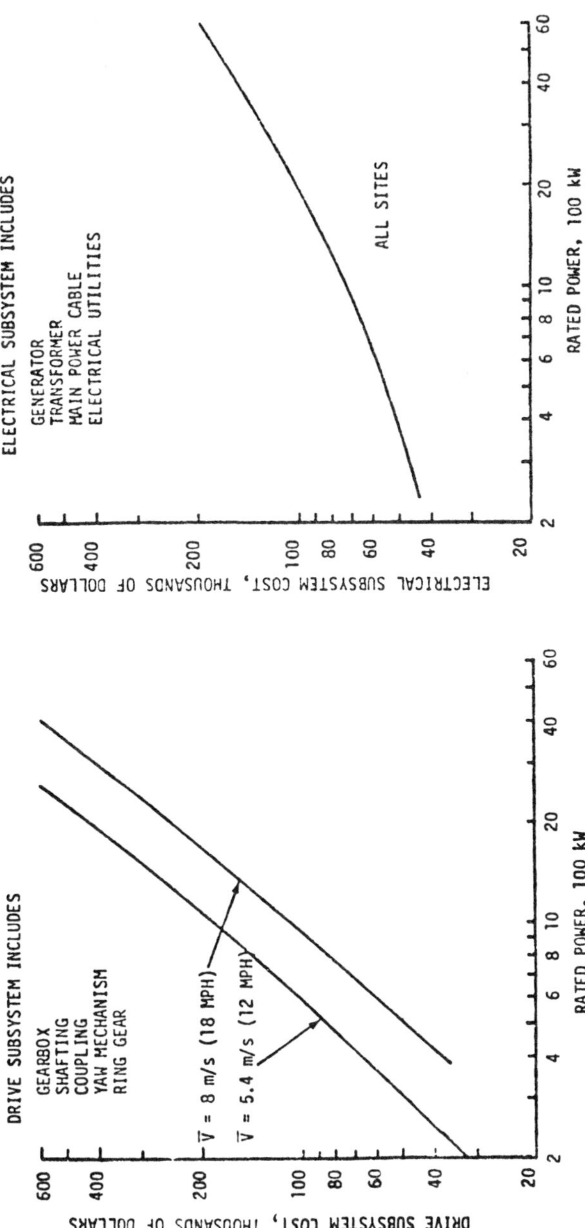

Figure 5.24: Drive Subsystem Cost

Figure 5.25: Electrical Subsystem Cost

Source: DOE/NASA-9404-76/1

Figure 5.26: Structure Cost

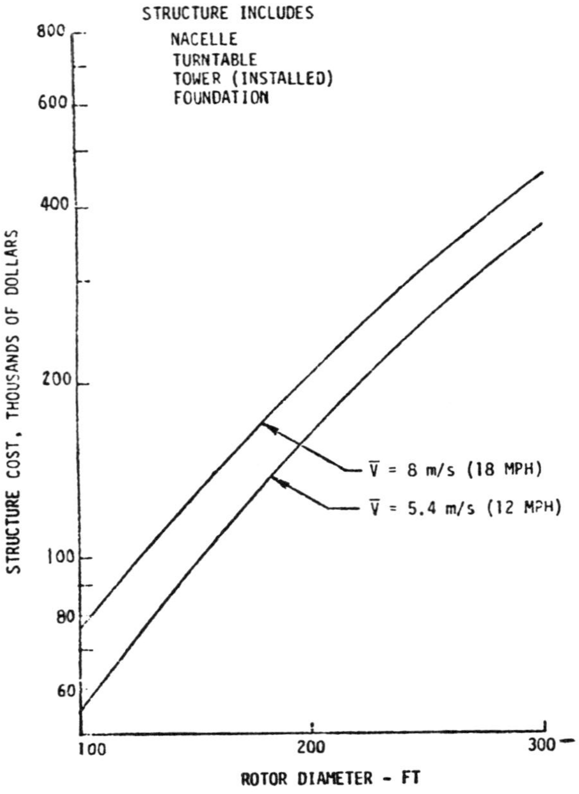

STRUCTURE INCLUDES
NACELLE
TURNTABLE
TOWER (INSTALLED)
FOUNDATION
STRUCTURE COST, THOUSANDS OF DOLLARS
800
600
400
200
100
80
60
V̄ = 8 m/s (18 MPH)
V̄ = 5.4 m/s (12 MPH)
100
200
300
ROTOR DIAMETER - FT

Figure 5.27: Other Costs

OTHER COSTS INCLUDE
SITE ACQUISITION
SITE PREPARATION
INSTALLATION OF COMPONENTS
OTHER COSTS, THOUSANDS OF DOLLARS
200
100
80
60
40
20
V̄ = 8 m/s (18 MPH)
V̄ = 5.4 m/s (12 MPH)
100
200
300
ROTOR DIAMETER - FT

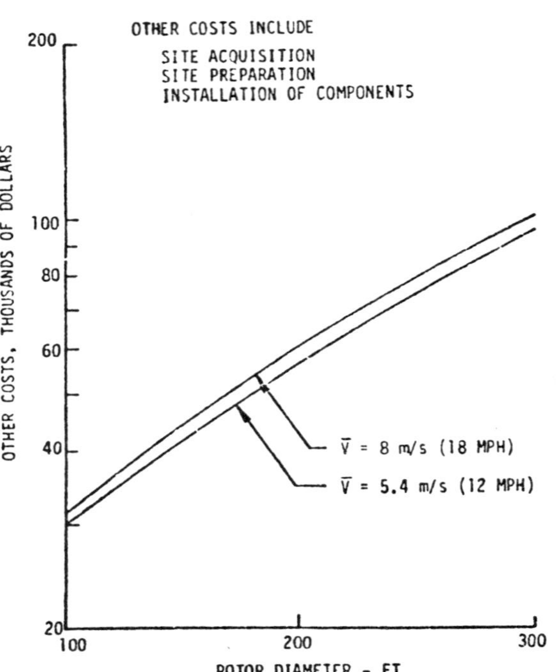

Source: DOE/NASA-9404-76/1

Figure 5.28: Energy Cost Variation with Rotor Diameter

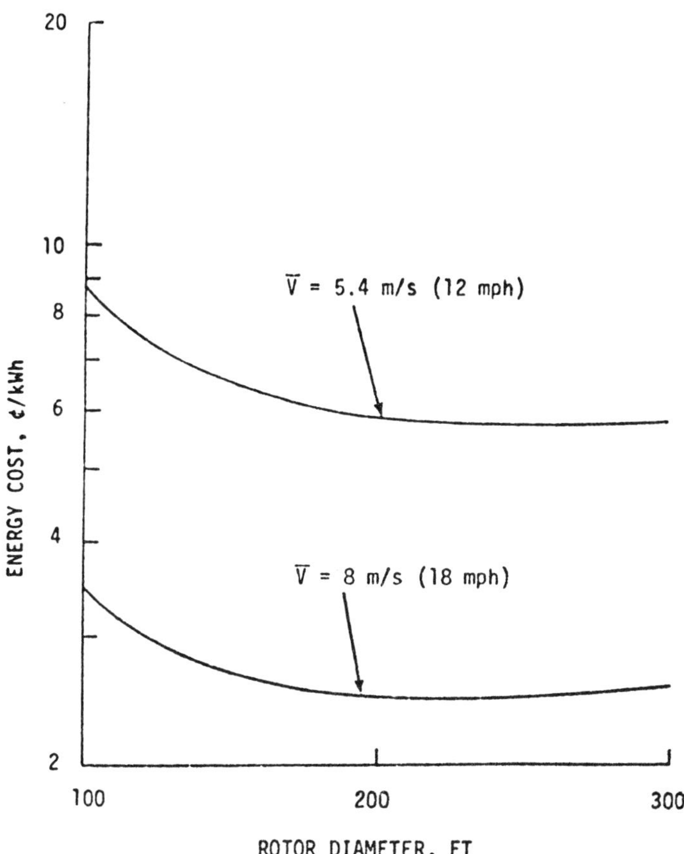

Source: DOE/NASA-9404-76/1

More specific definition of the WGS subsystem components, particularly the rotor subsystem, during preliminary design provided the basis for costs which were used in the preliminary design parametric model.

Component costs associated with two median wind speeds (12 and 18 mph) are shown, representing the influence of wind frequency distribution on optimum WGS sizing. Drive system costs are considerably higher for low wind speed applications because optimum rotors are larger in diameter and slower turning, thereby producing higher drive system torques, the primary factor in drive system cost. For structure subsystem costs, it was assumed that rotor ground clearance is 50 feet for all cases.

The control subsystem is not included among the curves relating costs to major parameters. Control subsystem cost is assumed to be constant, independent of both rotor size and rated power. The control system is described later in this summary, and includes only the electrical and electronic devices that govern rotor and yaw mechanism operation, telemetry and supervisory functions, and fault monitoring.

System Cost Trends: The preceding subsystem cost trends for the preliminary design parametric model were summed to obtain cost trends for the entire WGS. Direct capital cost as a function of rated power is shown in Figure 5.21, and energy cost as a function of rotor diameter and as a function of rated power is shown in Figure 5.28. Note that systems having minimum energy cost (¢/kWh) do not necessarily have the lowest direct capital cost ($/kW).

Requirements and Application in Utility Networks: For wind generators to assume a significant role in the production of electric energy in the future, they must be economically competitive with other forms of power generation and must be accepted by the utilities and the public. In addition to the economic question, utilities are concerned that wind generators be readily adaptable to the existing power grids and that they operate safely and reliably. Public acceptance will depend mainly on safety, environmental considerations, and how the wind generators affect monthly electric bills. There are many complex issues raised by the introduction of highly-visible wind generators into a utility network. Some of the more important are:

(1) Basic attitude of utilities toward wind generators.

(2) Integration into the financial and operational structure of utilities.

(3) Federal and state regulations, including licensing and safety.

(4) Environmental impact and public acceptance.

These questions had never been studied in detail previously. Therefore, a preliminary study was made to acquire a basic understanding of these issues from the point of view of the utility industry and to apply this knowledge in the analysis and preliminary design. Much of the information came from Kaman's utility consultant, Northeast Utilities Company. Colorado Springs Public Utilities also contributed. Providing assistance from the standpoint of government were the Connecticut Department of Environmental Protection, Connecticut Aeronautics Commission and OSHA.

Attitude of Utilities Toward Wind Generators: The two utilities consulted on the prospective use of wind generators gave insight into the question from two very different perspectives. Northeast Utilities is a large company with a total capacity of 5,500 megawatts (MW), including 1,100 MW of nuclear plant. Future expansion is expected to be mainly in the form of more nuclear capacity. Colorado Springs Public Utilities (CSPU), on the other hand, is a municipal utility with a modest 320 MW capacity. This company is converting the majority of its gas and oil-fueled capacity to coal. Most of CSPU's projected capacity expansion will also be coal-fueled. From Kaman's discussions with officials of these two companies, the following important factors were determined to affect utility views of wind generators and their future.

Fuel Costs – This is the major factor. Indeed, it is the impetus behind this research. However, it is felt more acutely by some utilities than by others. Those utilities heavily dependent on oil, particularly imported oil, are suffering most. Those who use low sulfur coal, hydroelectric and nuclear power are much less affected. Obviously, the former category is more likely to be attracted to wind power in the near future.

Environmental Problems – In response to public outcry, and now by law, utilities are giving increased attention to the environment, both from the standpoint of air and water pollution, and aesthetics. In the future, these considerations may preclude the construction of the optimum generating plants they prefer. As utilities consider ways to solve these problems, they may well become more attracted to wind generators, despite possible disadvantages in economy and efficiency.

Load Pattern and Growth – Utilities believe that until major improvements are made in energy storage systems the primary role of wind generators will be fuel savings. When wind is available, wind generators can be put on line to displace fuel-fixed units, thus reducing fuel costs. The value of the wind generator is obviously related to the cost of fuel saved and the comparative efficiency of the generating units displaced.

Further, because it is impractical to shut down and start up large base-load installations such as nuclear reactors, wind generators will have minimum value as fuel savers during periods of low demand. Alternatively, when they can replace the least efficient and most costly generators which are brought on line during periods of peak demand, their value will be greatest. Thus, it may well be that daily and seasonal variation in the demand pattern will be compared with the daily and seasonal wind pattern to determine the overall value of wind generators to a particular utility's needs. Ideally, of course, the most windy periods, daily and seasonally, would coincide with peak demand.

Anticipated load growth will also be a factor. Utilities with surplus capacity for the future would probably elect to bear high fuel cost rather than invest in more unneeded capacity. This is particularly likely if increased wind generator R&D is perceived as yielding cheaper, more efficient units in the future. Under such conditions, a utility with adequate near-term capacity would likely opt for postponement of wind generator capability.

In summary, there does not appear to be utility bias against wind generators. They are seen primarily as fuel savers. Those utilities facing critical fuel shortages and very high fuel costs, and those which need increased capacity soon, are likely to be the most receptive to a fuel-free energy source in the near future.

Utility Cost Estimating: Energy costs and direct capital costs to acquire and operate wind generators are based on standard utility cost procedures and the assumption of large-scale production. Table 5.12 shows the factors and assumptions supplied by NASA for the study.

Table 5.12: Economic Planning Factors and Assumptions

Parameter	Factor or Assumption
Useful life, dynamic components	30 years

(continued)

Table 5.12: (continued)

Parameter	Factor or Assumption
Useful life, structure	50 years
Financing	50% debt, 50% equity
Return on investment	9% on debt, 11.5% on equity
Depreciation	Straight line over 30 years
Corporate tax rate	48%

Source: DOE/NASA-9404-76/1

Direct capital cost includes procuring, transporting, erecting and readying the system on a site, the site and its preparation, supporting facilities and security. Average cost of the energy produced over the life of the system includes: recovery of capital; interest; taxes and operation and maintenance expenses.

Cost subroutines compatible with Northeast Utilities cost estimating procedures for annual carrying charges for capital recovery, interest and taxes were included in the parametric model developed earlier.

Operation and maintenance (O&M) costs are those established by Northeast Utilities for comparable generating plants and equipment with the exception of rotor maintenance cost which was based on Kaman's helicopter experience. A conservative estimate of rotor maintenance cost was assumed due to the lack of field experience with rotor of this size. Based on guidance from Northeast Utilities. yearly operating cost was estimated at 1.5% of direct capital cost. This covers the cost of operating personnel, supervision, and related indirect and overhead costs.

Public and Operating Safety – Neither public nor operating safety appears to be a significant problem. One concern is the possibility of the rotor shedding ice, which could be a hazard. Although this is a remote possibility, state and local zoning laws might require a buffer zone.

There is always the threat of vandalism or sabotage, especially since the sites are to be unattended. A peripheral fence, lighting and other common security measures minimize the threat, but can never absolutely preclude it. A shell tower with stairs on the interior would also contribute to security by keeping unauthorized persons out of the tower and hence, farther from the dynamic components.

The rotating blades will present a tempting moving target to would-be marksmen, but the construction of the blades makes them very tolerant of small arms bullet strikes.

Air traffic safety is not considered a problem. The wind generator will not likely be located close enough to airports to be affected by terminal area obstruction restrictions.

Operating safety becomes a consideration primarily during maintenance of the system. Requirements for personnel safety will be imposed on the design through compliance with OSHA and industry standards. Established utility practices prescribe standard safety procedures for personnel performing maintenance on various types of equipment. Doubtless such practices would be expanded to accommodate the unique requirements of the wind generator. Therefore, no significant operating safety problems are anticipated.

Maintenance: With the possible exception of the rotor, maintenance appears to present no unusual problems. The power train and generating subsystems and associated controls have counterparts in many of the facilities presently being maintained by the utility companies and are within their existing maintenance capability.

The wind generator will operate remotely and unattended for long periods. Therefore, reliability and long operating life are viewed as particularly important. Also important is the capability to detect critical faults and to initiate shutdown automatically and safely. Reliability and fail-safe provisions must be designed into the system.

Routine preventive maintenance and servicing should be kept simple and be required no more frequently than every 30 days. A major inspection, possibly involving some component tear-down and parts replacement, would be performed annually. Upkeep of the tower and associated structure would be scheduled at 10 year intervals.

The design must stress in-place repair of heavy components to avoid removing them from the tower. The need for special tools, equipment and skills not normally available to a utility is to be avoided.

Overhaul of major dynamic components off-site should be only on the basis of observed wear and deterioration, rather than according to a fixed operating hour schedule.

Environmental Impact – In recent years, environmental considerations have loomed large in utility development plans. Views on environmental impact of various utility facilities vary greatly with the locale, the relative influence of the organizations involved, and the weight of other priorities. Threats to the environment, real or imagined, tend to be measured against other problems of energy production, so that shortages and high energy cost have a mitigating influence on environmental concerns. These are exactly the conditions likely to spur the initial use of wind generators. Further, wind generators avoid the problems of air and water pollution with which utilities are presently contending. There are two possible environmental problems remaining, noise and aesthetics.

Noise – Rotor, bearing, and generator noise were considered. It is believed that this noise will be below objectionable levels, and that the units will be located far enough from population centers to eliminate noise as a major concern.

Aesthetics – Visual acceptability is the environmental issue most likely to spark controversy. It is unlikely that large wind generators in large numbers can be situated so as to avoid creating a displeasing visual effect. It is generally felt that public acceptance will decline with increasing numbers in a given locale. Unfortunately, the solution is not as simple as siting wind generators far from population centers. The population centers need the energy, whereas remote siting with accompanying transmission losses reduces the efficiency of wind generators. It will be desirable, therefore, to seek locations in sparsely populated areas and to subdue the presence of the generators where possible by spacing and use of natural cover. Again, unfortunately, the most windy locations may not be the most unobtrusive. Background blending may be of some help in mitigating the presence of these units. Colorado Springs Public Utilities has used this technique in some applications. They have used paint schemes, decorative plantings, rustic fencing and panels to disguise the base of transmission towers and transformer centers with some success.

Tower design is considered another important factor. The consensus is that the shell tower will be more appealing (or less objectionable) then the steel truss. Utilities will have to use every technique available to make the wind generators as unobtrusive as possible.

Licensing: Securing the necessary approval for a new generating plant can be a long and costly procedure for a utility. There are few, if any, precedents for licensing wind generators, so the requirements are unknown. Utilities agree, however, that wind generators will be licensed as generating plants and through state agencies. Whether each site will require a separate license or whether one will suffice for a total system of several sites is not known. Of course, if each site must be separately licensed, costs will be higher.

Applications: Ultimately, economics will decide what application, if any, will be feasible for wind generators. With this in mind, a preliminary analysis was made to evaluate the relative cost of wind generators when used in three typical applications. The basic approach was to calculate the break-even cost of the wind generator for replacing an existing fuel-fired unit.

Fuel Saver – This is the most obvious immediate application. The wind generator is connected directly into the grid during periods of available wind and when fuel-fired units of equivalent output are shut down. Any number of units can be used as fuel savers. Break-even cost of the WGS was computed as a function of the energy cost of the fuel saved, and is displayed in Figure 5.29. A 1,500 kW wind generator, at approximately $480/kW direct cost, is competitive with a gas turbine generator burning relatively expensive #2 oil, but not with coal or #6 oil-fired units. The wind generator would have to cost about $180/kW or $300/kW, respectively, to compete with coal and #6 oil.

With Base-Load Capacity – A wind generator could be credited with base-load capacity when some or all of its power output is available to meet the utility's daily base-load or basic energy demand with a high level of assurance. WGS use for base-load capacity necessarily assumes a number of wind generators disposed to benefit from wind variability over a large geographic area. Energy produced in excess of the rated base-load capacity could be credited toward fuel savings.

Analysis of wind generator break-even cost with base-load capacity with fuel savings was performed for a 1,000 MW system with 100 MW credited toward base-load at 0.7 plant factor. This means a 0.7 probability that at least 100 MW will be available to meet base-load requirements at all times. For a total system plant factor of 0.35, the system would have an effective fuel-saver plant factor of 0.315. This is obtained by subtracting the 10% of the 1,000 MW system's plant factor which is credited to base-load from the total plant factor. The analysis used a displaced base-load energy cost of 3.5¢/kWh and a displaced fuel cost of 2.35¢ per kWh.

Figure 5.29: Break-Even Costs–WGS vs Fuel Costs

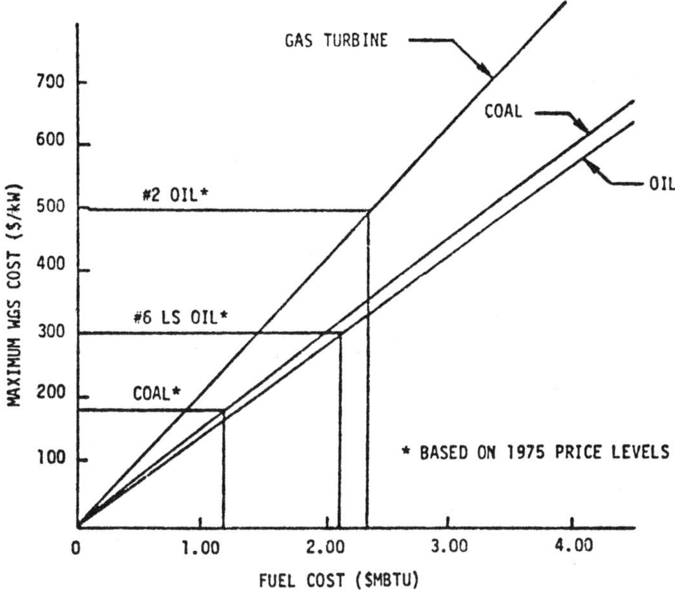

Source: DOE/NASA/9404-76/1

The results predicted a wind generator system cost of approximately $510/kW for base-load capacity plus fuel savings, which compares closely with the $480 per kW for saved gas turbine fuel in the fuel saving application alone.

With Storage – This application considers storing in some manner (not defined) energy generated with considerable wind variability, and gives capacity credit for peak load periods. Peak load capacity is the same as base-load capacity, except that wind generator power displaces the costly peaking units, such as gas turbines. In addition, surplus energy not needed for storage goes for fuel saving as before. Sufficient energy production to charge storage on at least 90% of available days is assumed. Besides the factors bearing on fuel savings, the analysis of the peak load application also considers the system yearly and daily capacity factors, storage efficiency and the displaced peak unit energy cost.

The break-even cost in this case has the cost of the storage system included. Figure 5.30 presents the results of the analysis. Break-even cost is shown as a function of system daily capacity for two storage efficiencies. Results indicate that both the daily plant factor and storage efficiency affect system break-even and cost significantly.

Results: The results of the economic analysis of possible utility company applications of wind generating systems are more meaningful when interpreted in light of the system cost analysis performed for the 500 kW and 1,500 kW WGS designs during the study program.

Figure 5.30: Value of WGS with Storage

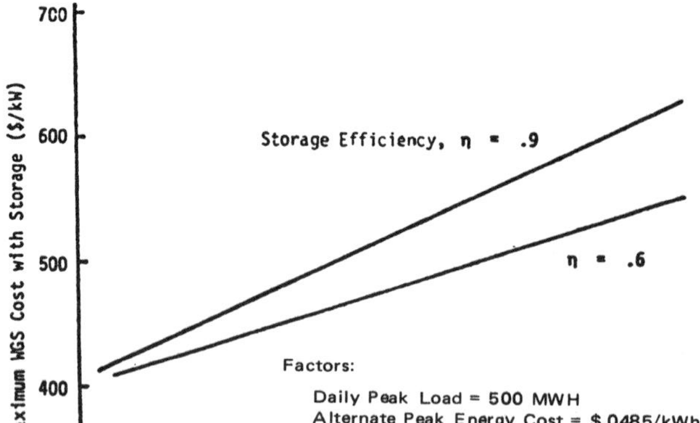

Source: DOE/NASA/9404-76/1

This detailed cost analysis predicts direct capital costs of approximately $900 per kW for the 500 kW wind generator and $480 per kW for the 1,500 kW system. Obviously, the larger WGS is more attractive economically. The $480 per kW capital cost of the 1,500 kW WGS compares favorably with the estimated $500 per kW break-even cost required in fuel saving applications for displacing gas turbine generating systems. However, compared to systems using coal and #6 oil as fuels, the 1,500 kW WGS is not competitive at present. Whether this system will be more competitive with the lower cost fuels in the future depends on how high fuel costs will rise and how much WGS costs might be reduced through continued development.

Fuel saving is judged to be the most promising application of wind generators at the present time. Systems for this application could range in size from a single wind generator to large multi-unit installations and, therefore, might find widespread use by both large and small utility companies.

For the fuel saving plus base-load capacity application, the predicted capital cost of the 1,500 kW WGS is slightly under the estimated break-even cost of $510 per kW. The assumptions made in the analysis are optimistic, however, even if the necessary wind conditions could be found, a very large number of WGSs would be required and transmission costs associated with such a system, which were not accounted for in the predicted WGS capital cost, would increase this cost further.

Despite the large number of wind generators required to satisfy the application, only about 10% of the total installed power, based on the example chosen, would be added to the user's base-load capacity. In effect, such a system is still primarily a variable output fuel saving system. Because of the large number of units required and the rather small base-load capacity afforded by the system, it appears that this application offers limited promise of being economically feasible for a typical utility in the foreseeable future.

The 1,500 kW WGS predicted capital cost is roughly equal to the break-even cost computed for the fuel saving plus peak load application. It should be noted, however, that the 1,500 kW WGS capital cost is optimistic for this application since allowances for system downtime are not included. The analysis of break-even cost assumed also that the stored wind energy would displace high-cost gas turbine-generated energy. If a less costly generating source is displaced, the WGS break-even cost would be correspondingly lower. The viability of the WGS in this application appears to be contingent upon the development of inexpensive and efficient storage systems.

The applications analysis has attempted to evaluate, in gross relative terms, the economic worth of a wind energy system to a typical electric utility. Because it has been conducted in the context of a utility operation, it did not explore applications requiring departures from present-day industry practice and regulation. Some of the regulations affecting utility operations, such as those governing rate structures and reserve capacity, do not easily accommodate variable or intermittent generating capacity, even though this type of power might be entirely feasible in specific applications. Using the wind generator as a variable energy source for selected segments of the market that could operate with anticipated power interruptions, and structuring the rates accordingly, might open up new areas of application.

Wind energy costs in the applications analysis have been based on capital, operations and maintenance, and fuel costs currently anticipated by a relatively large northeastern utility and on present-day projections of future fuel availability. These factors vary considerably among other utilities in different parts of the country and might change drastically in the future due to unforeseen circumstances.

Should the supply of oil and natural gas approach depletion more rapidly than predicted and the cost escalate accordingly, wind energy would become increasingly more competitive in the United States. In other areas of the world, primarily the nonindustrialized nations and even in some remote areas of the United States, conditions today may be such that wind energy is technically and economically a competitive alternative.

Preliminary Design

The optimized system characteristics formed the basis for the start of the preliminary design. They call for a 500 kW WGS for a median wind speed of 5.4 m/sec and a 1,500 kW system for a median wind speed of 8 m/sec.

Advantage was taken of the relative insensitivity of energy cost to rotor diameter in the 46 to 55 m range (Figure 5.28), and consideration was given to capital cost, plant factor, rotor model sensitivity and its adaptability to different sites, in choosing the rotor diameter. The 1,500 kW system's 54.9 m rotor diameter was chosen primarily to improve plant factor, but also to provide an eventual test unit with a singificantly larger rotor than that of the low-power system. Designing and evaluating two different size rotors also allows flexibility in interpreting and applying the results of the study. For instance, if due to the evolution of manfacturing techniques, rotor blade cost is reduced, the optimum rotor size will increase. Two distinct rotor diameters give an indication of effect of rotor diameter variation on the system.

System Description: A layout of the dynamic components of the 1,500 kW WGS is shown in Figure 5.31. The 500 kW system has essentially the same configuration. Both systems have a 2-bladed, variable-pitch, constant rpm rotor mounted downwind of the tower. The rotor consists of filament wound composite blades with hingeless attachment to a rigid hub. The hub is a rugged, welded structural steel assembly. Blade pitch is controlled by a linkage actuated by a hydraulic cylinder which rotates the blades on pitch bearings mounted on the hub.

The hub is supported by crossed roller bearings mounted on a fixed spindle. This allows the nonrotating spindle to carry the high bending moments produced by the rotor without large bearings to support a rotating shaft. The rotor torque is transmitted to a triple mesh commercial gearbox by a quill shaft running through the center of the static spindle. The gearbox is connected to a commercial AC synchronous generator of the type commonly used by utilities, with a parking brake/inching drive assembly located between the gearbox and the generator. This assembly is used to stop the rotor from low rpm and to position the rotor after the system is shut down.

The control system uses a microprocessor for all sequencing and data reporting functions, including WGS startup, shutdown, operational monitoring and failure reporting. Conventional electromechanical controls are used for blade pitch and positioning the rotor relative to the wind. The rotor is provided with a hub-mounted mechanical blade feathering capability so that if the control system fails or is overpowered by gusts, the blades are automatically feathered, preventing system overspeed and damage.

The generating machinery and controls are housed in a closed nacelle mounted on a structural steel turntable which, in turn, is mounted on top of the tower through a crossed roller bearing assembly. The assembly is oriented to the wind by a hydraulic motor driving a worm gear which engages a large ring gear on top of the tower (Figure 5.31).

The tower can be either a structural steel truss type or a precast, post-tensioned concrete shell type. The steel truss tower has a distinct advantage in small quantities. The precast, post-tensioned shell tower, while still slightly more expensive (about $5,000 to $6,000 for the 1,500 kW system) when 1,000 are produced, is considered to be better aesthetically, and hence could be the choice for a production run.

Another important tower effect must be considered. This is the aerodynamic blockage of the wind which causes a wake area behind the tower. Since the rotor is mounted on the downwind side of the tower, each blade passes through this wind wake, or "tower shadow" once per revolution and experiences a vibratory load input. Kaman selected a value of 30% wind velocity reduction in treating the effect of the tower shadow on the blades preliminary design.

(NOTE: Since this study was carried out, experimental findings from the NASA wind tunnel tests have quantified wind velocity reduction as a function of tower structural design. Although the downwind rotor placement continues to appear advantageous with open truss towers, the use of shell towers now is considered unlikely, since their tower shadow effects cannot readily be reduced to the values necessary for long blade fatigue life.)

Figure 5.31: Drive System Components 1,500 kW WGS

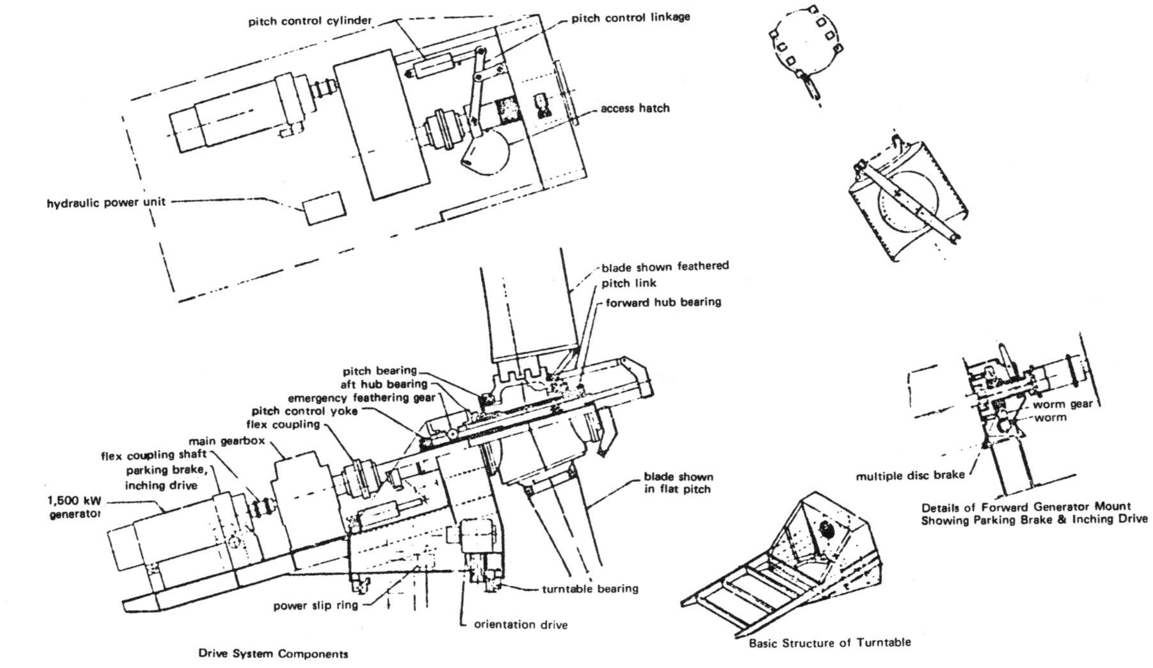

Source: DOE/NASA/9404-76/1

The choice of a foundation is highly dependent upon soil conditions and the tower type and size. This study assumed standard soil conditions. With this assumption, pile foundations are cheaper for the 1,500 kW system, but mass concrete is competitive with piles for the 500 kW system. However, the analysis is quite sensitive to changes in soil parameters. Therefore, the choice of foundation will likely be governed by soil condition at the particular site under consideration. Most of the electrical protective and power conditioning equipment is mounted on a pad at the base of the tower, along with the microprocessor, related control equipment, and the data recording and transmission equipment.

Preliminary Design Results: The characteristics of the low (500 kW) and high (1,500 kW) power WGS resulting from the preliminary design phase are summarized in Table 5.13. Table 5.14 lists weight breakdowns for both systems, assuming a steel truss tower. The costs of the two systems are shown in Table 5.15.

Table 5.13: System Preliminary Design Technical Characteristics

	Low Power System	High Power System
System		
Rated power, kW	500	1,500
Site median wind speed, m/sec	5.4	8
Rated wind speed, m/sec	9.3	11.5
Yearly energy output, kWh	1.3×10^6	5.7×10^6
Plant factor, %	29	43
Minimum wind speed, m/sec	4.5	5.4
Cut-out wind speed, m/sec	14	20
Design maximum wind speed, m/sec	54	54
Rotor subsystem		
Design shaft output power, kW	560	1,648
Rotor diameter, m	45.7	54.9
Rotor solidity, %	3	3
Rotor precone, degree	8	10
Rotor speed, rpm	32.3	34.4
Blade root chord, m	1.5	1.9
Blade tip chord, m	0.75	0.95
Drive subsystem		
Maximum orientation drive speed, rpm	⅓	⅓
Gearbox input torque, N-m	166×10^3	456×10^3
Electrical subsystem		
Generator rating, kW	510	1,522
Generator output voltage, kV	2.4	4.16
Generator output frequency, Hz	60	60
Generator speed, rpm	1,800	1,800
Control subsystem		
Rotor pitch rate, degree/sec	5	5
Orientation system yaw rate, degree/sec	2	2
Structure subsystem		
Tower height, m	33.5	38
Tower base span, m	9.5	10.7

Source: DOE/NASA-9404-76/1

Table 5.14: Preliminary Design System Weight–Steel Truss Tower

	Weight* (kg)	
	500 kW	1,500 kW
Rotor subsystem	8,020	17,430
Blades	(3,080)	(5,160)
Hub	(4,940)	(12,270)
Drive subsystem	16,780	35,180
Gearbox	(9,890)	(20,860)
Other	(6,890)	(14,320)
Electrical subsystem	2,820	6,950
Control subsystem	20	20
Structure subsystem	44,950	69,360
Structural steel	(37,470)	(52,210)
Other	(7,480)	(17,150)
Total weight on foundation	72,590	128,940

*Weight on foundation.

Source: DOE/NASA-9404-76/1

Table 5.15: Preliminary Design System Cost–Steel Truss Tower–(1,000 Units)

	Cost per Unit, $			
	500 kW		1,500 kW	
System integration	4,400		25,100	
Rotor subsystem	110,000		194,900	
Blades		(76,600)		(122,000)
Hub, including hub-mounted controls		(33,400)		(72,900)
Drive subsystem, including gearbox, shafting and couplings, ring gear and yaw mechanism	78,000		181,000	
Electrical subsystem, including generator, transformer, cable, utilities	43,800		64,800	
Control subsystem	31,400		31,400	
Structure subsystem, including bedplate, enclosure, ladders, tower (installed), foundation	97,400		134,600	
Total WGS unit cost	375,000		631,800	
Other capital costs, including site acquisition and clearing, shed, installation of components	75,670		89,000	
Direct capital cost, $	450,670		720,800	
Direct capital cost, $/kW	901		481	
Annual costs				
15% direct capital cost*	67,600		108,120	
Operation and maintenance	23,280		45,350	
Total yearly cost, $	90,880		153,470	
Yearly energy output, kWh	1.28×10^6		5.68×10^6	
Estimated energy cost, ¢/kWh	7.1		2.7	

Note: Subsystem cost trends are shown in Figures 5.23 through 5.27.

*Includes income taxes, debt service, return on equity and depreciation.

Source: DOE/NASA-9404-76/1

Rotor: The rotor subsystem, which includes the blades, hub and controls, is the largest single cost element of the WGS. It is the most technically demanding subsystem and unlike the others, is not available commercially. Further, little or no experience exists in design and fabrication of rotors as large as these. Finally, all other subsystem designs are influenced directly or indirectly by the rotor.

For these reasons, rotor design and analysis received great emphasis throughout all phases of the study. Particular attention was placed on evolving a rotor design that could be mass-produced at low cost with latest state-of-the-art technology, without neglecting the stringent technical requirements of the rotor.

Requirements – To be a competitive alternative for electrical energy production, the rotor must extract maximum energy from the available wind. It must operate in a wide range of weather conditions, including extremes of temperature and all types of precipitation. It must survive lightning strikes and hurricane winds. To be cost effective, the rotor must be economical to build, require little maintenance, and have a long operating life. With these requirements, the following design objectives were established:

(1) Maximize aerodynamic efficiency.

(2) Select the simplest design that will achieve desired aerodynamic performance objectives.

(3) Ensure that the rotor will operate satisfactorily in temperatures from -51°C to 49°C in precipitation, in salt spray, in winds up to the maximum anticipated gusts during operation, and survive 53.7 m/sec hurricane winds in the stowed position.

(4) Be impervious to lightning strikes.

(5) Survive foreign object damage such as bird strikes, stones, and small arms fire. Operate in sand and dust storms without leading edge erosion.

(6) Have 30-year operating life for blades, hub and grips.

(7) Use available technology to the maximum. Minimize production cost and development risk and reduce development costs for any new technology.

The design does not provide protection against freezing rain and structural ice. It is expected that the rotor will be shut down during those infrequent periods when heavy ice accumulates. However, operation during periods of light icing is permissible, though efficiency will be reduced.

Design Approach – The approach to the rotor design is similar to that of the other subsystems. During conceptual design and evaluation, several feasible rotor configurations were examined and one selected which offered lowest cost and risk while meeting technical requirements. The optimization of this concept included evaluation or rotor component options in greater depth. Finally, preliminary designs were prepared with drawings, weight estimates, cost estimates, specifications, and supporting analysis. Throughout the study, the latest state-of-the-art technology was selected to minimize cost and risk. Rotor configurations were proven concepts adapted from helicopter designs. When feasible, these designs were simplified to eliminate unnecessary elements. Standard commercial parts, such as pitch bearings, were selected whenever possible and low-cost fabrication was emphasized. Particular attention was paid to those operating conditions and regimes which have significant effects on rotor design. Starting operations, rotor-tower clearance, low rpm stability, transients during gusts, and load loss are examples. Emphasis was placed on avoiding operating the rotor near its natural frequencies to preclude high vibratory loads.

Candidate Concepts and Evaluation – A number of options for rotor blades, controls, and hub were examined during the conceptual design. The decisions made were reevaluated throughout the study and some were modified or changed as the work progressed.

Blades: Blades are the single largest contributor to rotor cost. Of primary importance to the economical design of the blades are the materials and technique for fabrication.

Conventional metal construction represents inexpensive, well-established technology. Metal blades, however, are not easily optimized for aerodynamic efficiency due to the difficulties of manufacture with optimum twist, taper and thickness distributions. Of equal importance is the length limitation of extruded metal spars. Current extrusion technology sets an upper spar length limit of 15 to 18 meters, considerably below the spar length required. Hence, had metal construction been chosen, spars would have had to have been fabricated in sections and joined in some manner. These joints would give rise to a number of problems, including adverse dynamic effects, stress concentrations and added weight. These considerations make the use of composite fabrication techniques and materials attractive.

In particular, filament wound composite construction can be used to fabricate blades of the size required for this application with little cost penalty for selecting optimum twist, taper and airfoil section. Additionally, use of filament wound composite construction facilitates achieving a close balance between the blade center of gravity (cg), feathering axis, and aerodynamic center (a.c.) without using balancing weights.

Experience has shown that serious dynamic instabilities can occur if the blade cg falls behind its a.c. Hence, blades balanced with cg and a.c. at the quarter chord (25% of distance between leading and trailing edge) and having the feathering axis near the quarter chord axis are traditional in helicopter blade design. These considerations were found to apply to WGS blades; hence a cg/a.c./feathering axis match was considered a design requirement. The result of these requirements was selection and retention of composite, filament wound blades.

Pitch Mode: Variable pitch wind-driven rotors can be operated in either of two control modes, positive pitch or negative pitch. The former produces high thrust loads along the rotor axis which are nonproductive, while the latter is primarily a torque-generating mode. The positive pitch mode at low wind speeds and low power levels is subject to rotor thrust instabilities familiar in the helicopter industry as the vortex ring state of rotor wake interference. The vortex ring state is a condition in which the descent velocity of a lifting rotor approximates the rotor downwash velocity, causing development of a large recirculation vortex around the periphery of the rotor disc. For the wind generator system rotor, a similar condition exists when the thrust-induced wake velocity is approximately equal to the wind velocity. Thrust fluctuations associated with the vortex ring state could cause vibratory loads at the top of the tower up to ±60% of the steady thrust load. Large blade tip deflections also occur, increasing the danger of blade tip intersection with the tower. For these reasons, the negative pitch control mode was selected for the WGS.

Controls: The rotor requires some method of torque and speed control. The variable pitch rotor was chosen over the fixed pitch rotor because its characteristics permit the blade to be rotated about its feathering axis from almost flat pitch to full feather. This facilitates startup and shutdown, permits effective torque control and stowing of the rotor in full feather in case of high winds. Also, the variable pitch rotor can be designed to avoid sustained operation at potentially hazardous resonant conditions during startup. As discussed earlier, these considerations dictated selection of the variable pitch rotor. The blade pitch, of course, must be controlled.

Blade pitch can be changed and controlled either by directly driving the blade root mechanically, or by using aerodynamic forces by moving a servo flap near the blade tip. The relative merits of the two systems will be discussed later.

Hub: A flex plate hub with low out-of-plane stiffness to give bending moment relief, and high in-plane stiffness to provide the needed characteristics to operate under relatively high gravity loads, was selected initially. During preliminary design, however, the out-of-plane natural frequencies were found to be too close to the operating speed, increasing vibratory response of the out-of-plane bending moments. A teetering type of articulation was then studied, but found not to be cost effective, as will be discussed later. The end result was a rigid hingeless configuration, which gave the necessary root end stiffness. During conceptual design and optimization, several major analyses were made to examine significant component and configuration alternatives. These investigations are summarized below.

Airfoil Section: Several airfoil sections were evaluated during the conceptual design and optimization phases. These included NACA 4412, 23012, 23018 and 63_2-615 sections. Characteristics used were from NACA Report No. 824 under standard roughness conditions at Reynolds number of 6 million, approximating that of the outer portion of the blades. Although lift-to-drag ratios of NACA 4412 and 63_2-615 airfoils are higher than the 23012, the rotor optimization study yielded very little difference in aerodynamic efficiency for the entire rotor.

Consequently, other considerations, such as lower aerodynamic pitching moments, satisfactory past performance, and especially better producibility, led to selection of the 230-series airfoil. This series is easier to make because of the absence of reflex curvature on the under surface, which would make filament wound composite fabrication more difficult. The 230-series airfoil was retained through the preliminary design. Blade tuning required the bending stiffness at the root to be increased, which led to increased airfoil thickness ratios inboard.

Rotor Size and Solidity: Early analysis indicated that minimizing rotor solidity (total blade planform area divided by rotor disc area) minimized the energy cost for large rotor diameters. A range of diameters at various solidities was examined to determine the limits of solidity and diameter. Results showed there were no inherent technical limits for solidity less than 0.02 and diameter less than 76.2 m. However, it was found that with solidity below 0.03, complex and costly construction methods and materials were indicated for the thick root sections required. It was also found that 0.03 solidity was near the lower bound for significant energy cost savings, and that below that solidity starting characteristics were affected adversely. Therefore, 0.03 solidity was tentatively selected and was carried through preliminary design. The rotor diameters considered during this analysis also bracket those selected for preliminary design.

Number of Blades: With the cost advantage of minimum solidity and the rotor diameter necessary to generate rated power at rated wind speed, consideration was given to a 3-bladed rotor. It was found that although a 3-bladed rotor reduces vibratory loads on the drive shaft, gearbox and tower, this was not an important enough advantage to offset the higher costs of the third blade. The 2-bladed rotor was selected and retained.

Blade Geometry: The filament wound composite construction permitted most of blade airfoil shape, twist and taper with minimal cost impact. Two geometry studies were made after final solidity and diameter were selected for the preliminary design. The studies covered planform taper, important for structural and tuning reasons, as well as cost, and twist distributions. Final planform was optimized with a 3:1 taper rate from mid-span to tip. It was found that a linear twist of from 10° to 12° results in a rotor efficiency only slightly less than that for an ideal twist. The linear twist reduces blade tooling complexity and difficulties that might arise in removing ideally twisted mandrels from the fabricated blades. Although optimum twist was retained, linear twist can be used with negligible effect on energy cost.

Blade Life: An important factor in rotor system design is selection of the most economical blade life for the particular application. The trade-off between blade life and initial cost is usually conducted on a total life cycle cost basis, where the cost of maintenance and replacement of the blade at periodic intervals is traded off against the unit cost of the blades. Such a trade-off study was conducted on the WGS blade, even though a 30-year life was a design specification. The results show that no savings in capital cost or energy cost will result from a blade with a life of less than 30 years.

Teetering vs Hingeless Hub: Although the flex plate hub selected during conceptual design offered blade root bending moment relief, some form of articulated blade could further reduce it, and possibly lead to weight and cost savings. A teetering hub was examined for possible advantages.

The teetering hub reduces out-of-plane vibratory bending moments substantially below those of the flex plate hub, but has no appreciable effect on in-plane bending moments. These results suggest that the teetering hub might offer some saving in weight and cost if the design requirements are dominated by out-of-plane bending. However, the out-of-plane static bending moments at the 54 m per sec maximum wind condition are the critical bending moments driving the blade design. Blade structure which meets the static maximum wind requirement results in low, noncritical fatigue stress levels under normal operating conditions, even for the higher bending moments imposed by the hingeless configuration.

When the additional complexity of the teetering hub was considered in this light, it was dropped from further consideration. Analysis conducted during preliminary design strengthened this conclusion. Blade stiffness distributions needed to tune blade natural frequencies imposed structural requirements on the blade more severe than those imposed by the hingeless rotor vibratory bending moments. Therefore, whether blades are sized for stiffness or high wind conditions, fatigue stress levels are low enough to achieve the required 30-year life.

Servo Flap vs Direct Pitch Control: The conceptual designs of the first phase of the study used a servo flap for blade pitch control. Subsequent analysis proved that although the servo flap alone would be very effective in alleviating wind loads and adequate for pitch regulation during normal operation, it would not serve for startup and shutdown. In these modes, the servo flap was found to lack adequate controllability and pitch resolution. To compensate for this shortcoming, auxiliary pitch control devices and special operational procedures would be required. Because direct blade root pitch control can accomplish the task across the operational spectrum unaided, it was incorporated into the preliminary design instead of the servo flap. However, if blade wind load alleviation becomes an important factor for cost or operational reasons, the servo flap offers a possible alternative.

Stability: Blade flutter and divergence boundaries were defined and analyzed. The rotor was free from flutter and divergence for the rated operating conditions examined. Similarly, rotor/tower stability was investigated and found to pose no threat over the operating regime of the system.

Overspeed: Rotor overspeed beyond the design rpm for normal operation can occur during large and rapid wind velocity changes associated with strong gust conditions, when the generator is abruptly disconnected from the network, or in the event of a rotor control system malfunction. Loss of the balancing torque load imparts a net accelerating torque on the rotor, causing overspeed. A moderate rate of change of blade pitch angle, 5° per second, was found adequate to provide the necessary overspeed and overtorque control. This insures that the maximum rotor overspeed under the worst combination of conditions is not greater than 150% of rated rpm, the structural limit of the generator. All other components have structural limits above 150%.

Tuning: The structural configuration and thickness distribution of blades were dictated largely by blade tuning requirements. Of particular concern in the overall blade tuning problem is the need to avoid lower order resonance crossings during rotor startup in marginal wind conditions. Under such conditions, rotor acceleration to normal operating rpm will be quite slow, particularly in the 80% to 100% rpm range, where accelerating torques are low. This requirement resulted in selecting blade configuration details, such as root-end airfoil thickness and blade stiffness, which tune the blade to eliminate these problems.

Fatigue Life: A potential for the accumulation of fatigue damage in WGS rotors exists in many regimes of operation. However, since 30 years is the required design life, almost all operation must occur below the blade endurance limit. Accordingly, a design goal was established which placed the endurance limit above all operating conditions, including startup, shutdown, and strong gusts in wind-up to cut-out wind speed, leaving the possibility of occurrence of fatigue damage only for extremely severe conditions, which are seldom encountered. The calculated fatigue strength for unlimited life for both rotor blade designs, compared with loads for the maximum gust above cut-out speed, shows large margins between allowable and expected loads. This indicates that the design fatigue life will be achieved.

Tower Shadow: The wind wake downwind of a tower is proportional to the drag coefficients and solidity of the structure. A dense tower will generate a sharp wind velocity reduction which produces a pulse input to the airload distribution of a downwind rotor as the blade passes through the tower wake. This pulse occurs at exactly 1/rev frequency and will, therefore, generate harmonic forces at all integral multiples of rotor speed, 1/rev, 2/rev, 3/rev...n/rev. Blade bending and torsion responses will occur at all of these frequencies. The magnitude of response will depend upon the amount of damping in the various bending and torsion modes, their proximity to multiples of rotor speed, and the strength of the tower wake.

The study reported herein used a 30% wind velocity reduction for tower shadow, based on available literature information. This value of tower shadow was not critical for fatigue at any wind velocity examined, from minimum wind speed to well beyond cut-out speed.

(NOTE: Subsequent information, obtained from NASA testing after completion of this study, quantified wind velocity reductions behind towers having high solidity and angular structural elements. The wind shadow behind such towers can produce substantially higher blade bending moments, with associated reductions in blade fatigue lives. Therefore, it is considered essential that tower design criteria include the requirement for minimum shadow effect by minimizing tower solidity and selecting structural elements having low drag coefficients.)

Preliminary Design and Analysis — Rotor preliminary design was supported by configuration, operational and structural analysis. Configuration analysis included blade geometry optimization studies which led to the selection of blade thickness, planform and twist distribution. The major operational analysis included pitch control mode selection, overspeed limit calculations, blade tuning analysis, determination of blade flutter and divergence boundaries, and whirl resonance analysis. The major structural analysis covered the blade design, fatigue analysis and hub design. Designs for both the 500 kW and 1,500 kW rotors are similar, hence only the 1,500 kW unit is illustrated in this summary.

Blades: In construction, three separate mandrels are successively wound with S-2 fiberglass at angles ranging from ±45° to ±60° to the blade centerline, with unidirectional layers interleaved spanwise during the winding process. The third mandrel is used to position aluminum honeycomb blankets and a trailing edge spline, after which the entire assembly is filament wound to form the complete structure. The outboard section of the blade leading edge is protected from erosion by a neoprene guard. Aluminum mesh screen embedded in the aft blade surface protects against lightning.

Head Assembly: The rotor head is shown in Figure 5.32 with a cutaway view including controls shown in Figure 5.33. The rotor head is a rigid, hingeless assembly supported by tapered roller bearings on a nonrotating hollow spindle shaft which reacts to out-of-plane bending moments.

Figure 5.32: WGS Rotor Head Assembly

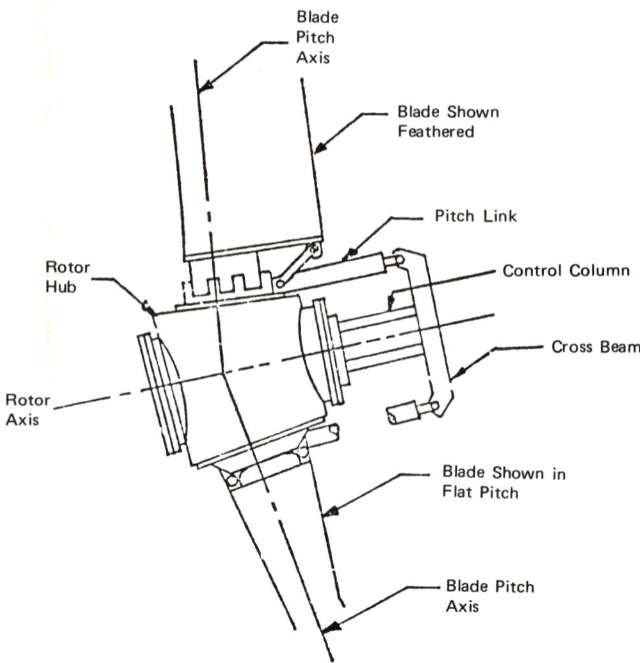

Source: DOE/NASA-9404-76/1

A central quill shaft transmits rotor torque to the gearbox. A linear-travel blade pitch control mechanism is supported by the rotating quill shaft within the spindle shaft. The pitch control system is linked directly to a clevis in each grip. Linear movement of the pitch control beam causes a blade to rotate about its pitch axis. Travel is sufficient for blade pitch control throughout the operating range, including full feathering. Multiple-lug blade grips and pitch control bearings complete the rotor head assembly.

Blade Grips: Forged aluminum grip fittings connect the blades to the hub. These fittings redistribute blade loads into the pitch bearings and pitch control linkage. Connecting the grips to the hub is a crossed roller pitch bearing, capable of reacting large moments as well as thrust. Blades are connected to the grips by a double-pinned multi-lug joint which picks up the main spar fitting. A truss to the blade trailing edge fitting carries trailing edge spline loads.

Hub: The hub is constructed as a weldment of three large rolled plate cylinders. The larger two cylinders have axes coincident with the rotor blade pitching axes and furnish the mounting for the blade pitch bearings. These cylinders are pierced by a smaller cylinder containing the hub bearings. The bearings are a tapered roller pair in which all thrust is carried by the aft bearing.

The welded structure was selected to reduce construction costs. Low stress levels are achieved throughout the hub by providing generous cross sections.

Figure 5.33: Rotor Head Assembly, Cutaway View

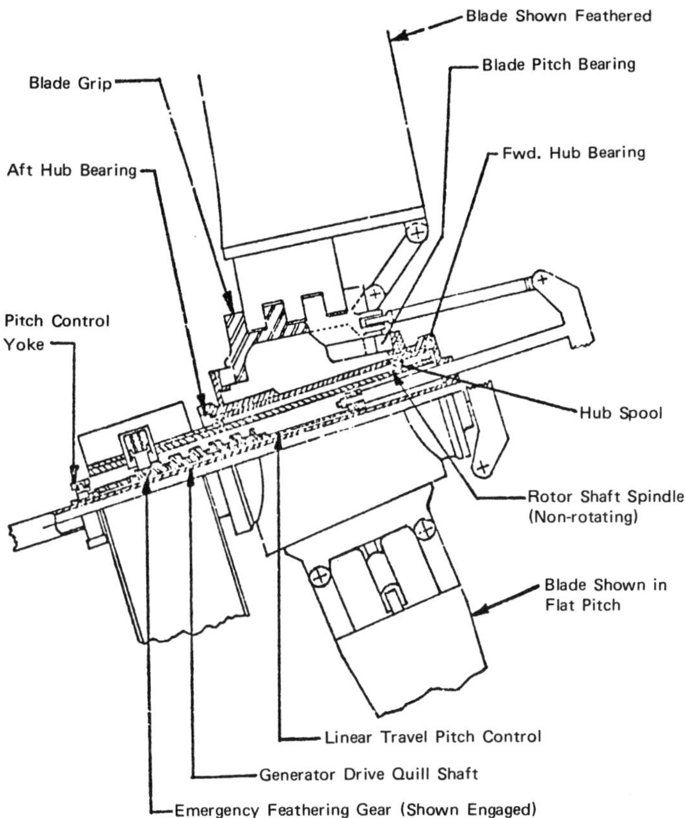

Source: DOE/NASA-9404-76/1

Controls: The controls subsystem conceptual design, optimization and preliminary design are discussed below. A primary tool for this process was a failure modes and effects analysis (FMEA).

The FMEA was used to guide the concept selection and design of the control system, based on the types of failures which might occur, the severity of their effect on the WGS, and the ability to detect and compensate automatically for such failures.

The control system uses a microprocessor for data telemetry and for startup and shutdown sequencing, with hydraulic servos and analog equipment for the primary rotor controls. A purely mechanical control backup is provided for emergency feathering and shutdown of the rotor.

Requirements – The controls subsystem must be designed for operation at a remote, unattended site. The subsystem must be fail-safe and self-monitoring; that is, it must be capable of detecting any failure within the WGS which may cause secondary damage to subsystems and take appropriate protective action. The controls must operate the system under extreme environmental conditions, such as wind gusting, and protective functions must be executed independent of the availability of external power. The controls must also be conservatively designed to maintain high reliability and be properly protected against induced transients from the power line or from lightning strikes. The control subsystem must do the following:

(1) Start up the WGS from rest to design rotor speed.

(2) Shut down and secure the WGS.

(3) Control blade pitch to regulate the rpm of the rotor when disconnected from the utility network.

(4) Control blade pitch to regulate power output when connected to the utility network.

(5) Control the yaw orientation of the nacelle

(6) Monitor operating parameters and telemeter information to a central control point

(7) Determine when a fault exists within the WGS or the utility line and take appropriate action to protect the system.

(8) Record significant data at the WGS site.

Many different functions are required of the control system, ranging from continuous fast-response proportional control to discrete fault detection and sequencing. Since the various control functions also have differing degrees of importance from a safety and reliability standpoint, the same equipment will not be optimum for all functions. Therefore, the control functions were categorized for purposes of concept selection and the optimum equipment was selected for each category.

The cost and weight of the electrical portion of the control subsystem are only a small percentage of the total subsystem cost and weight. Therefore, these were not major factors in the evaluation of control subsystem concepts. The major considerations in the evaluation of the candidate control concepts were reliability and safety. Other criteria included acceptability to electric utilities, compatibility with the specified environment, simplicity, life, and maintenance. In the case of critical functions, predictability and ease of detection of failure modes and the ability to initiate protective or backup action without external power were important considerations.

Design Approach – Types of equipment examined included fully-mechanical sensing and actuation devices, electrical devices, including ac and dc analog servo systems, standard digital logic, and digital microprocessors. Several actuation methods were also investigated, including aerodynamic, electrical, mechanical, pneumatic and hydraulic approaches. The sensing devices considered depended primarily on the parameter being sensed. However, preference was given to devices which do not require slip rings or commutators. Preference was also given to readily available, off-the-shelf components and concepts which have a firm and proven experience base in the rotary wing or power generation fields.

As the WGS design concept evolved through the conceptual design, parametric optimization and preliminary design phases, the control subsystem concept was developed and modified. The essential control subsystem concepts were selected at the outset of the preliminary design phase and then refined in concert with other subsystems.

Candidate Concepts and Evaluation – Critical Controls: Hydraulic servos supplied with pressure from a pump driven directly by the main gearbox were selected for the primary rotor controls. Standard analog servo amplifiers were selected to control the hydraulic actuators, primarily due to their large experience base in utility applications and their compatibility with automatic failure detection. The blade pitch control linkages from the actuator to the blade were selected as simple mechanical linkages to provide high reliability and safety. This linkage approach also permits use of the stored energy in the rotor to provide an emergency shutdown capability in the event of a power failure or control failure during operation.

Noncritical Components: A microprocessor-based computer system was selected to control the startup and shutdown sequencing of the WGS, and also to handle the data formatting and communications for telemetry and supervisory functions. This equipment is ideally suited to these functions, and is presently being used by utilities in a similar role.

Since the utility experience base with this equipment is not extensive, the microprocessor was limited to the functions described and was excluded from the primary rotor controls and the electrical protective and relaying equipment. Field experience with these devices indicates a satisfactory performance can be obtained in a remote site environment if proper precautions are followed, such as isolating inputs and outputs against transients, and special coding provisions to provide high noise immunity for the telemetry links. The microprocessor can also draw its power from a noninterruptible battery source.

Monitoring: Critical sensing, signal processing and actuation functions can be continuously monitored for failures, and appropriate action taken to prevent secondary damage in the event of a failure. This monitoring includes, as a minimum, the blade pitch servo control, the yaw servo control, the wind speed and direction sensing, the rotor rpm sensors and the critical portions of startup and shutdown sequences.

Preliminary Design and Analysis – The following description covers the electronic and electrical portions of the controls. The mechanical and hydraulic portions of the system are described in the drive subsystem section.

Yaw Control: The yaw servo keeps the shaft axis of the rotor aligned with the average wind when the rotor is turning. The yaw rate is limited to approximately ⅓ rpm to prevent sudden motions of the tower head which could result in large forces on the system due to gyroscopic effects. The yaw servo is only intended to trim the system to the average wind direction. A wind vane coupled to a synchro is used to sense wind direction error. The error signal is then amplified and used to operate hydraulic solenoid valves, which control the flow of hydraulic power to the yaw servo motor. Automatic fault monitoring of the yaw servo during operation is provided by independently monitoring the average wind direction error and checking for proper servo response to changes in the average direction of the wind.

Blade Pitch Control: The blade pitch servo amplifier provides a proportional dc output signal to operate the blade pitch hydraulic servo valve. The amplifier is automatically reconfigured for each mode of operation:

(1) Startup–programmed pitch change for rotor acceleration

(2) Standby/Synchronize–pitch controlled to regulate rpm

(3) Operate–pitch controlled to regulate power output to utility.

(4) Normal Shutdown–programmed pitch change for rotor deceleration

During both the standby/synchronize and operate modes, gross positioning of blade pitch is accomplished by providing the servo with a signal which is a function of the component of wind speed along the axis of the rotor shaft. For the operate mode, a power signal derived from the generator power output monitor provides a fine adjustment to hold the power output at the rated set point for wind speeds above rated. The rpm is used to control pitch for the standby/synchronize mode.

Continuous fault monitoring of the pitch servo during startup, standby/sync and operate modes is required to prevent possible overspeed/underspeed and/or reverse thrust on the rotor due to pitch control failures. The monitor must be capable of differentiating between pitch control failures and sudden or unusual motion of the controls due to wind gusts. This is accomplished by a monitoring device which checks for proper average blade pitch position via an independent blade pitch feedback sensor.

Several analyses of the control system were used to establish the rates and limits of the control components. The most important of these included the establishment of the rotor blade pitch rate, which is the basic system power control parameter. To establish the pitch rate required, an analysis of rotor response under gust conditions was conducted. The wind gust model used for this analysis was supplied by NASA.

When the generator is connected to the utility network, an increasing gust will cause increasing torque and power output, whereas a decreasing gust will cause decreasing torque and power output, and may also cause an undesirable thrust reversal on the rotor if the pitch control system is not sufficiently responsive. Using the static control characteristic of the rotor, in conjunction with the wind gust model, it is possible to estimate the effect of gusts on the system and to derive the control response requirements. The analysis indicated that the torque overload limit was the controlling requirement on pitch rate and showed the required pitch rate to be about 5° per second.

When the generator is disconnected from the network, the sudden loss of load, combined with a gust, can cause an overspeed condition. The overspeed analysis also showed a pitch rate requirement of about 5° per second to limit maximum overspeed to 150% of rated, the generator limit. Thus, analysis of both critical control conditions determined that a 5° per second pitch rate is adequate to control the rotor under extreme conditions. Ultimately, more detailed analysis may indicate somewhat faster rates are needed. However, the pitch rates should not be made faster than necessary, since excessively fast rates pose a greater hazard in the event of control system failure at the highest pitch rate position.

Structure: The WGS structure subsystem includes the tower and its foundation, and the turntable. Various tower configuration concepts were investigated in depth before selecting the final concepts for preliminary design. Preliminary designs of both steel truss and concrete shell towers were prepared for the 500 kW and 1,500 kW systems, since both types offer attractive combinations of cost, appearance and modification flexibility.

Requirements – The structure must support the power generating subsystems and controls, allow the rotor disk to be oriented normal to the wind and react the forces imposed by the rotor and by the wind acting on the tower itself. Fatigue strength of the tower must be great enough to withstand the rotor-induced vibratory loads, including the effects of startup and shutdown cycles, gust variations, tower shadow, and gravity for a 50-year service life. The stiffness of the tower must be selected so that the resulting tower natural frequencies avoid integral multiples of the operating frequency. The foundation must provide a firm anchor for the tower structure for all imposed loading conditions and certain earthquake conditions.

The turntable and associated connecting structure, orientation drive mechanism and protective shrouding must transmit loads developed by the rotor and power conversion machinery to the tower and protect the components from rain, snow, hail and lightning.

Many other detailed requirements, specific to each particular WGS concept, size and operating condition, were developed and used to guide the tower concept selection and design. The most important detailed requirements covered the subsystem structural criteria, including static strength, fatigue strength and stiffness.

The static strength requirement was established for three basic loading conditions:

> Blowover—For hurricane winds of 53.6 m/sec it was assumed that the rotor is parked with the blades vertical and that the wind direction was such that the wind impinges flatwise on the blades and broadside on the nacelle.
>
> Normal Operating plus Seismic Loads—The wind velocity for this condition is the rated wind speed of the WGS, which produces maximum rotor thrust. A horizontal load factor from the Uniform Building Code for a seismic disturbance was applied to produce inertia loads in the same direction as the rotor thrust.
>
> Maximum Operating Load—The normal operating loads at rated wind speed were multiplied by 2.0 (except that rotor torque was multiplied by 2.5) as conservative estimates of the worst transient loads, including dynamic response of the structure. The peak gust amplitudes were derived from the NASA gust model.

Fatigue strength requirements were established for two repeated loading conditions. For the high frequency vibratory hub moments existing in normal operation, infinite life was required. For the repeated loads incurred by startup and shutdown cycles (0 to 1.5 times maximum steady operating loads), a life of 50 years at five cycles per day was required. In order that structural natural frequencies avoid resonance with the rotor, it was judged to be necessary that the first mode bending and torsional natural frequencies be at least 1.5 and 2.5 times the rotor operating frequency, respectively, and that they should avoid integral multiples of normal operating rotor speed.

Design Approach – To minimize costs, existing commercial quality materials and parts and standard construction techniques were used for all tower concepts. The choice of materials was, therefore, limited to structural steel and concrete and the principal focus of the concept valuations was on the configuration of the tower and foundation.

Candidate Concepts and Evaluation – Several tower and foundation configurations were considered and evaluated earlier. Because large production run costs of both the truss tower and the pre-cast, post-tensioned concrete tower are reasonably close, and each has important advantages, both were carried forward to preliminary design.

A welded structural steel framework of standard sections was postulated for the turntable (Figure 5.31) as representative of the weight and cost of any feasible alternative.

A detailed study during the preliminary design phase examined the relative merits of foundations of mass concrete, piles, and piles with rock anchors, for both the 500 kW and 1,500 kW WGS. To establish strength properties for the foundation, average soil conditions and existence of bedrock at a reasonable depth were assumed. For the high power system, a combination of a truss tower and piles with rock anchor foundation showed a distinct cost advantage. Otherwise, no single foundation concept or tower-foundation combination was significantly advantageous. It was evident from the analysis that local soil conditions should be examined for a particular site to determine the most economical foundation.

Preliminary Design and Analysis – Since the tower concept studies showed that both the steel truss and pre-cast, post-tensioned concrete tower concepts have advantages, preliminary designs for both concepts were prepared for both the low and high power systems. Design configurations and system costs for all three foundation concepts for both tower types and power levels were also developed, assuming average soil conditions.

The 1,500 kW WGS steel truss and concrete shell towers are shown in Figure 5.34. The 500 kW system towers are similar. The 1,500 kW truss tower is a four-sided tower, constructed of standard grade structural steel H-beams and double angles. The width of the top of the tower was dictated by the bearing size required to react the rotor loads. Bending strength is primarily provided by the chords, or corner members, of the tower. The chords are constructed of wide-flange H-beams. Each chord is made up of three different sizes of H-beams, the heaviest at the base and the lightest at the top of the tower. They are constructed of various sizes of H-beams and double angles, again with heavier members used near the base of the tower.

Figure 5.34: 1,500 kW WGS Concrete Shell and Steel Truss Towers

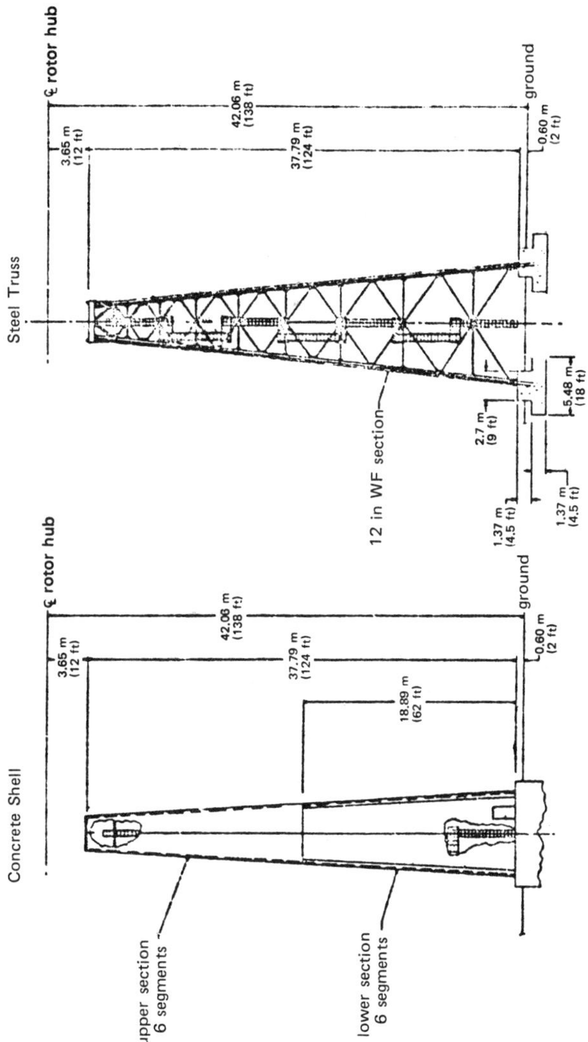

Source: DOE/NASA-9404-76/1

The purpose of the horizontal braces is to stabilize the diagonals and reduce their effective column length. The horizontal braces carry no primary loads.

The pre-cast concrete tower for the 1,500 kW WGS is a truncated circular cone with a constant cone angle and constant wall thickness, constructed of 12 factory match-cast segments, six upper and six lower, each covering 60° of circumference. The lower six segments are identical, except for a doorway in one, and the upper six are identical. Maximum size of each segment was determined by shipping considerations of weight and dimension. Access to the tower top is provided by caged ladders in the shell interior, protected from the environment.

Post-tensioning is accomplished during erection of the tower in two stages; the first post-tensioning is applied after the lower half is erected, while the second post-tensioning is applied after erection of the upper half. Post-tensioning strands are positioned at 12 locations within the shell wall which is thickened locally to provide protection to the strands. These steel strands, running through the shell wall from the foundation toward the top of the tower, are anchored to the foundation and are tightened during the post-tensioning process so that they compress the concrete. The amount of compression is sufficient to keep the concrete in compression for the maximum design loads. Shear ties between the segments are required along vertical joints to provide structural continuity.

For each tower design, the applied loads, member stresses, and natural vibration frequencies were analyzed in detail. Each structure was designed to satisfy the structural criteria described above. Of the three static loading conditions, the storm wind blowover condition was found to be most critical. Either this condition or the stiffness requirements governed the tower designs. The fatigue requirements were not found to be critical.

The relatively small difference in high production rate cost between the steel truss and post-tensioned concrete towers, leads to the conclusion that both tower types should be considered for production applications. The truss, however, is very clearly the most economical choice for demonstration and development programs.

(NOTE: As discussed earlier, due to the considerable importance of the excessive shadow effect produced by shell type towers, it now seems unlikely that shell towers will prove feasible.)

Drive: The drive subsystem, Figure 5.31 is composed of the mechanical components which support the rotor and transmit its torque to the electrical generator. In this study the control system power supply, the mechanical rotor controls, and the turntable orientation mechanism are defined as part of the drive subsystem.

Requirements — Table 5.16 lists the functional requirements of the drive subsystem.

Table 5.16: Functional Requirements of Drive Subsystem Components

Component	Function
Rotor Spindle or Shaft	Supports the rotor, allows its rotation

(continued)

Table 5.16: (continued)

Component	Function
Input drive	Transmits rotor torque to the gearbox
Gearbox	Provides rotor to generator speed conversion. Drives auxiliary equipment
Parking brake	Stops the rotor from low rpm and locks it
Inching drive	Repositions the parked rotor
Pitch control	Controls pitch setting of the rotor blades
Turntable control	Orients the rotor normal to the wind
Turntable bearing	Provides pivot between the turntable and tower
Hydraulic system	Provides power to pitch and orientation controls

Source: DOE/NASA-9404-76/1

Design Approach – In order to minimize cost, maximize reliability and provide long drive train component life, an approach emphasizing simplicity and design ruggedness was used. All drive subsystem concepts considered only off-the-shelf components operating well within their strength and performance capacities. Safe failure modes were designed into the drive subsystem where possible; e.g., feathering the blades and locking the turntable in the case of servo system failure.

Simple field assembly tasks were assured by assigning critical fitting tasks to shop fabrication steps and by allowing latitude for misalignment between field-assembled units. Maintainability was emphasized by the simple design approach, by designing good accessibility provisions and by including strategically placed diagnostic sensors.

The entire design approach to the drive subsystem was geared to achieve low initial and operating costs through the use of rugged components and straightforward, proven component integration techniques.

Candidate Concepts and Evaluation – A number of power transmission configurations were evaluated, including chain and belt drives, hydrostatic transmissions and fixed ratio gearboxes. Only fixed ratio gearboxes were found to be commercially available in sizes capable of handling rated rotor torque. Other candidate components would have to be combined, or staged, to meet the torque requirements. Since fixed ratio gearboxes also offer the best efficiency and highest reliability, they were the clear choice for the power transmission system.

The rotor support/drive shaft must take the form of either an overhung drive shaft supported on pillow blocks, or a static spindle with live bearings for rotor hub support and a quill shaft for torque drive. While simpler in concept, the pillow block shaft was not selected because of the higher weight of the shaft and bearings needed to handle the overhung rotating loads. The static spindle offers a much more efficient structure, not subject to large vibratory bending moments from the rotor.

Auxiliary drive system components, such as shafting, flexible couplings, bearings and clutches were found to be readily available as catalog items. Overall system cost and weight were generally not greatly influenced by these component choices, and most of the effort devoted to the evaluation of these devices concentrated on the selection of low cost designs.

One auxiliary function which received heavy emphasis was the provision for emergency shutdown. A primary brake, capable of stopping the rotor from an overspeed condition due to extreme gusts or network disconnect, was found to be impractical for the WGS. Brake size becomes unreasonably large when installed directly on the low speed rotor shaft, while gearbox strength and brake energy dissipation capability come into question with the brake installed on the generator high speed shaft. This problem was circumvented by providing a direct, fail-safe mechanical method for slowing the rotor by blade feathering, actuated by rotor inertia when overspeed exceeds safe limits. After slowing, the stop is completed with a small parking brake.

Two turntable support configurations were studied. In one, a system of three bearings was arranged on a pintle projecting from the bottom of the turntable. An alternative design used a single, large crossed roller bearing. Advantages leading to the choice of the second scheme were simpler assembly and lower cost, and a large clear center passageway from tower to turntable for personnel and power lines, an important design consideration.

Preliminary Design and Analysis: The preliminary design of the 1,500 kW high power WGS drive subsystem is shown in Figure 5.31. The drive subsystem for the 500 kW unit is similar. Both of these designs are described below, by major component:

Rotor Spindle: The rotor is supported on a tubular spindle which is fitted to a socket in the turntable structure shown in Figure 5.31. Attachment of the spindle to structure is by means of shear bolts. Stress levels in the spindle are sufficiently low to permit its designation as permanent structure with a life of 50 years. The spindle provides mounting for a pair of tapered roller bearings which support the rotor hub.

Quill Shaft: The driving connection between the hub and gearbox is a quill shaft which transmits torque only, since all other rotor loads are carried by the stationary spindle. In the 500 kW WGS, this tubular steel shaft is 0.2 m in diameter and, in the 1,500 kW system, the shaft diameter is 0.406 m. A flexible coupling connects the quill shaft to the gearbox, allowing for installation misalignment.

Gearbox: Rotor-to-generator speed conversion is by means of a triple mesh, parallel shaft gearbox. Candidate gearboxes are Philadelphia gear models for both the 500 kW and 1,500 kW WGS. These gearboxes are equipped with antifriction bearings throughout, and use case-hardened shaved, or ground, gears.

Pitch Control: The rotor pitch control system utilizes a central push-pull control column which is keyed to the rotating quill shaft (Figure 5.32). A rigid cross beam on the end of this column is linked directly to each of the two blade grip fittings by pitch links so that the push-pull motion is transformed into rotation of the blades about their pitch axes. The pitch column can be actuated by two independent means. Normal actuation is by means of a single, large hydraulic cylinder which is linked to a yoke on the inboard end of the column. A thrust bearing in the yoke allows transfer of the linear motion from the cylinder linkage to the rotating column. Motion of the hydraulic cylinder is controlled by an electro-hydraulic servo valve, and cylinder position is fed back by a linear displacement transducer.

Emergency actuation of the pitch column is accomplished when cam follower plungers drop into engagement with a helical cam slot in the column. Rotor shaft rotation will then move the column, very forcibly, in the feather direction. A deeper annular slot at the end of travel prevents motion beyond full feather, even if the shaft continues to rotate, and also prevents inadvertent reversing. The cam follower plungers are spring-loaded to the engaged position, and hydraulically disengaged. In the absence of hydraulic pressure, either through malfunction or actuation of a selector valve, the followers will engage and feathering will take place.

Orientation Control: Turntable orientation is accomplished by means of a hydraulic motor and gear train. The motor is mounted on, and drives through, a single stage vertical shaft worm gearbox of 60:1 ratio. Output of this gearbox drives a pinion gear which meshes with a large diameter internal gear. This large gear is integral with the inner race of the turntable bearing, which is secured to the tower. Ratio of the pinion-ring gear is 10:1. The worm gearbox is irreversible, so that the turntable will be rigidly held against the wind load, unless the wind load is assisting the hydraulic motor. Speed of the motor will be regulated by hydraulic flow control valves, so that turntable rotation will be held to ⅓ rpm, even with assisting wind. This is required to limit rotor gyroscopic forces.

Turntable Bearing: The turntable bearing is a single, large diameter crossed roller bearing. This type of bearing can withstand radial thrust and over-turning moment loads, separately or simultaneously. Balance of the complete turntable assembly with all equipment and nacelle is slightly offset, with the rotor on the light side. With the wind thrust, there is a constant moment tending to pitch the rotor down. The bearing is sized to handle all loads, including those generated by the maximum wind blowing broadside on the nacelle and feathered blades.

Parking Brake/Inching Drive: A parking brake/inching drive has been designed as a single unit. For this arrangement, the brake connects the generator shaft to a motor-driven worm gear, which is the inching drive. The purpose of this arrangement is to eliminate the need for a secondary clutch and actuator, and to bypass a hazardous sequencing step where the brake must be released when the inching drive is engaged. When the parking brake is engaged, it locks the generator shaft to the worm drive, which is irreversible to torque applied to the gear. The worm is driven by a gear motor so that the shaft can be slowly rotated to a selected parking position for stowing or maintenance. At no time is the brake released, until it is desired to start the WGS.

Electrical: Several types of electrical generation approaches were studied during the concept selection process, including variable rotor shaft speed and fixed rotor shaft speed configurations. Two fixed speed concepts, using either an induction or synchronous generator, were selected for the WGS preliminary design, since these concepts provide the highest system efficiency with the lowest system cost and complexity.

Requirements — The basic requirements of the electrical subsystem are to produce electric power at a voltage and frequency compatible with standard electric utility requirements and practices. The electrical generating equipment must produce this power using minimum cost equipment to achieve competitive energy costs.

The equipment must operate at a remote, unattended site. Therefore, automatic fault protection and synchronization of the WGS with the utility network must be provided. The equipment must operate over the required range of wind speed, temperature and other environmental effects, and maintain a stable interface with the utility network under expected wind gusts. This represents an additional requirement beyond those usually imposed on conventional utility generating equipment, since the power source, the wind, is a random variable. The WGS must, therefore, include provisions to limit the adverse effects of wind gusts on the WGS and the utility network. This is accomplished primarily through control of blade pitch and by control of generator field excitation.

Design Approach — In selecting the electrical subsystem concept and developing the preliminary design, primary emphasis was placed on minimizing the cost per kilowatt hour of energy delivered to the utility. Since the WGS spends a large percentage of its time operating below rated power, good partial power efficiency was an important consideration in the selection of the generating and interface equipment. Emphasis was also placed on the selection of equipment commercially available and conforming to standard utility practices. Other decision criteria included equipment cost, total weight of the electrical subsystem which affects tower cost, electrical subsystem reliability and maintainability, and utility equipment preferences and control considerations.

Candidate Concepts and Evaluations — Generators: There are two fundamentally different approaches for generating electric power with the WGS. One approach is to use an electrical subsystem which can accept variable shaft rpm, allowing the rotor to operate at a speed proportional to the wind. The second approach operates the electrical equipment at a fixed rpm and requires a fixed rpm shaft drive from the gearbox to the generator.

Both approaches must provide constant frequency output. Although operation at variable rpm complicates the electrical equipment, it does permit the rotor to operate at its most efficient speed over a range of wind speeds. For variable speed systems, a means must be provided to convert the variable shaft rpm into the constant frequency required by the utility network.

In the study, three basic concepts were considered for generating dc power from the variable shaft rpm and then converting to ac power at constant frequency for delivery to the utility network. These three concepts were:

(1) A dc motor driving an induction generator

(2) A dc motor driving a synchronous generator

(3) A three-phase solid state inverter

Two concepts were considered for fixed rpm electrical systems. One using an induction motor operated as a generator, and the second using a synchronous generator of the type normally used by electric utilities. These approaches are straightforward concepts using standard equipment, and very similar to small conventional utility generation installations. The evaluation of these alternatives determined that the variable rpm electrical system configurations were in all cases considerably more complex, more costly, and less efficient than the fixed rpm schemes. This is shown in Table 5.17, which gives the results of a study for a 1,000 kW WGS in terms of system efficiency and total equipment cost. As might be expected, the slightly higher efficiency of the rotor when operated at variable rpm is overwhelmed by the poor efficiency and high cost of the variable rpm electrical systems.

Table 5.17: Electrical Subsystem Efficiency and Cost*

	. . . . Load Net Efficiency, %.				
Type	**Full**	**¾**	**½**	**¼**	**Total Cost, $**
Constant rpm/induction generator	94.3	94.0	92.2	87.0	59,305
Constant rpm/synchronous generator	95.0	94.4	92.2	87.5	86,815
Variable rpm/MG set (induction generator)	82.6	81.3	79.4	70.7	208,350
Variable rpm/MG set (synchronous generator)	83.2	81.6	79.4	71.7	235,860
Variable rpm/inverter	78.3	77.0	75.5	44.4	364,325

*For 1,000 kW high power WGS.

Source: DOE/NASA-9404-76/1

Because of the small cost difference between them, either the induction or the synchronous generator can be used with the fixed rpm system, depending on the preference of the using utility. To provide this option, protective equipment and switch gear specifications were developed to be compatible with either type of generator. Both were carried forward to preliminary design.

Protective and Relay Equipment: Although there are several new developments in the field of generator protective and relaying equipment which hold promise for the future, they were not adopted for the WGS. Included in this category is the microprocessor which was suggested to replace the standard electromechanical and solid state devices commonly used by utilities. The microprocessor was not selected for this role because of the limited experience utilities have with it. However, it was selected for noncritical sequencing, monitoring and housekeeping.

The failure modes and effects analysis provided further evidence that a microprocessor is suitable for use in the sequencing and supervisory control functions of a wind generator system. The microprocessor can detect and take the necessary corrective action in the event of failures in the remainder of the system that could lead to hazardous conditions. In most cases, the microprocessor can even detect its own failures, or at least their effects that might create hazardous conditions, and then initiate proper corrective action.

In one respect, the microprocessor based control system is quite different from a control system using conventional techniques. This is in the interface with the operator or maintainer. A control system, using conventional techniques, generally has a number of indicator lamps, annunciators, or other directly observable devices that indicate the conditions within the control system at all times. The microprocessor based system, on the other hand, carries out its logical decision making within the microprocessor circuits themselves, where it is invisible to the operator or maintainer. Diagnostic aids should, therefore, be a part of the microprocessor based sequencing and supervisory control system. This not only provides the maintainer with a starting point for his maintenance procedures, but also provides a higher degree of confidence that the trouble he finds and fixes is, in fact, the one which caused the system to shut down.

Interface Equipment: Options for the interface equipment to tie the WGS into the utility network are limited. The required transformers may be either air- or oil-cooled, but the oil-cooled unit was selected because of its lower cost and suitability to outdoor operation.

Circuit Interrupts: Standard circuit breakers and recloser breakers were considered. Since the recloser was at first thought to be cheaper, it was evaluated. It was found, however, that the standard recloser lacks repeatable closing times needed to assure accurate synchronization between the WGS and the power grid. Therefore, standard breakers were selected.

Emergency Power Supply: In the event station service power to the WGS is interrupted, emergency power must be available to shut down and secure the system without damage. The two alternatives considered were a gasoline or diesel powered auxiliary generator, or batteries with a charger, constantly floating on line. The higher reliability of the batteries and the preference of the utilities for them were the governing factors in their selection. However, the small amount of power available from batteries dictated that all emergency sequences be designed for low power consumption. This requirement influenced the control system design significantly.

Connections from Tower to Ground: Two methods were considered for getting the electrical power and control wiring from the tower to the ground. The first was slip rings and the second was cable. The cable would be allowed to twist up some predetermined amount before it was untwisted by turning the tower head.

Allowing the cable to twist up is a technique that has been used on other large systems with success. Usually, the net twist accumulated over a period of time does not amount to more than a turn or two. This approach has lower cost and higher reliability than slip rings and is more suitable for the large numbers of wires that might be required for the control and signal cables. It does require a long length of cable to absorb the twist and a sensor for the yaw servo during the shutdown sequence to remove the twist.

The major disadvantage of this approach with respect to the power cables is lower reliability; manufacturers of these cables recommend against twisting or flexural motions. It was, therefore, decided to run the power wiring through slip rings and to use direct connections only for the control and signal wiring with the yaw servo being used to untwist the cable as necessary during each shutdown of the system.

Preliminary Design and Analysis — After selection of the electrical subsystem and component concepts, the system preliminary design process evolved the detailed electrical subsystem configuration and supporting analyses. Descriptions of the major system components are given below and apply to both the 500 kW and 1,500 kW units.

Generator: The generator is a high speed induction or synchronous machine operated at 1,800 rpm. Voltage is either 2,400 or 4,160 volts, selected to minimize cable weight and cost, and the cost of the slip ring used to bring the power cables around the rotating joint at the tower head. Cooling is provided by forced ambient air.

Protective and Control Equipment: The protective and control equipment performs the following functions:

(1) Protects the WGS against damage from electrical faults
(2) Protects the utility network against failures in the WGS
(3) Differentiates between faults internal to the WGS and faults in the utility system.

The protective equipment is designed so that electrical faults within the WGS equipment initiate shutdown and lockout of the WGS until repaired. External faults, on the other hand, trip the main breaker and the WGS is then allowed to resynchronize with the network after the fault has been cleared.

Lightning protection for both systems is accomplished by providing good paths to ground for direct strikes along the outside surfaces of the WGS so that the equipment is inside an effective shield. Lightning arrestors are provided at critical locations to protect against lightning transients.

Utility Interface: The WGS is equipped with an oil filled step-up transformer to match the generator to the voltage of the distribution system, and to limit fault currents. The breaker is located on the high side of the distribution transformer so that, when the breaker is open, the system is completely isolated from the utility network, except for the station service supply which bypasses the breaker. Large currents can be generated by the grid feeding back into faults within the WGS, therefore, the breaker has been sized to interrupt these currents. Reliability considerations require that emergency power at the WGS site be provided by a trickle charged battery source, rather than a gasoline or diesel driven generator. The WGS has been designed to be compatible with this type of emergency power source.

The final design of the electrical subsystem must be tailored to the needs of the particular user utility company. The design will also be influenced by the location within the network system where the WGS is installed. Its location in the system can affect the type of generator, induction or synchronous, the size of the breakers required, and the type and settings of the required relaying and protective equipment. The electrical subsystem design developed for the WGS provides for this tailoring, and is compatible with standard utility requirements.

Transient Analysis: When the synchronous wind generator unit is connected to the electrical distribution system, the interaction must be analyzed to ascertain the effect this unit will have on the operation of the distribution system, and the requirements for protective and control equipment for the electrical system and the wind generator itself. Because of the importance of this issue, a study was conducted to determine the general response of the WGS unit to faults and switching operations on the distribution system and to changes in unit input torque due to wind gusts. The stability analysis was performed by Northeast Utilities for the 1,500 kW WGS. The 1,500 kW system was studied for the preliminary design analysis, since stability questions on the 500 kW system will be less critical than the 1,500 kW unit when connected to the same network.

The modeling of the generator and network was done by use of a transient stability program which is a standard tool used by electric power system engineers in the study of dynamic performance of power systems. The distribution feeder was modeled as a 13 km long feeder with 7 MW of distributed loads along the feeder. The WGS was connected to the feeder through a dedicated 1.6 km line. Modeling of the wind generator portion was based on a quasi-static approximation of the rotor and controls. Although a more comprehensive model should be developed for analyzing the final detailed system design, the current analysis results are representative of a typical WGS/utility interface.

Results of the study indicate that the wind generator should be disconnected from the distribution feeder for any disturbance which causes the normal supply to the feeder to open. This will prevent the wind generator from attempting to supply the load on the feeder in an isolated mode, and will assure interruption of power on the line to allow clearing time for faults. Voltage variations on the distribution system due to wind gusting conditions are more severe if the synchronous generator is connected to the feeder at a point remote from the source (substation). They are also more severe for decreasing gusts (loss of wind) than for increasing gusts of the same magnitude.

With the generator connected to the system at the substation, voltage variations on any feeder supplied from that substation are less than 0.5% for even the most severe wind gust studied. With the generator connected near the end of the feeder, the same wind gust causes distribution voltage to vary 2.2%.

The acceptability of these more severe voltage variations depends upon their frequency of occurrence and upon the standard established by the particular utility company to which the generator is connected. Ultimately, optimization of the design of the pitch control system and the generator regulator could reduce these variations. However, wind gusts up to the maximum design intensity are not likely to cause the synchronous generator to pull out of synchronism with the system, for the typical feeder investigated in this study.

While the results of this study are typical, giving general characteristics of system performance, each specific installation will have to be studied separately to establish the operating conditions and electrical requirements particular to that installation. A detailed transient analysis of the WGS connected to the network should also be performed for each installation to optimize the installation and assure stability of the system under all operating conditions at the selected site. Since this is normal practice for utility generating installations, it does not present any unusual demands on the user utility.

Conclusions and Recommendations

(1) The WGS concept selected for preliminary design consisted of a 2-bladed variable pitch rotor downwind of the tower driving an ac synchronous generator at constant rpm through a fixed-ratio gearbox. The dynamic components are mounted on either a steel truss or concrete shell tower. Preliminary design confirmed that this concept offers lowest capital investment and energy cost, highest efficiency and reliability, least maintenance and lowest technical risk. This system is recommended, with steel truss tower, for development, test and demonstration.

(2) Two WGSs were designed for sites with median wind speeds of 5.4 m/sec and 8 m/sec, a low power system of 500 kW and a high power system of 1,600 kW, respectively. For any site with median wind speed between 3.5 m/sec and 6.3 m/sec, the 500 kW system will yield energy at a cost comparable to a system optimized for that site. Similarly, the 1,500 kW WGS can be economically employed for sites with median wind speeds between 6.3 m/sec and 9 m/sec. Therefore, these two designs can be economically used at most feasible wind power plant sites in the United States, and are recommended for development.

(3) For rotors of the size required for the WGS, 40 to 60 m in diameter, technical considerations strongly favor composite construction for the blades to meet the demanding structural and dynamic requirements. It is recommended that automatic filament wound fabrication techniques be used for blade construction.

(4) The rigid, nonarticulated rotor hub design minimizes cost, complexity and potential dynamic problems and is recommended for the WGS.

(5) Rotor torque control by changing blade pitch minimizes blade operating loads and permits operation and control in fluctuating aerodynamic flow regimes. Torque control through pitch variation is recommended. Since the blade pitch control rate is determined by gust requirements, it is further recommended that a design gust spectrum be defined prior to detail design of the control system.

(6) Brakes capable of preventing rotor overspeed due to control system failure or extreme gusts are not available commercially. Therefore, a mechanical fail-safe emergency blade feathering mechanism is recommended for incorporation into the rotor design to prevent excessive overspeed.

(7) Conventional electro-mechanical control of rotor yaw orientation and blade pitch is the method most acceptable to electric utility companies. The mechanical emergency feathering feature recommended above can be economically integrated into such controls. Electro-mechanical control for rotor yaw orientation and blade pitch is recommended.

(8) Digital microprocessors offer significant advantages for normal operations, such as sequencing and supervising control of startup, shutdown, operations monitoring, failure detection, and data transmission and recording. Microprocessors for control of these operations are recommended.

(9) Standard off-the-shelf electrical generating equipment such as generators, transformers and switchgear can meet all technical requirements of the WGS and is recommended. Either synchronous or induction generators are suitable for

the WGS. Proper design of protective devices and electrical interface equipment will make both compatible with the WGS. Therefore, it is recommended that the detail design include provisions for either generator, with the choice left to the utility company.

(10) On a typical utility network, analysis shows that the WGS remains stable and synchronized under most operational fault and wind gust conditions. It is recommended, however, that the physical and operational characteristics of the utility distribution network be defined to permit selection of breaker and relay ratings, and generator and regulator characteristics for the WGS detail designs. These network characteristics should be defined to minimize adjustments for a specific installation.

(11) The steel truss tower is recommended. Although the pre-cast, post-tensioned concrete shell tower is competitive in large production runs and is considered preferable aesthetically, wind shadow effects, emphasized by NASA's experimental findings subsequent to completion of this study, will probably preclude use of shell towers.

(12) Analyses made during this study show that a WGS can be competitive with other energy sources. It is recommended that more detailed analyses be performed during the detail design of the WGS to determine specific applications and utility interface requirements.

(13) Capital costs for the 500 kW and 1,500 kW systems were derived during this study as $901/kW and $481/kW, respectively, for quantity production. Energy cost of 7.1¢/kWh and 2.7¢/kWh for the 500 kW and 1,500 kW systems were derived. Yearly rotor maintenance costs, which were a major influence on the energy cost figures, were conservatively estimated due to the lack of actual operating experience for such rotors.

(14) Operational and institutional issues, including utility attitude, public acceptance, environmental impact, licensing, and safety appear to present no insuperable barriers to the introduction of WGS. However, it is recommended that the question of visual acceptability of large numbers of units be explored to guide tower design and siting.

REFERENCES

(1) Putman, P.C., *Power from the Wind,* D. Van Nostrand Co., Inc., (1948).
(2) Jayadeviah, T.S., and Smith, R.T., "Generation Schemes for Wind Power Plants," paper presented in the 10th Intersociety Energy Conversion Engineering Conference (IECEC), Newark, Delaware, (August, 1975).
(3) *NSF/NASA/UTILITY Wind Energy Conference Proceedings,* NASA-LEWIS Research Center, Cleveland, Ohio, (December, 1974).
(4) Owen, T.B., "Variable-Speed Constant-Frequency Devices: a Survey of the Methods in Use and Proposed," *AIEE Transactions* (Part II. Applications and Industry), Vol. 78, pp. 321-326, (November 1959).
(5) Chirgwin, K.M., and Stratton, L.J., "Variable-Speed Constant-Frequency Generator System for Aircraft," *AIEE Transactions* (Part II. Applications and Industry), Vol. 78, pp. 304-310, (November 1959).
(6) Rauch, S.E., and Johnson, L.J., "Precision Power Frequency with Variable Speed Generators," AIEE Conference paper No. CP 59-772.

(7) Hoard, B.V., "Constant-Frequency Variable-Speed Frequency-Make-Up Generators," *AIEE Transactions* (Part II. Applications and Industry), Vol. 78, pp. 297-304, (November 1959).

(8) Reitan, D.K., "Wind-Powered Asynchronous AC/DC/AC Converter System," *NSF/NASA Wind Energy Conversion Systems Workshop Proceedings,* NSF/RA/W-73-006, pp. 109-114, (December 1973).

(9) Jesse, R.D., and Spaven, W.J., "Constant-Frequency a-c Power Using Variable Speed Generation,"*AIEE Transactions* (Part II. Applications and Industry), Vol. 78, pp. 411-418, (January 1960).

(10) Chirgwin, K.M., Stratton, L.J., and Toth, J.R., "Precise Frequency Power Generation from an Unregulated Shaft," *AIEE Transactions* (Part II. Applications and Industry), Vol. 79, pp. 442-451, (January 1961).

(11) Bird, B.M., and Ridge, J., "Amplitude-Modulated Frequency Changer," *Proceedings IEE,* Vol. 119, No. 8, pp 1153-1161, (August 1972).

(12) Wickson, A.K., "A Simple Variable Speed Independent Frequency Generator," AIEE Conference paper No. 59-915.

(13) Bernstein, T., and Schmitz, N.L., "Variable Speed Constant Frequency Generator Circuit Using a Controlled Rectifier Power Demodulator," AIEE Conference paper No. 60-1053.

(14) Ramakumar, R., Allison, H.J., and Hughes, W.L., "Description and Performance of a Field Modulated Frequency Down Converter," *1972 SWIEEECO Record,* Dallas, Texas. IEEE Catalog No. 72 CHO 595-9 SWIECO, pp. 252-256, (April 1972).

(15) Allison, H.J., Ramakumar, R., and Hughes, W.L., "A Field Modulated Frequency Down Conversion Power System," *IEEE Transaction on Industry Applications,* Vol. IA-9, No. 2, pp. 220-226, (March/April 1973).

(16) Oklahoma State University progress reports: NSF/RANN/SE/GI-39457/PR/74/1 dated April 18, 1974 and NSF/RANN/SE/GI-39457/PR/74/3 dated October 15, 1974.

(17) Savino, J.M., "A Brief Summary of Large W.G.S. in the U.S.," NASA Tech. Memo. X-71605, (1974).

(18) Reitan, D.K., "Wind Powered AC/DC/AC Converter System," *Wind Energy Conversion Systems Workshop Proceedings,* 109, (1974).

(19) Ramakumar, R., Allison, H.J., and Hughes, W.L., "A Self Excited Field Modulated Three-Phase Power System," C74 318-2 IEEE-PES Summer Meeting, (July 1974).

(20) Golding, E.W., *The Generation of Electricity by Windpower,* E & F.N. Spon, London (1955).

(21) Kostenko, M.P., "A.C. Commutator Generation with Frequency Regulation Independent of Rotational Speed," *Electrichestuo, No. 2, (1948).*

(22) Kostenko, M.P., and Piotvousky, L., *Electrical Machines,* Vol. II, MIR Publishers, Moscow. (1969).

(23) Adkins, B., and Gibbs, W.J., *Polyphase Commutator Machines,* Cambridge University Press, (1951)

(24) Wall, T.F., "Large Wind Driven Synchronous Generators," *Engineering,* 155, 4037, p 532.

(25) Meyer, V.M., "Uber die Unter Synchrone Stromrichter Kaskade," *ETZ-A,* (September 1961).

(26) Paice, D.A., "Speed Control of Large Induction Motors by Thyristor Converters," *IEEE Trans.,* IGA, 5, 545, (October, 1969).

(27) Linke, S., Teshome, A., and Yehsakul P.D., "A Study of Transmission and Protection Elements for Wind Energy Generating Systems," Brookhaven National Laboratory, (unpublished) (December 1975).

(28) Jorgensen, G.E., Lotker, M., Meier, R.C., and Brierly, D., "Design, Economic and System Considerations of Large Wind-Driven Generators", IEEE Power Engineering Society, Winter Meeting, (January 1976).

(29) Jayadev, T.S., "Windmills Stage a Comeback," *IEEE Spectrum,* p.45 (November 1976).

(30) "Wind Energy Mission Analysis", Lockheed California Company, Report #LR27611, (September 1976).

(31) *Design Study of Wind Turbines 50 kW to 3,000 kW for Electric Utility Applications,* General Electric Company, Valley Forge Space Center, ERDA/NASA-19403-76/2, (February 1976).

(32) *Design Study of Wind Turbines 50 kW to 3,000 kW for Electric Utility Applications,* Kaman Aerospace Corporation, ERDA/NASA-19404-76/2, (February 1976).

(33) "Operational, Cost and Technical Study of Large Wind Power Systems Integrated with an Existing Electrical Utility, "Southwest Research Institute, (1976).

(34) Todd, C.J., Eddy, P.L., James, R.C., and Howell, W.E., "Cost-Effective Electrical Power Generation from the Wind," *Proceedings of the 1977 Annual Meeting, American Section International Solar Energy Society.*

(35) "Wind Energy Mission Analysis," General Electric Company, Valley Forge Space Center, Report #76, SDS4267, (February 1977).

(36) "Potential Pumped Storage Projects in the Pacific Southwest," Federal Power Commission, Washington, D.C., (1975).

(37) "Pumped Storage in the Pacific Northwest, an Inventory," Corps of Engineers, North Pacific Division, (1976).

FARM AND RURAL USE SYSTEMS

The material for this chapter has been based upon a report by the Institute of Gas Technology, Chicago, Illinois (PB 259 318) and papers presented at the Second Workshop on Wind Energy Conversion Systems (NSF-RA-N-75-050).

WIND POWERED HYDROGEN ELECTRIC SYSTEMS

The objective of a study by the Institute of Gas Technology was to determine the current technology of a wind-energy conversion system (WECS) of minimal cost for rural applications. Specifically, IGT evaluated available methods for converting shaft horsepower from a wind turbine to electricity and hydrogen. A workable mix of these two energy forms with storage that can support the energy needs of selected farming operations and the rural home was sought.

Energy load patterns of several farming operations were examined for interfacing with the energy storage and delivery systems that are supplied by wind turbines dependent on the prevalent winds. Several preliminary designs have been completed so that a follow-up program for developing detailed designs and demonstration units can be implemented if desired.

Methodology

The work was divided into four major task areas to ensure an ordered progression throughout and to provide a means for measuring progress. Those tasks were:

(1) Determine energy load patterns
(2) Establish preliminary system sizing and design
(3) Select equipment
(4) Optimize system design

Task 1 required identifying both the selected loads' service energy requirements and the wind energy available. Using data obtained through personal contacts, from the available published literature, and data available from the Economic Research Services branch of the United States Department of Agriculture, the energy requirements for the following four farm types were characterized: 40.5 ha (100 acre) cash grain operation; 2,000-head/yr swine operation; 10,000-layer poultry operation; and 100-head dairy operation.

In addition, using unreported in-house data on energy use for two homes in Canton, Ohio, and the available climatological data, the energy requirements for a medium-sized, ranch-style home were characterized in three regions: Midwest (Chicago, Ill.), Southeast (Atlanta, Ga.), and Southwest (San Diego, Cal.).

Wind-energy availability was established using data from Argonne National Laboratory and from the National Climatic Center. Both the wind energy available and the energy requirements of the load were characterized on a monthly basis. A shorter interval would have been more desirable in order to examine the effects of daily and weekly variations of the load and wind on the system. However, accurate data for the loads and, to a large degree, for the wind are not available, thus precluding the possibility of pursuing this more desirable approach in this study.

During Task 2, system designs were conceptualized and analyzed for possible component interface and operational problems. Where major problems seemed imminent, designs were reworked or rejected. Also, during this task, decisions were made concerning the quality of the systems' outputs, and the number of days of energy storage for the systems was set at 3. This storage period is assumed to be conservative; however, it has not been verified by actual wind data.

After the preliminary designs were established, information on the equipment needed in the wind-energy systems was gathered. Interfacing, maintenance, and reliability characteristics were evaluated along with the equipment cost and availability. From this data set equipment types were selected for use in the optimized systems.

The concluding phase of the study was to identify the optimum wind-energy conversion system designs and to evaluate the system economics. Wind availability, storage capacity, load to be satisfied, and WECS design were all considered during this phase.

Load Factors for Farms

Agricultural production can be classified into two basic groups: crop production for cash or internal use (primary), and feeding operations for animal or animal-product production (secondary). Such a classification is meaningful from an economic as well as a management point of view and is useful here because the energy-conversion devices and load profiles in each group are quite different.

In recent years, several factors have increased the degree of specialization in agricultural production and, in particular, the separation of crop production from animal husbandry. So, field-rearing systems are being replaced, to a large extent, by confinement or intensive systems in which all the variables affecting conversion efficiency can be standardized and controlled. Moreover, productivity in such systems depends on specialized managerial techniques that are distinctly different from those required for crop production. At the same time, the introduction of cultural energy has substituted tractor fuel for animals, and fertilizers for crop rotation. Thus, land also tends to become specialized in the production of those crops for which the economic return is greatest. The integration of crop and animal production is economic (to spread risk) rather than technological. The price and supply of feed for the animal-husbandry operation tend to be stabilized through integration with crop production.

Furthermore, because of the large capital requirements for both crop production and animal production, it is difficult to optimize two separate operations and still achieve a balance. As a result, mixed farm operations vary widely in size, in the ratio of crop to animal production, and in the degree of automation of the confinement system. On this basis, crop production and each type of animal production were considered separate operations, each with its own characteristic energy-use profile. The applicability of wind-energy systems was evaluated for each of three types of animal-production and one crop-production operation. The mixed-farm operation was not evaluated due to current unfavorable prospects of powering the crop-production operation with a wind-energy system.

Crop Production: For this group a 40.5-ha operation consisting of equal acreage in corn and soybeans was selected. The tillage method for each crop was considered the same, except that the corn required drying from a moisture content of 26.5 to 13.0% (1). Energy-conversion devices for these operations mainly consist of mobile tillage equipment (tractors, combines, etc.) and stationary grain-drying equipment.

Seasonal Energy Requirements – The annual mobile fuel requirement for tillage operations (plowing, planting, etc.) is 70 x 10^{12} J on the 40.5-ha cash-crop operation. The energy consumption for the tillage operations is shown in Table 6.1 and assumes total fall plowing. These data are from the University of Nebraska as reported by the Doanes Agricultural Report (2)(3).

Table 6.1: Energy Consumption of Tillage Operations

	..................Fuel Use (diesel)..................			
	Corn	Soybeans	Corn Ensilage	Hay
Operation/Crop	................l/ha (gal/acre)................			
Chop stalks	–	6.74 (0.72)*	–	NA**
Plowing	14.23 (1.52)	14.23 (1.52)	14.23 (1.52)	NA
Disc and drag	9.07 (0.97)	9.07 (0.97)	9.07 (0.97)	NA
Plant	3.00 (0.32)	3.00 (0.32)	3.00 (0.32)	NA
Spray and fertilize	3.74 (0.40)	3.74 (0.40)	0.75 (0.08)	NA
Cultivate twice	4.86 (0.52)	4.86 (0.52)	4.86 (0.52)	NA
Harvest	6.08 (0.65)	6.08 (0.65)	78.56 (8.40)	NA
Total	41.0 (4.38)	47.7 (5.10)	110.5 (11.8)	18.7 (2.0)

*Corn stalks from previous year due to crop rotation.
**Not available.

Source: PB 259 318

The corn-drying operation was assumed to require 5.21 x 10^9 J/ha, based on the removal of 128.7 kg H_2O/m^3, production of 8.7 m^3/ha, and an energy requirement of 4.65 x 10^6 J/kg H_2O (3)(4). For the 20 ha of corn, this amounts to the consumption of 1.04 x 10^{11} J over a period of 3 months.

Figure 6.1 shows that even without drying, the energy needed for crop production is consumed during two separate 90-day periods. The energy storage required for this rate and pattern of energy use is currently uneconomical, with batteries or hydrogen. If the plowing operation was performed in the spring, as is many times necessary, about 0.27 x 10^{12} J of energy use would be shifted to the month of March. However, this would have little effect on the storage requirements of the system.

Figure 6.1: Cumulative Energy Consumption of 40.5-ha (100-Acre) Operation: 20.25 ha in Corn and 20.25 ha in Soybeans

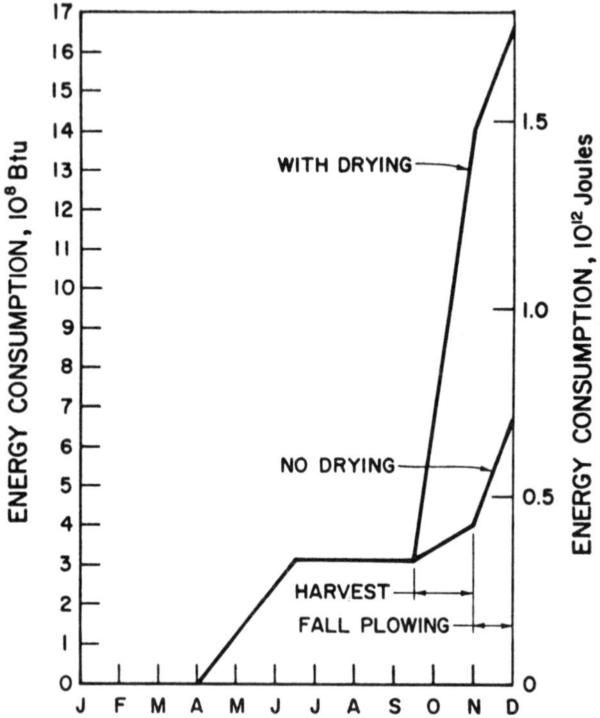

Source: PB 259 318

The energy profile of hay and/or haylage production was compared with that for corn and/or soybeans. In this case, the total energy expenditure is a function of the number of cuttings and acreage. Thus, the fuel-use rate over the growing season tends to approach a constant when the time span of the harvest cycle approximates the growing time between cuttings. The production of hay, though limited in use, would somewhat alleviate the large energy-storage requirement. However, a major problem still exists with providing energy derived from a WECS to the mobile equipment necessary for the tillage operations.

Tractor-Fuel Storage – While many specialized self-propelled farm machines are available for use in crop production, the farmer's work horse is the tractor. Most tractors in use today fall in the 70 to 110 horsepower range; for this reason, a 90-horsepower unit was selected for comparison in this study. A 90-horsepower tractor has a fuel capacity of 113 to 151 liters with a consumption rate of about 22.7 l/hr at full load; this yields a running time of 5 to 7 hours (1).

Ballast weights up to 748 kg are attached to the tractor for some operations and often left on year round. This ballast weight, combined with the weight of a full fuel load, amounts to about 862 kg. On this basis, a 907-kg energy-storage system (fuel tank) was chosen as an upper limit for use in this study.

The use of current-technology, lead-antimony batteries, rechargeable from a WECS, could only supply about 10 kWh of energy (based on a 1-hour discharge rate) with the 907-kg weight restriction. This battery pack would be about 1.53 m^3 and only provide 1 hour of full load operation. Much work is being carried out on advanced battery concepts which could improve the viability of battery-powered tractors; however, with this study being primarily limited to current technologies, no further investigations into battery-powered tractors were made.

Hydrogen as a fuel produced by the WECS was also considered for tractor use. This application of hydrogen would require some modifications to the tractor's engine. Although combustion engines operating on hydrogen cannot be considered off-the-shelf items, the technology for this operating mode can be considered current. Several researchers have demonstrated satisfactory operation of internal-combustion engines on hydrogen in recent years (5).

The problem of on-board fuel supply is yet to be solved. Liquefied hydrogen or improved (lighter) metal hydrides seem to be required before the hydrogen tractor will become viable, as illustrated by the following examples:

High pressure gas storage tanks*		
Total tankage volume	0.5 m^3	(17.63 ft^3)
Total tankage weight	909 kg	(2,004 lb)
Storage pressure	24.9 mPa	(3,600 psi)
Hydrogen stored	10.24 kg	(22.59 lb)
Full load run time (approximately)	1.4 hr	
FeTi hydrid storage		
Reaction	$FeTiH_{1.5} \rightleftharpoons FeTiH_{0.2} + 0.65H_2$	
Container weight (estimated)	640 kg/m^3	(40 lb/ft^3)
Bed density	3,299 kg/m^3	(206 lb/ft^3)
Total weight	907 kg	(2,000 lb)
Total volume	0.23 m^3	(8.16 ft^3)
Hydrogen stored	9.49 kg	(20.93 lb)
Full load run time (approximately)	1.3 hr	
Liquefied hydrogen		
Tanks	two 35" diameter spherical tanks	
Total weight (+ fuel)	331 kg	(730 lb)

(continued)

Total volume	378.5 l	(100 gal)
Hydrogen stored	26.76 kg	(59 lb)
Full load run time (approximately)	3 to 4 hr	

*Four 323.8 x 1,752.6 mm o.d. (12¾" x 5'9" o.d.)
one 323.8 x 1,143 mm o.d. (12¾" x 3'9" o.d.)

The fuel requirements of a tractor coupled with the energy-use pattern for cash-crop production make the application of a WECS to this type of farming operation impractical. This is due to the high harvest-time fuel consumption necessitating large energy-storage requirements and the difficulty in providing the tractor with an adequate fuel reservoir.

Animal and Animal-Product Production: Three confinement-type, mass-production, animal-husbandry operations were selected for study: dairy (milk production), poultry (egg production), and swine (meat production). Energy-conversion devices in confinement operations are predominately stationary, consisting of electric motors, lights, and space heaters. The relatively unchanging use pattern for these devices results in a predictable energy-demand rate. The use of climate-control equipment is the largest contributor to demand-rate fluctuations.

Energy requirements for different agricultural operations vary extensively. Current U.S. animal-husbandry operations range from labor-intensive pasture rearing to energy-intensive, highly automated, total confinement systems. Although the trend is definitely toward the use of confinement-type systems, most operations have stopped short of the ultimate in automated agricultural operations. Therefore, confinement operations have been selected in this study that can be considered representative of automated systems now in the industry, in terms of both size and degree of automation.

Although a considerable amount of data on various farming operations can be found in the literature, it is highly scattered and not well systematized for characterizing the various animal-production operations. The Economic Research Services (ERS) branch of the USDA is, however, currently in the process of assembling such information. Consequently, it has been elected to characterize the energy requirements of selected operations by engineering calculations based on reasonable assumptions and on the available seasonal-energy-use profiles for the various operations. These estimates have been checked for general consistency with data from one or all of the following: data obtained by personal contact, available published literature, and available ERS data.

Dairy Operation — A 100-head dairy farm, with cold free-stall barn and four-on-a-side herringbone milking parlor, was selected for this study. The assumed milk production was 6,350 kg/yr with twice-a-day milking and every-other-day milk pickup.

The energy consumption rate (on a monthly basis) has a maximum variation of 20%. This is a result of the winter parlor heat and water-warmer (keeps animal water tanks ice-free) energy consumption being offset by the parlor ventilation and increased milk-cooling requirements during the summer. The months of April and October, which require neither heating nor ventilation, are the lowest energy-consuming months. The highest energy-consuming months are January in the winter and July and August in the summer. Thermal loads, such as water heating and parlor heat, represent as much as 45% of the monthly energy requirement during the winter and decrease to about 30% of the monthly energy requirements during the summer. The energy budget for the 100-head dairy operation is shown in Table 6.2 (6).

Swine Operation – The energy requirement for a 2,000 head/yr operation has been characterized. Animal houses used in the operation consist of a farrowing-nursery house, a growing-finishing house, and a gestation house. The farrowing cycle assumed allows for a 1-week period per cycle during which no animals would be contained in the farrowing house. This practice permits periodic, thorough cleaning of the farrowing house to reduce the possibilities of disease. The total monthly energy consumption rate varied by as much as 250% over the year. This results from the additional heat required during cleaning operations in the winter, and the large ventilation requirements during the summer.

Table 6.3 shows the energy budget calculated for the swine operation. The months of February, April, October, and December require additional heat during the cleaning operation because of the farrowing cycle assumed. Different farrowing-cycle assumptions will change the amount of heat and the specific period of time this heat is needed. This has little effect, however, on the WECS component sizes, since the wind energy available in the Midwest is considerably higher in the winter months than in the summer.

Poultry Operation – A cage-type layer operation with a steady state population of 10,000 layers producing 2.4 million eggs/yr was selected for this portion of the study. It was assumed that feed was purchased or custom-ground and compounded off the premises. It was also assumed that eggs were collected manually several times a day, washed, and cooled to 12.8°C for delivery to an offsite assembler-packer for distribution. On this basis, components of the energy budget are ventilation (as much as 62%), lighting, automatic feeding, water pumping, and egg-cooling. It was found that for steady state operation at the population densities assumed, no auxiliary heat is required during the winter months. The energy budget for the 10,000 layer operation is shown in Table 6.4.

Energy-Load Patterns of the Home

The energy requirements for residential climate control and appliance usage vary widely, depending on the size, design, and quality of construction of the structure itself as well as the lifestyle of the occupants. It is difficult, therefore, to define a typical or average residential energy-use profile, particularly with so little published data. IGT does have, however, extensive in-house data on the detailed energy-use profiles in two lived-in homes built by East Ohio Gas Co. in Canton, Ohio, in 1965 (which are used as a base for establishing the home energy requirements). These data were collected over 2 years while the homes were lived in by two typical families of four in a suburban setting, and thus reflect the effects of door openings and vent fan usage on air infiltration (7).

Home Design and Construction: A brief description of the home is as follows.

Design

3 BR single-level, 1½ baths, attached 2-car garage
First floor heated area: 118 m^2 (1,270 ft^2)
Partial basement, heated: 91.7 m^2 (989 ft^2)
Total heated area: 210 m^2 (2,259 ft^2)

Table 6.2: Estimated Energy Demand for 100-Head Dairy Operation

Major Operation	January	February	March	April	May	June	July	August	September	October	November	December	Total
	kWh												
Silo unloading	400	361	400	264	272	264	243	243	235	310	300	310	3,602
Feed conveyor	136	123	136	132	136	132	136	136	132	136	132	136	1,603
Bank feeder	102	92	102	99	102	99	102	102	99	102	99	102	1,202
Vacuum pump	611	552	611	592	611	592	611	611	592	611	592	611	7,197
Milk cooling	1,006	908	1,006	1,398	1,444	1,398	1,253	1,253	1,214	1,061	1,028	1,061	14,030
Lighting maintenance	275	249	275	244	252	244	236	236	228	256	248	256	2,999
Pumping water	207	186	207	198	209	198	236	236	238	202	196	202	2,515
Ventilation (parlor)	0	0	0	0	100	200	437	437	126	0	0	0	1,300
Water warmer	384	320	96	0	0	0	0	0	0	0	47	352	1,199
Parlor heat (electric)	600	585	272	0	0	0	0	0	0	0	110	580	2,147
Water heating (electric)	1,332	1,203	1,332	1,288	1,332	1,288	1,332	1,332	1,288	1,332	1,288	1,332	15,679
Parlor heat (gas)	863	843	390	0	0	0	0	0	0	0	158	834	3,088
Water heat (gas)	1,936	1,748	1,936	1,873	1,936	1,873	1,936	1,936	1,873	1,936	1,873	1,936	22,792
Total all electric	5,053	4,579	4,437	4,215	4,453	4,415	4,586	4,586	4,152	4,010	4,040	4,942	53,468
Total gas/electric	5,920	5,382	5,159	4,800	5,057	5,000	5,190	5,190	4,737	4,614	4,673	5,800	61,522

Table 6.3: Estimated Energy Demand for 2,000 Swine per Year Operation

Major Operation	January	February	March	April	May	June	July	August	September	October	November	December	Total	Percent
	kWh													
Ventilation	236	219	313	487	1,071	1,404	1,507	1,504	1,382	710	341	251	9,425	52.0
Feed grinding	476	430	476	460	476	460	476	476	460	476	460	476	5,602	30.1
Feeding	108	100	108	106	108	106	108	108	106	108	106	108	1,280	7.1
Water	68	62	68	67	68	67	68	68	67	68	67	68	806	4.4
Heat	22	22	22	22	22	0	0	22	22	22	22	22	220	1.2
Auxiliary heat	–	1,590	0	541	0	0	0	0	0	241	0	1,521	3,893	0
Lights	82	75	77	71	56	44	41	43	71	77	79	82	798	4.4
Total	992	2,498	1,064	1,754	1,801	2,081	2,200	2,221	2,108	1,702	1,075	2,528	22,024	99.2

Source: PB 259 318

Table 6.4: Estimated Energy Demand for 10,000 Layer Cage House in Chicago

Major Operation	January	February	March	April	May	June	July	August	September	October	November	December	Total
								kWh					
Ventilation	270	256	377	916	1,516	1,962	2,178	2,178	1,825	1,196	420	296	13,390
Feeders	275	249	275	266	275	266	275	275	266	275	266	275	3,238
Water	33	30	33	32	33	32	33	33	32	33	32	33	389
Egg cooler	30	27	30	29	30	29	30	30	29	30	29	30	353
Lights	953	860	953	921	953	921	953	953	921	953	921	953	11,215
Total	1,561	1,422	1,668	2,164	2,807	3,210	3,469	3,469	3,073	2,487	1,668	1,587	28,585

Table 6.5: Estimated Energy Demand for Model Home in Chicago

System, Major Appliance	January	February	March	April	May	June	July	August	September	October	November	December	Total
								kWh					
All electric													
Furnace	4,960	4,284	3,651	1,969	864	196	0	0	331	1,336	3,088	4,865	25,544
Water heater	604	546	604	585	604	585	604	604	585	604	585	604	7,114
Dryer	109	99	109	106	109	106	109	109	106	109	106	109	1,286
Range and oven	182	164	182	176	182	176	182	182	176	182	176	182	2,142
Air conditioning	0	0	–	–	76	275	433	398	120	29	0	0	1,331
Other miscellaneous	1,019	916	1,019	986	1,019	986	1,019	1,019	986	1,019	986	1,019	11,993
Total	6,874	6,009	5,565	3,822	2,854	2,324	2,347	2,312	2,304	3,279	4,941	6,779	49,410
Gas/electric													
Furnace	7,631	6,591	5,617	3,029	1,329	302	0	0	509	2,055	4,751	7,023	38,837
Water heater	867	784	867	840	867	840	867	867	840	867	840	867	10,213
Dryer	109	99	109	106	109	106	109	109	106	109	106	109	1,286
Range and oven	364	329	364	352	364	352	364	364	352	364	352	364	4,285
Air conditioning	0	0	0	0	76	275	433	398	120	29	0	0	1,331
Other miscellaneous	1,019	916	1,019	986	1,019	986	1,019	1,019	986	1,019	986	1,019	11,993
Total	9,990	8,719	7,976	5,313	3,764	2,861	2,792	2,757	2,913	4,443	7,035	9,382	67,945

Source: PB 259 318

Table 6.6: Estimated Energy Demand for Model Home in Atlanta

System, Major Appliance	January	February	March	April	May	June	July	August	September	October	November	December	Total
								kWh					
All electric													
Furnace	2,875	2,297	1,817	591	111	0	0	0	33	562	1,674	2,736	12,696
Water heater	604	546	604	585	604	585	604	604	585	604	585	604	7,114
Dryer	109	99	109	106	109	106	109	109	106	109	106	109	1,286
Range and oven	182	164	182	176	182	176	182	182	176	182	176	182	2,142
Air conditioning	0	0	17	42	243	505	635	610	357	90	0	0	2,499
Other miscellaneous	1,019	916	1,019	986	1,019	986	1,019	1,019	986	1,019	986	1,019	11,993
Total	4,789	4,022	3,748	2,486	2,268	2,358	2,549	2,524	2,243	2,566	3,527	4,650	37,730
Gas/electric													
Furnace	4,423	3,534	2,795	909	171	0	0	0	51	865	2,575	4,209	19,532
Water heater	867	784	867	840	867	840	867	867	840	867	840	867	10,213
Dryer	109	99	109	106	109	106	109	109	106	109	106	109	1,286
Range and oven	364	329	364	352	364	352	364	364	352	364	352	364	4,285
Air conditioning	0	0	17	42	243	505	635	610	357	90	0	0	2,499
Other miscellaneous	1,019	916	1,019	986	1,019	986	1,019	1,019	986	1,019	986	1,019	11,993
Total	6,782	5,662	5,171	3,235	2,773	2,789	2,994	2,969	2,692	3,314	4,859	6,568	49,808

Table 6.7: Estimated Energy Demand for Model Home in San Diego

System, Major Appliance	January	February	March	April	May	June	July	August	September	October	November	December	Total
								kWh					
All electric													
Furnace	1,288	972	898	591	324	213	25	0	66	176	574	1,054	6,181
Water heater	604	546	604	585	604	585	604	604	585	604	585	604	7,114
Dryer	109	99	109	106	109	106	109	109	106	109	106	109	1,286
Range and oven	182	164	182	176	182	176	182	182	176	182	176	182	2,142
Air conditioning	12	0	0	20	34	87	192	259	211	100	18	0	933
Other miscellaneous	1,019	916	1,019	986	1,019	986	1,019	1,019	986	1,019	986	1,019	11,993
Total	3,214	2,697	2,812	2,464	2,272	2,153	2,131	2,173	2,130	2,190	2,445	2,968	29,649
Gas/electric													
Furnace	1,982	1,495	1,382	909	498	328	38	0	102	271	883	1,622	9,510
Water heater	867	784	867	840	867	840	867	867	840	867	840	867	10,213
Dryer	109	99	109	106	109	106	109	109	106	109	106	109	1,286
Range and oven	364	329	364	352	364	352	364	364	352	364	352	364	4,285
Air conditioning	12	0	0	20	34	87	192	259	211	100	18	0	933
Other miscellaneous	1,019	916	1,019	986	1,019	986	1,019	1,019	986	1,019	986	1,019	11,993
Total	4,353	3,623	3,741	3,213	2,891	2,699	2,589	2,618	2,597	2,730	3,185	3,981	38,220

Source: PB 259 318

Construction

Walls–Brick facing
Insulated sheathing
92 mm Hagan Super-R blown cellulose insulation
19 mm lath and plaster
Basement–103 mm concrete blocks
Ceiling–19 mm lath and plaster
152 mm Hagan Super-R cellulose insulation
Floor–No insulation over basement
76 mm foil between family room and crawl space
Windows–Weather-stripped with storm windows and doors.

Estimation of Seasonal Heating and Cooling Loads: From the above construction data and the IGT-Canton home measurements of heating and cooling energy needs, the characteristic heat-loss (or heat-gain) rate was derived, and the seasonal heating and cooling loads for various climates estimated. A heat-loss rate of 2.66×10^7 J/°C-day was used for this study. This is a relatively low heat loss, but is well within the range economically achievable by current energy-conservation methods (8)(9).

Appliance loads were taken directly from the IGT-Canton home measurements, except that the cooking load was increased and a freezer added, based on the assumption that rural families do more cooking and raise a portion of their own food. Energy budgets for homes in the three different regions are shown in Tables 6.5 to 6.7.

Peak Demand and Power Factor: Calculation of the expected peak demand for each load (shown in Table 6.8) is required to determine the proper size of the ac-output components in the WECS designs. The calculations are based on the work of L.B. Altman and L.F. Charity for sizing distribution transformers and secondary services in rural areas (10). The general validity of these estimates was then confirmed by the Transformer Section of Commonwealth Edison (11).

Power factor determination for each load was less straightforward. Using published data and assumptions on end-use equipment in each load, the power factor at peak demand was assumed to be 0.8 or greater (12)(13).

Table 6.8: Expected Peak AC Power Demand and Power Factor

Load	Demand, kW	Power Factor
House (Chicago)	13.3	0.85
House (Atlanta)	13.7	0.85
House (San Diego)	12.8	0.85
Dairy	23.7	0.80
House and dairy	32.7	0.80
Swine	13.8	0.80
House and swine	26.7	0.80
Poultry	16.6	0.80
House and poultry	25.6	0.80

Source: PB 259 318

Wind Regime

Data Used: For the Midwest region, wind data collected by Argonne National Laboratory from 1950 through 1964 were used. The Argonne data were collected at elevations of 5.7 and 45.7 m (19 and 150 ft). A 30.5-m (100-ft) level was selected as a practical height for the wind turbine in this system, given the need to clear surrounding buildings and trees which may be near the desired installation site. To accommodate this decision, the Argonne data at 5.7 m was shifted to reflect the wind speeds at 30.5 m using the power law:

$$V/V_o = (H/H_o)^N$$

where V = wind speed at height H; V_o = wind speed at observed position; H = selected height; H_o = height of observation; and N = 0.25 for wind speeds of 0 to 12.8 km/hr (0 to 8 mph), 0.11 for wind speeds above 12.8 km/hr (8 mph). [The values used for N are typical of the values calculated by Argonne National Laboratory in correlating the 5.7-m data with the 45.7-m data (14).]

Wind data (for examining regional effects) in the form of surface-weather observation summaries for the Southeast and Southwest were obtained from the National Climatic Center, Asheville, N.C. The Southeast data were collected at Dobbins Air Force Base, Ga., from 1946 to 1972. During the measurement period, the wind equipment's height and location were changed four times. IGT used the weighted average (21.3 m) of these heights (9.7, 16.2, 22.8, and 31.3 m) as the height of observation for the data.

Wind data collected at the Fleet Weather Office in San Diego, Cal., were used for the Southwest location. These data were collected from 1945 to 1972 at three different heights (9.1, 26.5, and 29 m). Again, the weighted average (19.4 m) was used as the height of observation.

As with the Argonne data, the Southeast and Southwest data were also shifted to reflect wind speeds at the 30.5-m elevation. Monthly mean wind-speed profiles for the three areas are shown graphically in Figure 6.2.

Correlations with Other Locations: To insure that using Argonne wind data would not overly limit this study to a specific site, additional wind data was surveyed. These data, from eight other Midwestern cities, were compared with the Argonne data. Examination of the mean wind-speed profiles (Figures 6.3 to 6.6) showed that all locations recorded their lowest mean wind speeds in August and highest mean wind speeds in March or April. This correlated well with the Argonne data and was considered sufficient justification for concluding that the general shape of the mean wind-speed profile for the Argonne data was not site-specific.

Calculations: Energy Available — Monthly percentage frequency distribution information was used to calculate the total energy available for each month. The total hours for each wind-speed range were divided equally among each increment in the range in order to approximate data in 0.45 m/s (1 mph) increments. After modification by the power law, the energy available each month was calculated as

$$\sum_{i=1}^{\ell} KAV_i^3H_i$$

Figure 6.2: Mean Wind-Speed Profiles

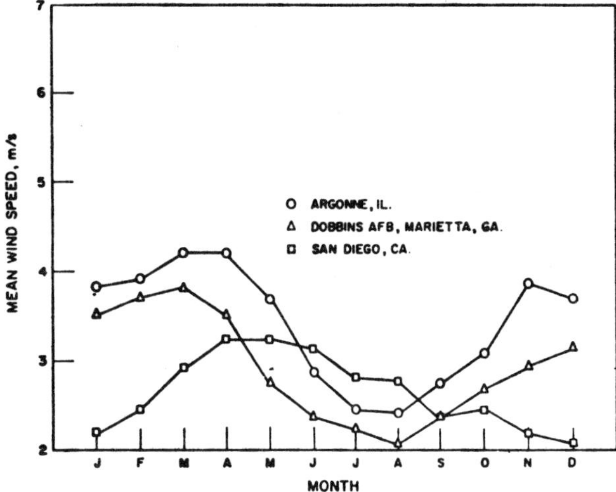

Approximate Instrument Heights
Argonne: 5.7 m (19 ft)
Marietta: 21.3 m (70 ft)
San Diego: 19.4 m (63.71 ft)

Figure 6.3: Mean Wind-Speed Profiles

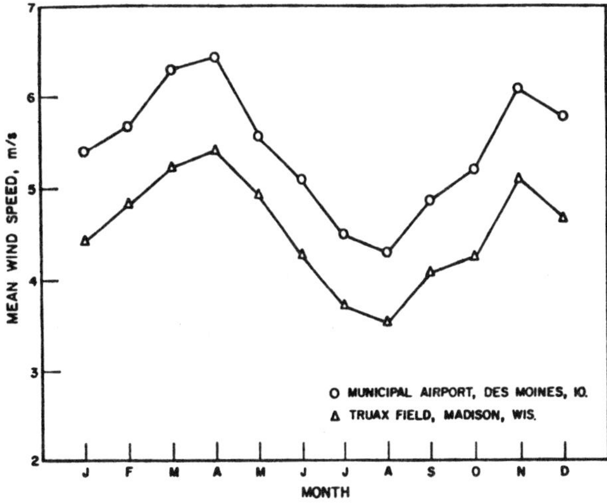

Approximate Instrument Heights
Des Moines: 21.0 m (69 ft)
Madison: 11.6 m (38 ft)

Source: PB 259 318

Figure 6.4: Mean Wind-Speed Profiles

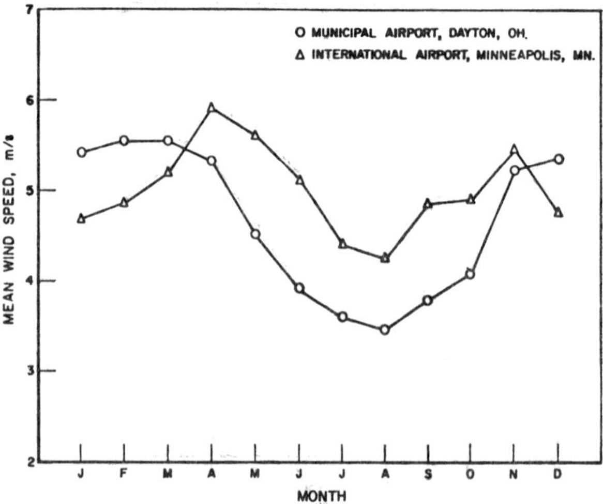

Approximate Instrument Height
Dayton: 16.7 m (55 ft)
Minneapolis: 22.8 m (75 ft)

Figure 6.5: Mean Wind-Speed Profiles

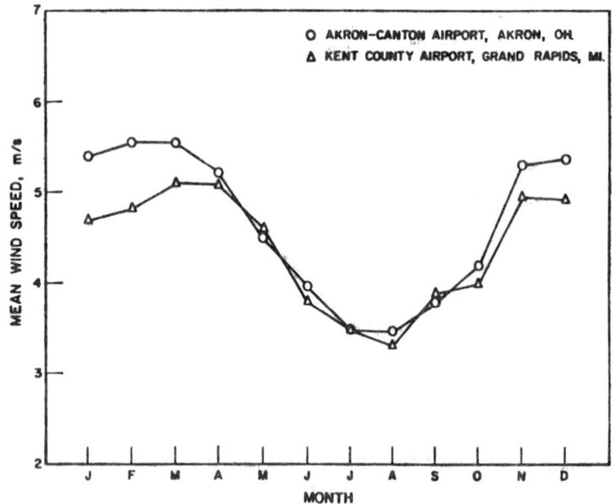

Approximate Instrument Height
Akron: 6.1 m (20 ft)
Grand Rapids: 19.5 m (64 ft)

Source: PB 259 318

Figure 6.6: Mean Wind-Speed Profiles

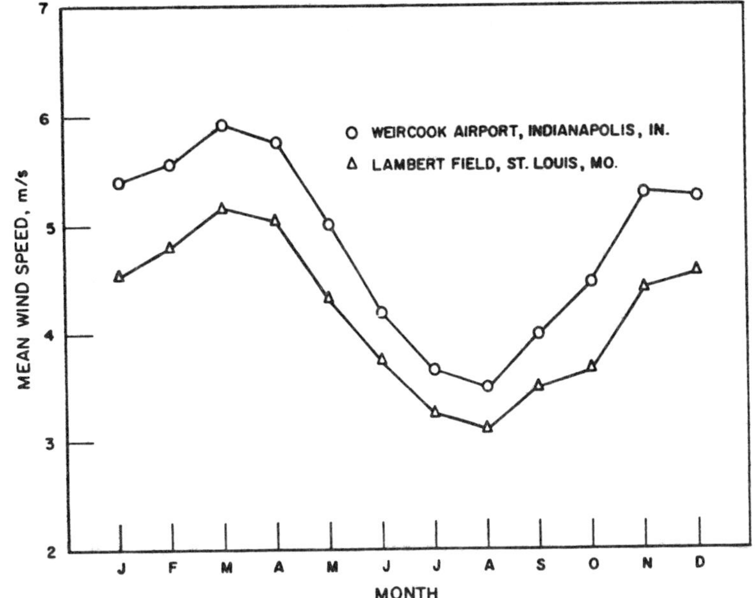

Approximate Instrument Height
Indianapolis: 18 m (59 ft)
St. Louis: 25 m (82 ft)

Source: PB 259 318

In the above expression,

> K is a constant affecting average wind-turbine efficiency, air density, and unit-conversion factors;
>
> A is wind-turbine sweep area;
>
> V_i is wind speed;
>
> H_i is amount of time where wind speed = V_i; and
>
> ℓ is number of increments for V_i and H_i.

Annual Mean Wind-Speed Adjustments – Separate wind data for testing the sensitivity of the WECS designs in higher annual mean wind speeds were not specifically obtained. Instead the Argonne data were elevated while maintaining the same mean wind-speed profile. To accomplish this, the number of hours (H_i) for each wind speed (V_i) was increased. This method of increasing the wind speed by essentially increasing the time the wind blows at each speed produces the results shown in Figure 6.7.

Limitations – The methodology for utilizing the Argonne wind data results in two possible problems. The first is inherently due to the one month sampling period. Wind variations within the month are relatively unknown, and, in extreme cases of low wind periods of long duration, the required storage system capacity could exceed the 3 day requirement imposed on these WECS designs.

The second area of concern is the annual pattern of monthly, mean-wind-speed, peak-to-peak variations. This method of elevating the annual mean wind speed maintained the variations exhibited in the lower wind regime in the higher mean wind speeds. If changes in annual mean wind speed additionally resulted in significant changes in monthly variations, such as increasing the difference between the highest and lowest monthly mean wind speed, optimum WECS component sizes and costs could be affected.

Figure 6.7: Mean Wind-Speed Profiles for Base Argonne Data and Elevated Cases

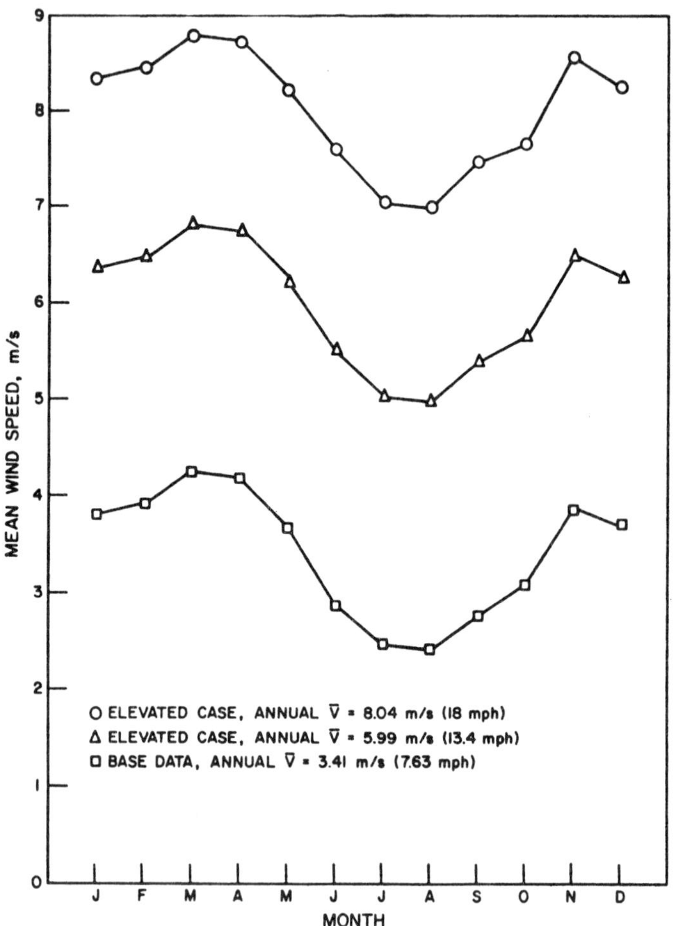

Source: PB 259 318

System Sizing and Design

Sizing: Component sizes within any of the WECS designs are functions of the load requirements, the wind energy available, and the efficiency of one or more of the systems' components.

Output – The output of the WECS must be able to meet the demands of the load. On this basis, the output components within the system that produce ac-power (inverter, engine generator, motor generator) and hot water must be sized.

Though the ac-output component's characteristics of durability, dependability, efficiency, cost, and input energy form must be considered from a system interface perspective, the foremost requirement of this component is to provide constant voltage and constant frequency ac-power at the rate and power factor demanded by the load. If this requirement is not met, judicious load management will be required to prevent overloads resulting in system and/or component failures. Required ac-output component sizes ranged from 13 to 33 kW with power factors of 0.8 and greater at full load.

Components for supplying domestic hot water and space heat are assumed to operate on dc, hydrogen, or a combination of the two. This enables the ac-power-output components to be sized somewhat smaller by reducing the expected peak ac-power demand. These heating units would differ very little from those in present use including the sizes required.

Storage – The capacity of the systems' energy-storage units are first determined by the number of hours or days of energy reserve desired and the efficiency of the output components. A 3-day reserve was arbitrarily chosen for the designs, resulting in minimum storage capacities ranging from 273 to 1278 kWh. Increases in the minimum storage capacity can result when long-term storage is advantageous (i.e., storing energy in January to be used in July) and, in the case of batteries, when this storage capacity limits the output from the wind turbine due to the batteries' maximum acceptable charging rates. Table 6.9 shows battery-bank capacity and the maximum respective charging rates (wind turbine size) based on five times a finish charge rate of 7 A/100 Ah of capacity (15).

Table 6.9: Minimum Battery Cell Size and Energy Stored for Selected Wind-Turbine Output

Energy Stored in 120 V Battery Bank (kWh)	Rating of Battery Cells (Ah)	Maximum Wind-Turbine Output (kW)
28.5	238	10
85.7	714	30
171.4	1,429	60
285.7	2,381	100

Source: PB 259 318

Input – The wind turbine and the components necessary to interface it with the storage system (i.e., electrolyzer) are the last components of the system to be sized. Interface components are simply sized to accept the full output of the required wind turbine. Failing to do this could result in insufficient energy available to the output components.

The exact size of the wind turbine (primarily sweep area) is determined by the amount of energy required and the amount available from the wind, specifically the energy density of the wind. For a 69% efficient wind turbine, output ratings of 12 kW up to several hundred kW were identified for various wind and load combinations.

Design: Preliminary Designs – Several preliminary WECS designs incorporating various combinations of storage-system and output devices were considered. Many combinations were rejected because significant development work is required on one or more of the system components. Other candidate systems, though technically feasible, utilized components that were cost prohibitive.

Four preliminary designs (Figures 6.8 to 6.11) were chosen for a more detailed equipment selection and optimization. In addition, the use of short-term storage via hot water coupled with some load management is examined for its effect on the optimum system. This subsystem is shown in Figure 6.12.

Figure 6.8 shows the basic configuration for a WECS design operating with battery storage. The required wind turbines produce dc electricity for battery charging and for direct use (not shown) by the ac-output component via their associated rectifiers (power-conditioning units). The ac-output component for this design may be either a solid-state inverter or a dc motor/ac generator set.

Figure 6.8: WECS Design with Battery Storage

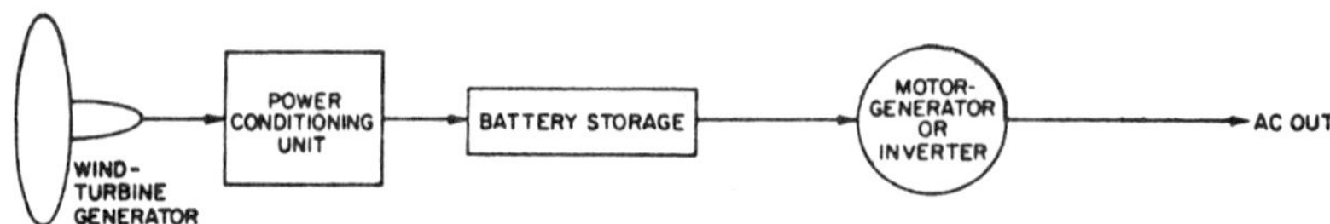

Figure 6.9: WECS Design with Hydrogen Storage

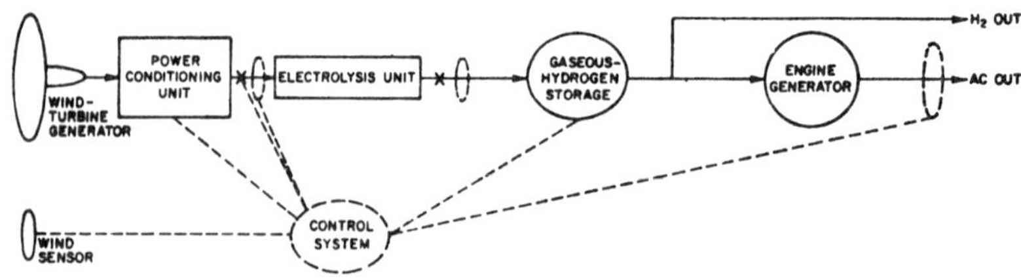

Source: PB 259 318

Figure 6.10: WECS Design with Hydrogen-Fueled Battery Charger

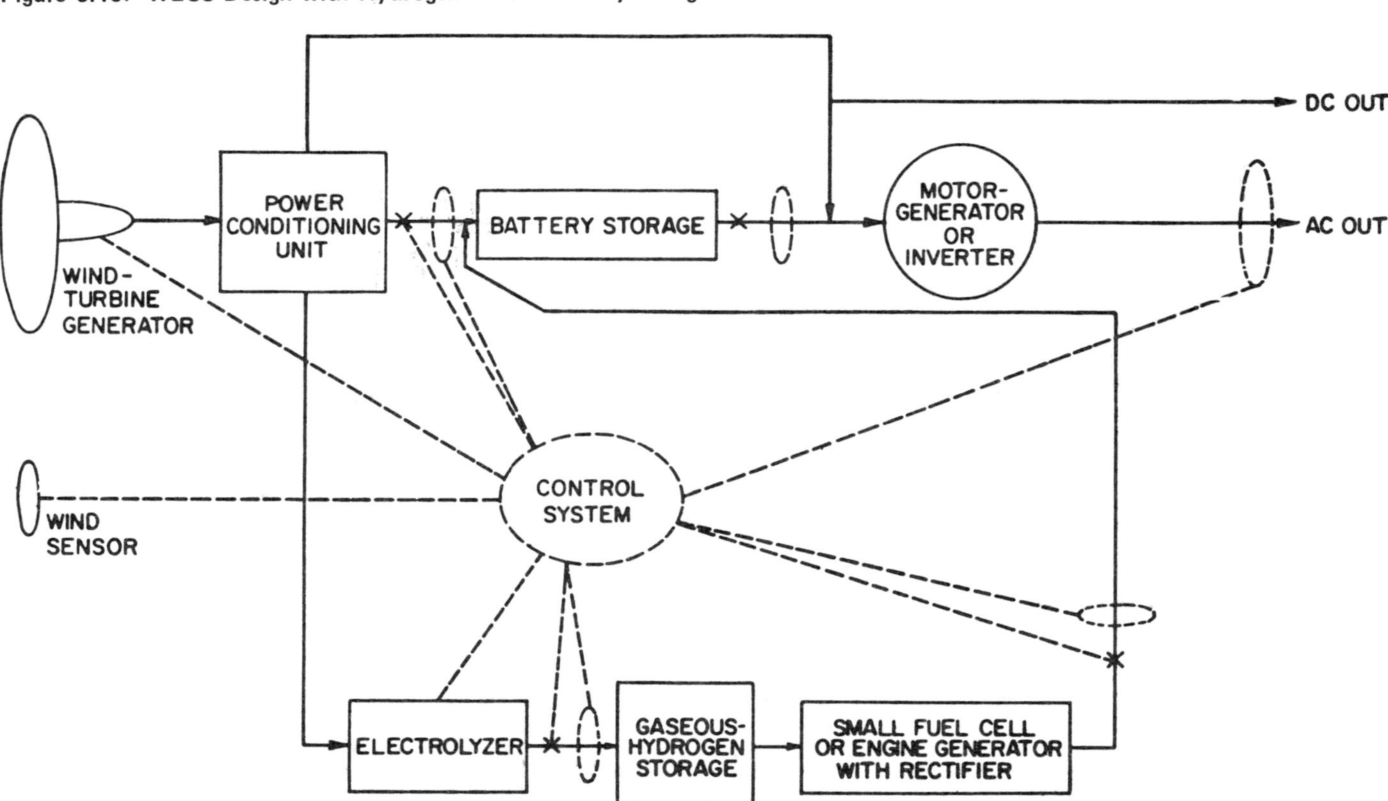

Source: PB 259 318

Figure 6.11: WECS Design with Hydrogen and Battery Storage

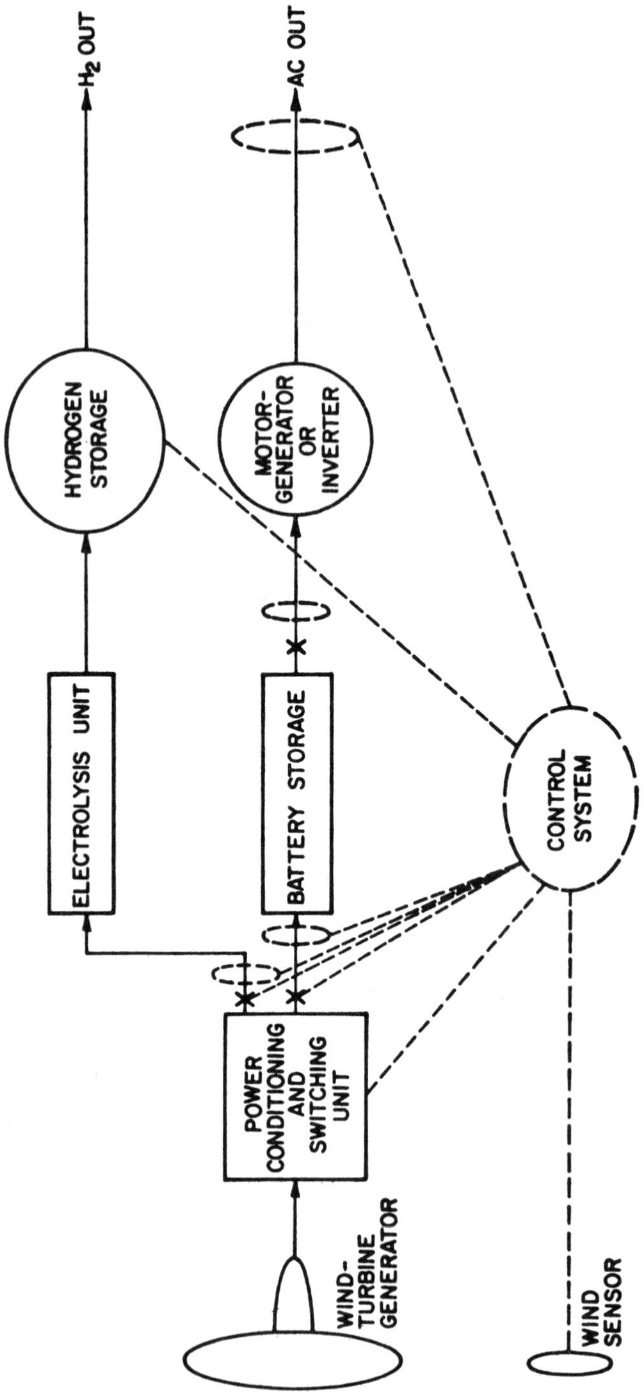

Source: PB 259 318

Figure 6.12: Short-Term Hot Water Subsystem

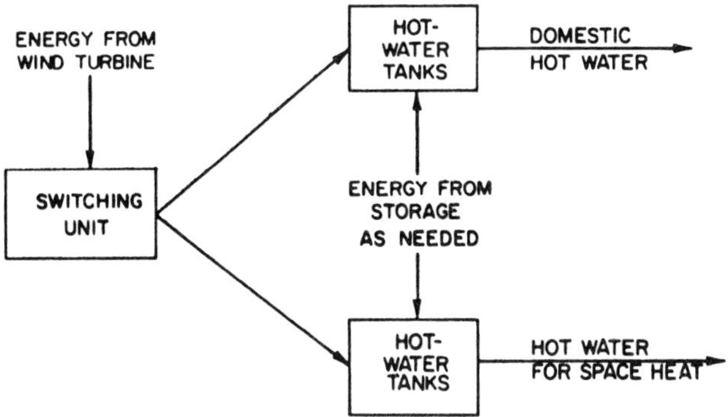

Source: PB 259 318

Figure 6.9 shows the basic configuration for a WECS design employing hydrogen storage. Again, the wind-turbine output is dc electricity. In this system, however, the direct use of this energy by the ac-output component can be achieved only if a fuel-cell/inverter combination is used. Engine generators, which are the other alternatives for converting hydrogen to ac power, cannot utilize a dc input.

Figure 6.10 shows basically the same design as Figure 6.8, with the addition of a hydrogen-fueled, battery-charger storage system and necessary controls. The addition of this component provides a means for equalizing the charging of the batteries. Such a step helps prolong the operating life of the batteries and has the added potential of being more economic when large amounts of stored energy are desirable.

In Figure 6.11, a combination battery/hydrogen system is shown. This system was originally selected for application with the grain farm where hydrogen gas is a preferred fuel. It was subsequently eliminated from consideration after reviewing the energy budget and mobile fuel requirement for the grain operation. In addition, the use of this design for the other farm operations resulted in increased costs due to the separation of subsystems. Excess capacity in one subsystem could not easily supplement the other during peak demand periods.

Control Systems – With WECS designs (Figures 6.10 and 6.11, for example) having multiple loads relative to what the wind turbine sees, the addition of a logic-centered control system (possibly a mini-computer) appears desirable. This control system, operating on a predetermined algorithm, would control the distribution of energy from the wind turbine to the various loads. Using as inputs the energy available (via a wind sensor), the instantaneous load demand, and the state of each storage element, the system would direct switching functions to ensure (1) system reliability, and (2) maximum energy efficiency.

If a mini- or micro-computer is used for the heart of the system, the algorithm governing the switching decisions could be continuously updated by the control system itself. To perform this internal housekeeping, the computer would merely monitor its own performance and then use hindsight, so to speak, to update its basic algorithm. Systems of this form could also monitor the energy systems' components to give advance warning of failures and to implement emergency procedures when necessary.

Complete control packages of this nature are not readily available, but the necessary hardware and software expertise can be developed. Several hardware/software firms were contacted and the equipment requirements and costs for this type of control system were discussed. For a one-of-a-kind system, a cost of $3,000 to $10,000 was estimated. If production run quantities of similar systems were to be required, a total control-system cost of less than $1,000 may be realized (16).

Equipment Selection: State-of-the-Art Components — The major system components required for the candidate WECS designs are described here.

Input Components: Wind Turbines — Four wind turbines in the 8- to 15-kW power range were identified for use in farm and rural-home application. There are other commercial units available, but their output power levels are too low for this application.

Only one of the four machines is a vertical-axis type, which has a rotor assembly consisting of a vertically mounted central shaft with three curved airfoil blades. The other three wind turbines are of the three-bladed, horizontal-axis type, operating downwind. All of the units employ some form of rotor over-speed control to prevent physical damage to the equipment in high winds. The most sophisticated unit has an electromechanical pitch-control mechanism actuated by a wind-pressure sensor circuit for feathering the blades. Centrifugally actuated air brakers will also deploy in the event of electrical failure. One wind turbine has servomotors to turn the rotor blades out of high winds. Two other units have centrifugally operated spoiler flaps or weights in the rotor-hub assembly to feather the blades. Most units have a manual brake for the rotor so that inspection and maintenance can be performed safely.

The vertical-axis machine is installed on a free-standing, steel, truss-type tower with three guy wires attached to the top, bearing rotor-shaft housing. For the other machines, three support-tower types are available: (1) free-standing, steel, truss frame; (2) guyed steel truss, and (3) prestressed, centrifugally cast concrete.

All the wind turbines surveyed produce some form of ac electricity; none of them, however, produce constant voltage/constant frequency, 60-Hz electricity. Instead, the output, depending upon the particular unit, was either constant voltage/variable frequency or variable voltage/variable frequency.

This ac output is then sent to the rectifier which, in one case, is an integral part of the alternator. With the other units, appropriate rectifiers can be supplied with the wind turbines at nominal cost.

For systems that produce variable voltage/variable frequency electricity, the alternator must be tuned (modification of the electrical characteristics of the alternator) to match the characteristics of the load (batteries or electrolyzers).

For systems that can produce constant voltage/variable frequency, this internal tuning will not be required because the output voltage can be controlled externally.

Costs for these wind turbines ranged from \$204 to \$484/m^2 of sweep area. On a 1976 \$/kW basis, the wind-turbine costs ranged from about \$550 to \$1,470. It should be noted, however, that the \$204/m^2 unit is not the \$550/kW model; this cost results from the manufacturer's selection of rated wind speed and alternator size.

To illustrate this, assume the existence of two rotor assemblies, each having the same efficiency, size, cost, and wind-loading capability. One rotor is fitted with a 15-kW alternator and can obtain this output in a 13.4 m/s wind. The second rotor is fitted with a 10-kW alternator and produces full output in an 8.9 m/s wind. If it is assumed that both alternators are approximately the same price, the 10-kW wind turbine will cost about one-third more when expressed on a \$/kW basis.

Furthermore, it cannot be assumed that the 15-kW wind turbine is preferable over the 10-kW unit based on their \$/kW cost. If both units were operated in a low wind regime, the 10-kW unit would provide more energy out because it would operate nearer its rated output more often. (Most alternators are more efficient when operated near their rated output.) If, on the other hand, both wind turbines were operated in a much higher wind regime, the 15-kW unit would provide the most energy, since the 10-kW machine cannot increase its output to utilize the higher wind speeds.

Because of the ambiguity of the \$/kW wind-turbine rating, until specific site information and additional wind-turbine data are known (at which time energy cost can be determined), it has been elected to characterize the wind-turbine cost as a composite of the rotor-assembly and alternator costs. Thus, the total wind-turbine cost equals the sweep area multiplied by the rotor cost in \$/unit sweep area plus the alternator size multiplied by an alternator cost of \$50/kW (value assumed in study).

Using a rotor cost of \$269/m^2 and rated wind velocities of 8.9 and 15.6 m/s, the resulting total cost of the wind turbine, expressed in \$/unit sweep area, ranged from \$277.7 to \$316.5/m^2. This value was selected to correlate with the costs of the units surveyed and is assumed to include the tower costs and rectifiers. In addition, since individual, identified, wind-turbine characteristics were not used, the wind turbines in this study were assumed to have an average efficiency of 69% (approximately equivalent to a C_p of 0.41).

Electrolyzers – To effect an acceptable system interface in WECS designs, more technical modifications will be required for the use of electrolyzers than for other components discussed. Although the electrolysis process utilizes dc electricity, the input-voltage-control, system-monitoring, and control-valve operations are all dependent on the 60-Hz ac for which existing units are designed. System-monitoring and control-valve operations, if isolated from the voltage-control portion of the electrolyzer's power-conditioning/control unit, could still be ac-operated via the ac output of the WECS. This would eliminate a costly redesign of that portion of the electrolyzer package.

Although control of the input dc voltage for the electrolysis process may be somewhat more difficult, it appears to be feasible by any of the following three methods. First, if the WECS design employing the electrolyzer has no additional loads for direct use of the wind-turbine-produced energy (such as water heating), the necessary voltage control can be achieved at the wind turbine. Wind turbines having alternators with permanent magnetic fields can be internally modified so that the voltage (which varies with the wind-speed/power-out relationship of the wind turbine) matches the power requirement of the electrolyzer at the voltages produced.

The voltage outputted from turbines using alternators that have wire-wound fields can be externally controlled via their field current. Control of this field current and of the resulting output voltage with a logic-centered control system, as previously discussed for wind turbines encountering multiple loads, appears to be the most efficient method. Using this control method would allow for additional control of the wind-turbine output to avoid overpowering the electrolysis unit during warm-up periods and to provide the capability for minor adjustments to match specific operating conditions.

Second, in systems where multiple loads are encountered by the wind turbine, the use of wind turbines with permanent magnetic alternators appears undesirable, because supplying energy to additional loads simultaneously would unbalance the wind-turbine/electrolyzer match. Thus, either a predesigned mismatch or discrete switching techniques would have to be employed neither of which are desirable from an energy-efficiency standpoint. The use of an externally controlled wind turbine with rectifiers and logic-centered control system is best suited here.

Third, a wind turbine that supplies constant-voltage/variable-frequency output can be used. The standard power-conditioning/rectification unit could then be modified to accept a variable-frequency input and to function in its normal manner with one exception: signals for controlling the unit's dc output voltage must be supplied from an external source, because the energy demand of the electrolysis module may exceed that which is available from the wind turbine.

For hydrogen-based WECS designs presented within this report, the use of wind turbines with externally controllable alternators combined with a logic-centered control system was assumed. Cost for this control system was set at $4,000. No modification costs were charged to the electrolyzer, assuming that capital-cost reduction from the elimination of the rectification portion of the power-conditioning/control unit would offset these costs. Costs for electrolyzers considered in this study ranged from $2,612 to $416/kW. When the electrolyzer required was over 100 kW, a somewhat reduced cost of $300 to $350/kW was used. This $66 to $116/kW cost reduction, when compared to the 80-kW unit, reflects the estimate of the cost reductions obtainable with increased capacity.

Storage Components: Hydrogen Storage – Compressed-gas storage was chosen over metal-hydride and liquid storage based upon the results of a previous IGT study (7). Hydrogen can be stored by chemically combining it with various metals and alloys as a hydride, but a commercial storage unit of this type is still in the developmental stages and may not be available for some time. In addition, the current costs for hydride-hydrogen storage systems are at best speculative, because they are not based upon a reliable plant design. The capital investment and operating costs encountered with hydrogen liquefaction and storage facilities cannot be justified for the small-quantity, short-turnover storage conditions found in farm and rural applications.

The most economical hydrogen-storage pressure for this system was determined by examining the storage costs at several different pressures, with hydrogen being supplied from electrolyzers outputting at 104 kPa and 791 kPa. In all cases, as shown in Figure 6.13 and Table 6.10, the 791-kPa storage pressure is optimum. It should be noted, however, that, if land area becomes a critical problem, the 9,030-kPa storage pressure would be preferred. This result was previously obtained in an IGT study (7) in which huge quantities of hydrogen were assumed to be produced by electric utility off-peak power.

Hydrogen-storage costs at 791 kPa were determined for storage volumes ranging from 14 to 1,139 scm. The results plotted in Figure 6.14 show storage costs ranging from $85 to $35/scm for the high-pressure electrolyzer, and from $1,440 to $53/scm for the low-pressure electrolyzer.

Figure 6.13: Hydrogen-Storage Cost Versus Pressure

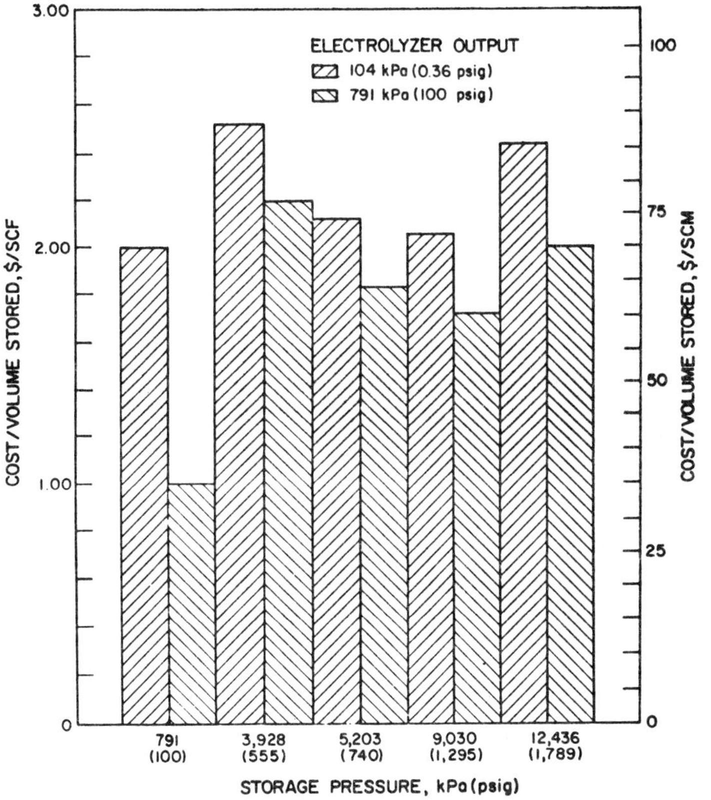

Source: PB 259 318

Figure 6.14: Hydrogen-Storage Cost Versus Volume

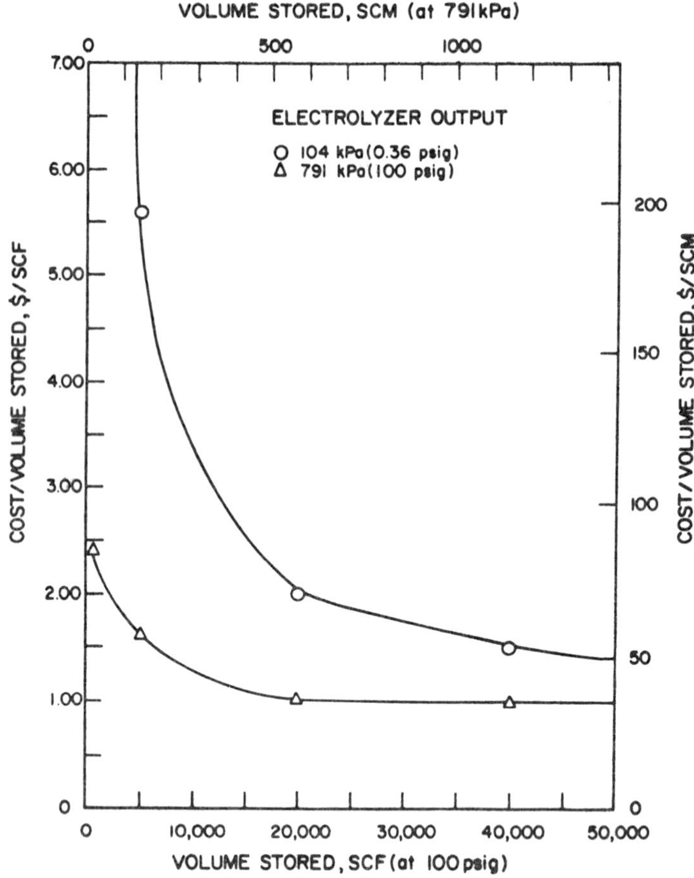

Source: PB 259 318

Table 6.10: Hydrogen-Storage Systems

	Storage Pressures, kPa (psig)				
	791 (100)	3,928 (555)	5,203 (740)	9,030 (1,295)	12,436 (1,789)
Vessel type	steel blimp	seamless-steel pressure bottles, 24" o.d.			
Number required	1	5	4	2	2
Volume stored, scm (scf)	566.4 (20,000)	561.4 (19,825)	610.6 (21,560)	490.1 (17,306)	475.8 (16,800)
Tank cost, $	20,000* **	25,000***	20,800***	11,000***	11,000***

(continued)

Table 6.10: (continued)

	Storage Pressures, kPa (psig)				
	791 (100)	3,928 (555)	5,203 (740)	9,030 (1,295)	12,436 (1,789)
Compressor type					
A†	none	diaphragm/single-stage			
B††	diaphragm/single-stage or reciprocating/two-stage	diaphragm/two-stage			
Compressor cost, $					
A†	0	18,900†††	18,900†††	18,900†††	22,600†††
B††	20,000††† §	25,000†††	25,000†††	25,000†††	30,000†††
Total cost, $					
A†	20,000	43,900	39,700	29,900	33,600
B††	40,000	50,000	45,800	36,000	41,000
Cost/volume stored, $/scm ($/scf)					
A†	35.31 (1.00)	78.20 (2.21)	65.02 (1.84)	61.01 (1.72)	70.62 (2.00)
B††	70.62 (2.00)	89.06 (2.52)	75.01 (2.12)	73.45 (2.07)	86.17 (2.44)

*Imperial Steel Tank Co.
**Chicago Boiler Co.
***U.S. Steel Corp.
†Electrolyzer output at 791 kPa (100 psig).
††Electrolyzer output at 104 kPa (0.36 psig).
†††Autoclave Engineers, Inc.
§Ingersoll-Rand Co.

Source: PB 259 318

The cost differential between the storage systems mated to the two electrolyzers results from the compressor costs associated with the low-pressure electrolyzer. Compressors rated for the required volumes and pressures were found to be specialized such that their cost, over the range of gas volumes, remained the same; the compressor cost then becomes a significant portion of the storage-system cost as the required volume decreases. The implementation of a high-pressure electrolyzer was assumed in calculating the hydrogen-storage costs used in this study. Also, a pressure regulator may be required at the hydrogen-storage output to effect proper output-component performance.

Battery Storage – The lead-antimony type of lead-acid storage cell has been recommended for this application (17)(18). Cells of this type have a good record of reliability and long life in many current industrial applications.

For inverters (an output component) that require a nominal input voltage of 120 v, 60 lead-acid series-connected cells will be required. For inverters or motor generators that require higher or lower input voltages, appropriate cell combinations can be effected.

Given the storage requirement that has been imposed upon the system (3 days), it is anticipated that the battery bank will operate close to the full float mode. For cells operating in this mode, a low full-charge specific gravity (1.210) is recommended to ensure long life (19).

Assuming proper selection and installation, the lead-antimony-type battery life expectancy can be 15 years or more in stationary applications. (A 15-year battery life was assumed in this study.) This life will depend upon the depth of discharges, frequency of deep cycling and site temperatures. The best battery life expectancy is obtained in the 15° to 25°C temperature range (19).

Many of the control techniques and/or modifications discussed earlier, which are required to interface the wind turbine and electrolyzer, will also be required with battery systems. The variable input characteristics of the battery bank (which is a function of its state of charge) are analogous to the electrolyzers' input characteristic. Costs for the lead-antimony storage cells ranged from $0.16 to $0.18 A-hr based on their 8-hr rating. On an energy-stored basis, the cost is $80 to $90/kWh.

Output Components: Internal-Combustion Engine-Generator Sets – Engine-generator sets in the 12 to 55-kW output power range were selected as one method of converting hydrogen to 60-Hz ac power. The four- or six-cylinder engines normally employed are designed to operate on gasoline, natural gas, LP-gas, or diesel fuel. Conversion to hydrogen operation, although not a standard practice, has been demonstrated and should not cause significant technical problems (20).

To ensure continuous power availability, it is recommended that dual-engine generator sets be employed (21)(22)(23). The use of dual engine generator sets will increase the WECS reliability though not increase total costs. Hydrogen-based WECS designs will use a total of 4 engine-generator sets over a 20-year life, using dual sets and alternating their duty will ensure availability and improve maintainability.

Routine maintenance for the engine-generator units consists primarily of engine oil and filter changes, tune-ups, and compression checks. It is estimated that both engines will require overhaul at 40-month intervals if the engine generators are operated as recommended. The engines will have a 10-year life span, on the basis that each engine may be rebuilt twice during its life (22).

Inverters – The use of solid-state inverters is one method of converting the dc electricity from the battery bank (or other sources) to 60-Hz, ac power. There are many types of inverters; some employ transistors, and others use thyristors to generate the required wave form. The power dissipation characteristics of available transistors limit their use to 15-kW and lower units. Thyristors are being used in the higher power units. With all units, filtering is required to eliminate unwanted harmonics.

The output wave form and the resulting harmonic content is an important point to be considered when selecting an inverter. Some electronic equipment and at least one line of currently manufactured household refrigerators will perform unsatisfactorily when operated on ac power having a poorly approximated sine wave form (24).

It is also important that inverters selected for this application employ some type of magnetic voltage stabilizer circuit and be capable of handling inductive loads. Without these features, load surges resulting from motor starting could cause inverter failure. Inverters having the proper requirements are available. Several units in the 8 to 60 kW size range were identified during the course of this study.

Motor Generators – Given the short time period over which large solid-state inverters have had to prove their reliability, the prospects of using a dc-powered motor-generator set was investigated. It was found that while these sets are not off-the-shelf per se, the components needed are. Motor-generator sets designed to operate with a range of dc input voltage, as would be provided from a discharging battery bank, can be assembled (25). Estimated specifications for these units are as follows (25): frequency stability ±1.5%; voltage regulation ±3%; overload capability of 15% for 1 hour, 10% for 2 hours; and efficiency of 86% at full load. A cost estimate for a 30-kW system was obtained. After development costs of about $15,000, the estimated cost of this unit was quoted at $25,000 (25).

None of the system costs presented later in this report are based on the use of a motor-generator set. It is acceptable, however, to assume their use, because motor-generator and inverter costs are similar.

Fuel Cells – The discussion of hydrogen-air fuel cells in this section is primarily provided for completeness. Although fuel cells are available in the 12 to 750 W range and can be assembled in larger units, their cost is currently prohibitive for this application. These cells are of the acid-matrix-type and have a predicted stack life of only 5 years. Continued developments in this area will undoubtedly improve this picture.

System Housing – A utility building is required to house the control system and all WECS equipment except the wind turbines and the hydrogen-storage tanks. Depending upon the system employed, the area under roof varies from approximately 18.6 to 55.8 m^2. Modular buildings with corrugated sheet steel or rigid-rib aluminum sheathing built on a concrete pad are the most economical structures suitable for these purposes. These buildings may have 2 x 4 pine-wood or bolted-angle iron frames. Costs for these structures (26)(27)(28), including the concrete pad, range from $129 to $258/m^2. A representative figure of $162/m^2 was used in the system-cost computations.

Future Technology – In order to make some projections on future WECS costs and designs, information on anticipated component improvements was obtained. This information primarily consists of cost and efficiency projections.

Input Components: Wind Turbines – No projections on wind-turbine costs were obtained from existing manufacturers. Instead, a future cost of $129/m^2 of sweep area was assumed, based on assumptions of cost reductions resulting from production-run quantities and reported data (29)(30).

Electrolyzers – Cost reductions for electrolyzer systems in the 30 to 100 kW range are not anticipated to be as substantial as those predicted for major electrolytic installations (31). Most cost improvements for these large installations result from technological advances which permit larger, more efficient, module construction.

Electrolyzers for WECS applications should benefit in efficiency from the technological gains made on large units. However, this is expected to have only minor effects on cost. The major cost reduction for the WECS electrolyzer system will be the elimination of the currently expensive power-conditioning unit.

On this basis, future electrolyzer costs of \$250/kW have been assumed, representing a reduction of approximately \$150/kW.

Storage Components: Hydrogen Storage – Reductions in the cost of storing hydrogen as a gas in steel tanks are not anticipated. Steel handling and fabricating technology is mature, with major cost-reducing breakthroughs appearing unlikely. However, gaseous-hydrogen storage at low pressure in concrete vessels may offer some cost reductions if this methodology were to be developed (32). The magnitude of these cost reductions cannot be quantified at this time.

Hydride-storage systems appear to offer little advantage in stationary applications where space is not a problem. If significant reductions in the present hydride costs are not achieved, gaseous-hydrogen storage will remain not only simpler, but cheaper.

With iron titanium costs of about \$6.61/kg and a usable hydrogen-storage density of 1.6% by weight, only about \$0.22/kg could be allotted for tankage and heat exchangers before exceeding the cost of storing gaseous hydrogen at 791 kPa. While some lower cost projections have been made in the past, no cost projections were assumed in this study.

Battery Storage – No projections on improved battery storage systems were specifically considered in this study. However, the system cost presented for a battery system having 1-day storage can be interpreted as 3-day storage for systems using advanced batteries. The sodium/sulfur batteries with projected costs of one-third to one-half of the cost of lead-acid cells (\$90/kWh) and having similar usable life expectancies (33)(34) would fit this rough approximation.

Output Components: Inverters and Fuel Cells – The Power Systems Division of United Technologies Corp. in South Windsor, Conn. provided the following cost and efficiency information for advanced inverters and fuel cells (acid-matrix-type) in sizes applicable to this use:

Component	Cost, \$/Kw	Efficiency, %
Inverter	100	86-90 (10-100% of rated load)
Fuel Cell	300-500	45-47 (25-100% of rated load)

System Operation

In order to firm up the energy-flow, equipment, and control-system interface relationship, a description of system operation is presented here. The WECS design shown in Figure 6.15 will be used for illustration. This design consists of both hydrogen and short-term hot water storage. For the discussion of this system, the following equipment and/or control techniques are assumed:

(1) The wind turbine has a variable-frequency, externally controllable voltage alternator.

(2) The power-conditioning unit is nothing more than a rectifier circuit. Control of the electrolysis system itself is assumed to be powered by the WECS ac output and is not shown in Figure 6.15.

(3) The engine generator runs continuously and is rpm-governed for 60-Hz output. Switching from one engine-generator set

to another (for the system having two engine-generator sets) is not discussed because the operation is straightforward, using existing transfer switches.

(4) Hot-water heating units consist of multiple elements that can be switched on and off separately via the power flow control unit.

(5) The WECS has been in operation long enough such that all equipment is at full operational status, and the hydrogen storage tank is partially full.

Figure 6.15: WECS Design with Hydrogen Storage and Short-Term Hot-Water Subsystem

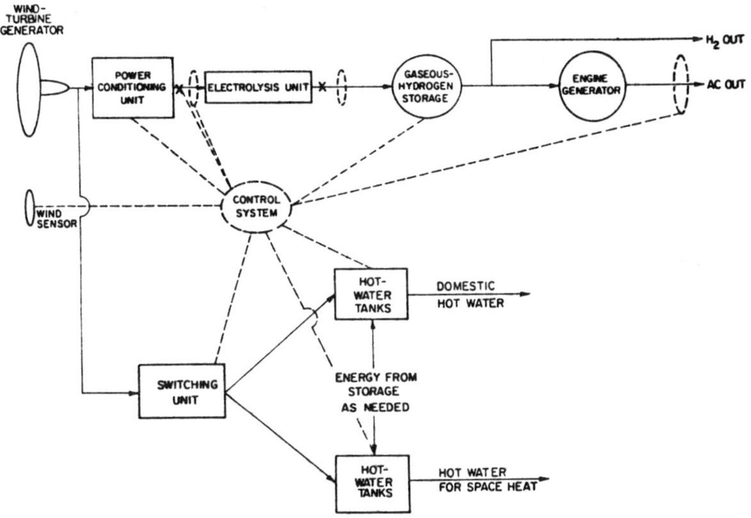

Source: PB 259 318

With the system in a steady-state condition, where the available wind power is sufficient for electricity generation and power is being supplied to both the electrolyzer and to the hot-water subsystems, the control system determining this power distribution performs in conjunction with the other system components as follows: using a sampling period of once a minute, for example, the available wind energy is calculated from the wind-speed data provided by the wind sensor. The correct distribution of this energy is determined using: (1) a preset algorithm that anticipates demand on the various subsystems, and (2) data on the state of each subsystem, such as the temperature of the hot water and the volume of hydrogen stored.

Once this determination is made, the output voltage of the wind turbine is set to effect the desired energy flow into the electrolyzer. After this has been accomplished and checked by monitoring the input current and voltage to the electrolyzer, the remaining energy is fed to the hot-water subsystem by switching on the appropriate number of heating elements. The system updates itself, with each sampling interval establishing new energy-flow quantities as needed. After a long calm period, it may be necessary to purge the electrolysis system with nitrogen (or another inert gas) prior to continued operation. In this event, the ac-powered electrolysis system control would signal the main WECS control to initiate automatic purge procedures before switching power to the electrolyzer.

Outputs from the system, in this case 60-Hz ac and hot water for space heating (domestic hot water has been omitted in this example), are provided upon demand. Control of the engine generator is accomplished by its rpm governor, and requires no assistance from the main WECS control system.

The hot-water, space-heating system will require some control in addition to the demand-actuated circulating pump. This control governs the amount of additional energy inputted from the storage system. During a space-heating demand period, for example, additional energy must be supplied when the water temperature falls to a preselected lower limit. While this can be accomplished with simple thermostat control, the decision to continue this energy input once the demand is over (circulating pump off) cannot. It is here that the use of a logic-centered control system appears advantageous.

By estimating the next demand period and the wind energy available, the decision whether to discontinue the additional energy input can be made. The fail-safe for this methodology is the placement and size of the additional heating element(s). If the element is large enough or is placed near the water outlet, rapid recovery could be achieved without significant discomfort within the home.

In summary, wind energy is converted to electricity by the wind turbine and is distributed to the electrolyzer and the hot-water subsystems. The hydrogen produced by the electrolyzer is stored until needed by the engine generator or hot-water subsystem. The outputs (60-Hz ac and hot water) are available on demand.

System Optimization

In previous sections, the various components of the WECS (input, output, and storage equipment) have been described. The way in which all these components can be linked together into a unified system has also been presented. With this sytem structure in mind, an important issue is the optimal sizing of system components, in order to meet given energy demands (both ac and thermal) from available wind energy at least cost.

Linear Programming Model: The design of a minimum cost system has been accomplished through the use of linear programming. The total cost of the WECS is simply the sum of the costs of individual components. This total cost is the objective function to be minimized in the linear programming (LP) formulation.

Since the energy system is a dynamic system, that is, energy is being generated and used continuously in time and at time-varying rates, a major modeling assumption is necessary with respect to the length of a discrete time interval, which will serve as the basis of the analysis. For the purposes of optimization of system capacity, the basic time period selected is 1 month. While 1-month time periods may be quite long for many purposes, it is felt that for the purpose of an initial investigation of component sizing, monthly periods are appropriate.

Furthermore, data availability on shorter time periods for many important variables is quite limited, so that optimization based on shorter periods with estimated data may be an empty exercise.

Constraints on the objective function to be minimized are provided by system structure, available wind energy, and required energy to meet demand. There are seven basic types of constraints on the system optimization:

(1) For given wind and wind-turbine characteristics, wind-turbine sweep area determines the maximum amount of energy available for direct use or for storage in each time period.
(2) Exogenously specified requirements for ac power in each time period must be met, either from energy generated directly in that month, or from energy generated in previous months and stored.
(3) Exogenously specified requirements for thermal energy in each time period must be met, as in the case of ac power either from directly generated energy or from energy previously generated and stored.
(4) Flow of energy in the system must be conserved. For example, total energy put into a storage component in a time period must be equal to energy taken from storage in that period, plus the change in the amount stored in energy, e.g., energy into storage = energy out of storage + Δenergy in storage.
(5) The amount of energy put into storage in any given period is constrained by the capacity of the equipment required to convert the output of wind-turbine to a form suitable for storage in batteries, as hydrogen, or as hot water.
(6) The amount of energy in storage at any time must be nonnegative. That is, energy cannot be drawn from storage which has not yet been stored.
(7) The capacity of the storage components must be sufficient to accommodate the energy stored at any given time.

Each of the constraints listed is a linear function, so the problem of optimization of system capacity is one of minimizing a linear function subject to a set of linear constraints. However, the component costs associated with many of the linear constraints are not linear. Thus, in order to implement the LP model, component sizes were estimated and the approximate linear cost for each component within a size range were entered.

Table 6.11 shows the equipment costs and efficiencies used for the optimization of current technology systems. Table 6.12 shows the equipment costs and efficiencies that were used to make projections of the cost of future WECS designs.

Optimum WECS Designs: Midwest Loads — Base Cases: For the seven Midwest loads (home, dairy, swine, and poultry operations; and home-plus-dairy, home-plus-swine, and home-plus-poultry operations), several initial computer optimizations were performed using the basic battery and hydrogen WECS designs (Figures 6.8 and 6.9). These analyses assumed that all energy used passed through storage. Each load/WECS-design combination was examined in three wind regimes, with annual means of 3.4, 6, and 8 m/s (7.63, 13.4, and 18 mph), and with wind turbines having different rated velocities (V_R).

Table 6.11: Equipment Cost and Efficiency, Current Technology

Component	Cost	Efficiency, %
Wind turbine	\$269/m² (\$25/ft²) + \$50/kW (for alternator)	wind to electricity, 69 ($C_p \approx 0.41$)
Rectifier	assumed included in wind turbine cost	85
Batteries	\$0.16-\$0.18/Ah (\$80-\$90/kWh)	80
Electrolysis system	\$2,600-\$300/kW	68*
Hydrogen storage	\$84.75-\$35.31/scm (\$2.40-\$1.00/scf)	99
Inverter, dc to ac	~\$1,200/kW	85
Engine generator	\$400-\$190/kW	34
Hydrogen to thermal energy	—	70
Electricity to thermal energy	—	99

*Does not include rectifier efficiency.

Source: PB 259 318

Table 6.12: Equipment Cost and Efficiency, Future Technology

Component	Cost	Efficiency, %
Wind turbine	\$129/m² (\$12/ft²) + \$50/kW (for alternator)	69 ($C_p \approx 0.41$)
Rectifier	assumed included in wind turbine cost	90
Batteries	\$0.16-\$0.18/Ah (\$80-\$90/kWh)	80
Electrolysis system	\$250-\$300/kW	86*
Hydrogen storage	\$84.75-\$35.31/scm (\$2.40-\$1.00/scf)	99
Inverter, dc to ac	\$100/kW	90
Fuel cell	\$400/kW	46
Hydrogen to thermal energy	—	70
Electricity to thermal energy	—	99

*Does not include rectifier efficiency.

Source: PB 259 318

Only with the hydrogen-based system in the 3.4-m/s wind regime was it economic to store energy from 1 month to the next. In all other wind regimes and with the battery-based WECS design, the wind-turbine size selected was large enough to eliminate costly month-to-month energy storage. When the WECS design is sized in this manner, the system constraints are reduced to a single constraint: For given wind and wind-turbine characteristics, wind-turbine sweep area determines the energy available, and this energy must equal or exceed the combined exogenously specified ac and thermal energy requirements with appropriate considerations for equipment efficiencies during each time period.

Each month, the constraint is applied, and the required sweep area (A) is determined. In only 1 of the 12 months will this constraint be binding on the system (the month requiring the largest sweep area). Though this binding or critical month varied with load type and wind regime, the months of December, January, or August most often determined the wind-turbine size, December and January for loads such as the home (and combinations with farms), which has a high winter thermal requirement; and August, a low-wind energy month, for loads such as the poultry operation, which has a large summer ventilation requirement.

The results of these initial optimizations, expressed in terms of total installed system cost, are shown in Table 6.13. As can be seen, the hydrogen-based system for all combinations of loads and wind regimes is at least 1.4 times as costly as the battery-based system.

The use of an assumed average wind-turbine efficiency (69%) prevents finding an optimum wind-turbine rated velocity (V_R) for each wind regime. The inefficiencies resulting from operating a wind turbine with high V_R are equal to or greater than any wind speed available. It is for this reason that Table 6.13 shows decreasing system cost with increasing V_R in all but four entries.

In the 8-m/s (18-mph) wind regime, increased costs for the hydrogen system are shown for V_R = 15.6, 17.9, and 20 m/s. At these points, the increase in energy collected does not offset the higher costs resulting from the increased wind-turbine kW rating and the larger electrolyzer required. The fourth entry is again in the 8-m/s wind regime, but this time for the battery system with V_R = 20 m/s. Again, this is due to increased wind-turbine kW rating. In terms of the equations that produce these results,

$$\text{kW rating of wind turbine} = kAV_R^3$$

$$\text{energy available, kWh} = \sum_{i=1}^{\ell} kAV_1^3 h_i$$

such that

$$\text{if } V_1 < V_C \quad V_1 = 0$$

$$\text{if } V_1 > V_R \quad V_1 = V_R$$

$$\text{if } V_1 > V_O \quad V_1 = 0$$

$$\text{cost of wind turbine} = X_1A + X_2kAV_R^3 = A(X_1 + X_2kV_R^3)$$

where

- k = a constant consisting of average wind-turbine efficiency, air density, and unit conversion factors
- A = wind-turbine sweep area (determined by LP model)
- V_1 = wind speed
- V_C = minimum wind speed for wind-turbine output
- V_R = rated wind velocity of wind turbine
- V_O = maximum wind speed for wind-turbine output
- h_i = amount of time wind speed was V_1
- X_1 = cost/unit sweep area of wind turbine
- X_2 = cost/kW of alternator
- ℓ = number of increments for V_1 and h_i

Table 6.13: Installed System Cost for Loads in Midwest

Load	$\bar{V}$ (Annual Mean Wind Speed), m/sec (mph): 3.4 (7.63)		6 (13.4)			8 (18)					
	V_R (Rated Wind Speed of Wind Turbine), m/sec (mph): 8.9 (20)		11.2 (25)	13.4 (30)	15.6 (35)	8.9 (20)	11.2 (25)	13.4 (30)	15.6 (35)	17.9 (40)	20.1 (45)
	Installed Cost, $ x 10^3/kW										
Home											
Hydrogen system*	279.0	163.0	–	–	143.0	136.8	–	–	109.7	–	–
Battery system**	163.0	120.1	–	100.6	96.7	106.8	–	90.1	86.9	–	–
Swine operation											
Hydrogen system*	283.3	140.7	–	–	104.3	115.1	–	–	80.9	–	–
Battery system**	129.2	70.8	–	57.5	55.0	61.4	–	51.2	50.4	–	–
Dairy operation											
Hydrogen system*	482.9	236.3	–	–	163.4	147.3	–	–	123.9	–	–
Battery system**	236.6	119.8	–	93.5	88.4	100.5	–	81.7	78.3	–	–
Poultry operation											
Hydrogen system*	432.2	200.8	–	–	155.4	149.4	–	–	110.4	–	–
Battery system**	180.5	88.5	–	67.8	63.9	73.1	–	58.8	55.8	–	–
Home and dairy											
Hydrogen system*	713.7	343.7	289.8	267.0	259.0	236.9	210.5	164.3	195.6	202.2	231.5
Battery system**	383.9	214.8	192.3	183.3	172.5	191.4	171.6	161.1	155.5	153.0	153.4
Home and swine											
Hydrogen system*	537.7	159.6	–	–	184.4	179.5	–	–	137.8	–	–
Battery system**	249.2	131.0	–	–	112.2	111.4	–	109.9	98.4	–	–
Home and poultry											
Hydrogen system*	630.8	314.9	–	–	222.1	210.2	–	–	157.7	–	–
Battery system**	298.0	146.9	–	–	106.6	121.7	–	97.7	93.4	–	–

*Figure 6.8.
**Figure 6.9.

Source: PB 259 318

If increasing V_R does not result in an offsetting decrease in A to obtain the same amount of energy, wind-turbine costs will increase, which is what occurred when V_R equalled 20 m/s in the 8-m/s wind regime. In Table 6.14, the installed capital cost has been divided by the annual energy requirement (in kWh) of its respective load. Each load/WECS combination can now be compared on an installed cost/annual kWh required basis ($/kWh-yr). These amounts may be used to estimate the cost of WECS installations for loads of different magnitude, provided the energy-use pattern and the consumption per annum is about the same.

Reduced Costs: In Table 6.14, reduced installed costs on the $/kW-yr basis can be seen from the combination of the home with the farming operations. The energy-use patterns of the midwestern home and the farming operations tend to be complementary in that their combination produces an energy-use pattern with increased similarity to that of the energy available.

Figure 6.16 shows the energy profiles for the home, poultry operation, and their combination and the energy available from a 9-m wind turbine in a 6-m/s wind regime. Additionally, the home/farm combination yields improved equipment utilization. Thus, components within the WECS supplying the home/farm combination are generally smaller than the sum of corresponding components in the individual WECS. Table 6.15 shows the sizes of the major components for battery-based WECS serving the midwestern loads in a 6-m/s wind regime.

Further WECS cost reductions are obtained through (1) the use of short-term (1 to 2 days) hot-water storage to satisfy thermal loads, and (2) the direct use of energy from the rectifier (power-conditioning unit) by the inverter (battery system only) for a portion of the ac load. WECS designs incorporating one or both of these features are shown in Figures 6.15 and 6.17. Cost reductions are obtainable with this system due to an increase in system efficiency. The base-case battery and hydrogen WECS designs have the following efficiencies:

	Efficiency, %*
Battery system	
Wind to ac	40
Wind to thermal	46
Hydrogen system	
Wind to ac	13
Wind to thermal	28

*100% = 59.3% of energy contained in the wind

If 30% of the ac load and 75% of the thermal load in each month can be supplied directly from the wind turbine or the rectifier (power-conditioning unit) bypassing the storage system, the following system efficiencies can be obtained:

	Efficiency, %
Battery system	
Wind to ac	43
Wind to thermal	63
Hydrogen system	
Wind to ac	13*
Wind to thermal	58

*No improvement can be obtained because the ac output component (engine generator) cannot use an electrical input

Table 6.14: Installed Cost on a $/kWh-yr Required Basis

Load	$\bar{V}$ (Annual Mean Wind Speed), m/sec (mph)										
	3.4 (7.63)		6 (13.4)				8 (18)				
	V_R (Rated Wind Speed of Wind Turbine), m/sec (mph)										
	8.9 (20)		11.2 (25)	13.4 (30)	15.6 (35)	8.9 (20)	11.2 (25)	13.4 (30)	15.6 (35)	17.9 (40)	20.1 (45)
	Installed Cost, $/kWh-yr Required										
Home											
Hydrogen system*	5.68	3.32	–	–	2.91	2.78	–	–	2.23	–	–
Battery system**	3.32	2.44	–	2.05	1.97	2.17	–	1.83	1.78	–	–
Swine operation											
Hydrogen system*	12.86	6.39	–	–	4.96	5.22	–	–	3.67	–	–
Battery system**	5.87	3.21	–	2.61	2.50	2.79	–	2.32	2.28	–	–
Dairy operation											
Hydrogen system*	9.03	4.42	–	–	3.06	2.75	–	–	2.32	–	–
Battery system**	4.42	2.21	–	1.75	1.65	1.88	–	1.53	1.46	–	–
Poultry operation											
Hydrogen system*	14.80	7.02	–	–	5.44	5.23	–	–	3.86	–	–
Battery system**	6.31	3.10	–	2.37	2.23	2.55	–	2.04	1.95	–	–
Home and dairy											
Hydrogen system*	6.96	3.35	2.83	2.60	2.52	2.31	2.05	1.60	1.91	1.97	2.26
Battery system**	3.74	2.09	1.87	1.78	1.68	1.87	1.67	1.57	1.52	1.49	1.50
Home and swine											
Hydrogen system*	7.56	3.65	–	–	2.59	2.52	–	–	1.94	–	–
Battery system**	3.50	1.84	–	–	1.58	1.56	–	1.54	1.38	–	–
Home and poultry											
Hydrogen system*	8.12	4.05	–	–	2.86	2.70	–	–	2.03	–	–
Battery system**	3.83	1.89	–	–	1.37	1.56	–	1.26	1.20	–	–

*Figure 6.8.
**Figure 6.9.

Source: PB 259 318

Table 6.15: Major WECS Components Sizes for Midwestern Loads in 6-m/s Wind Regime

Load	. . . . .Wind Turbine* . . . . . Diameter, m	Rating, kW	Storage (3 day), kWh	Inverter Size, kW
Home	9.8	77	716	12
Dairy operation	9.4	71	552	24
Swine operation	6.7	36	273	15
Poultry operation	8.4	56	408	16
Home and dairy	13.2	139	1,268	30
Home and swine	11.6	109	950	24
Home and poultry	10.7	86	900	24

*V_R (rated wind speed of wind turbine) = 15.6 m/sec (35 mph).

Source: PB 259 318

Figure 6.16: Energy Profiles of Selected Loads and Wind Turbine

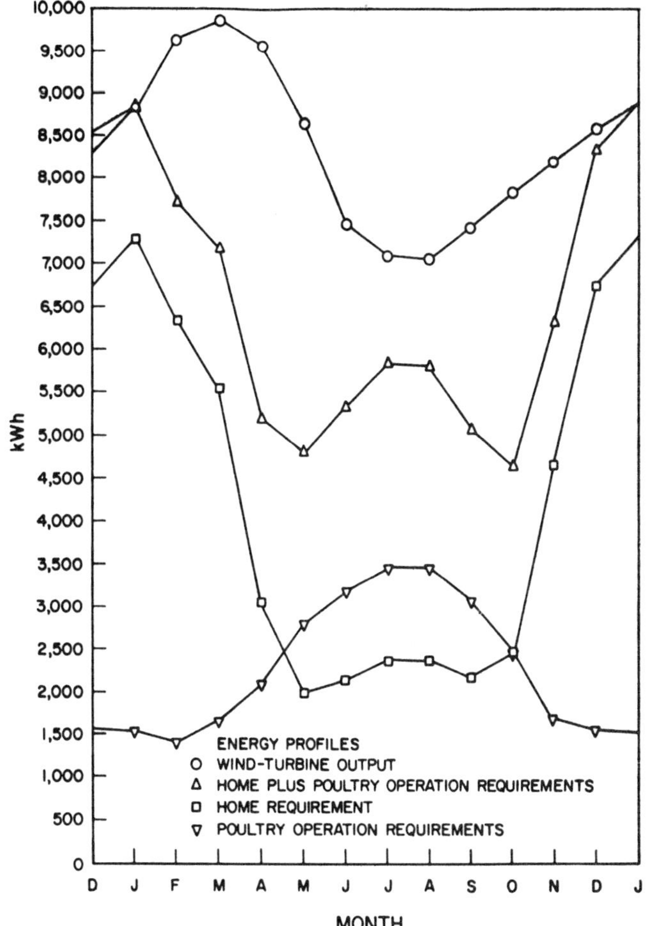

Source: PB 259 318

Figure 6.17: WECS Design with Battery Storage and Short-Term Hot-Water Subsystem

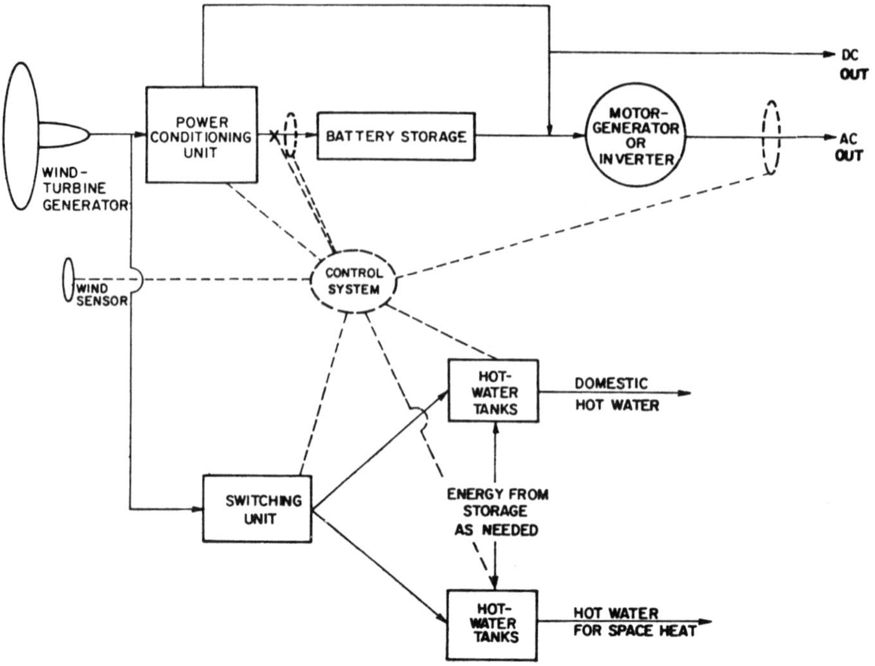

WIND-TURBINE GENERATOR
POWER CONDITIONING UNIT
BATTERY STORAGE
MOTOR-GENERATOR OR INVERTER
DC OUT
AC OUT
WIND SENSOR
CONTROL SYSTEM
HOT-WATER TANKS
DOMESTIC HOT WATER
SWITCHING UNIT
ENERGY FROM STORAGE AS NEEDED
HOT-WATER TANKS
HOT WATER FOR SPACE HEAT

Source: PB 259 318

The 30 and 75% figures are assumed to be the maximum reasonable values for direct energy use (bypassing storage) in a WECS design. In months where considerable excess energy is available (Figure 6.16), these values could be exceeded. It would seem over-optimistic, however, to assume higher values for the month where the energy available is equal to the energy used (the critical month).

The Argonne wind data shows that the wind energy available is about evenly split between the day- and nighttime hours (6:00 AM to 6:00 PM and 6:00 PM to 6:00 AM) (14). With the majority of the ac power demand of the load occurring during the daytime, 30% of this demand could equal as much as 60% of the energy available (assuming no thermal load) during this daytime period. Satisfying the thermal load with 75% direct energy appears to be less constraining because the short-term storage system allows for delayed energy use. On this basis, without data on the real-time correlation between the wind energy available and that demanded by the load, higher percentages for direct energy use were not justifiable.

Here it has been assumed that the Argonne 12-hour daytime period matches a hypothetical 12-hour period when the load's total ac demand occurs. While in actuality this is not the case, it allows for the presentation of a worst case argument. The improved efficiency resulting from this direct energy use yields a smaller wind-turbine requirement and, for the hydrogen-based system, a smaller electrolyzer. Other WECS components (storage and output) are unaffected. Table 6.16 shows wind-turbine size and installed-system cost information for WECS designs with and without direct energy use in a 6-m/s wind regime.

Table 6.16: WECS Designs With and Without Direct Energy Use Compared*

	Battery-Based System			
	Without Direct Energy Use (Figure 6.8)		With Direct Energy Use (Figure 6.17)	
Load	Wind Turbine Diameter, m	Installed Cost, $	Wind Turbine Diameter, m	Installed Cost, $
Home	9.8	96,700	8.6	90,900
Swine operation	6.7	55,000	6.3	53,500
Poultry operation	8.4	63,900	7.8	61,500
Dairy operation	9.4	88,400	8.6	84,600
Home and poultry	10.7	106,600	7.2	88,900
Home and swine	11.6	112,200	7.2	88,800
Home and dairy	13.2	172,500	11.9	163,700
	Hydrogen-Based System			
Home and dairy	19.6	259,000	17.9	224,400

*$\bar{V}$ (annual mean wind speed) = 6 m/sec (13.4 mph); V_R (rated wind speed of wind turbine) = 15.6 m/sec (35 mph).

Source: PB 259 318

One additional method for reducing system costs was identified, the use of hydrogen storage coupled with a battery-based WECS (Figure 6.13). When the desired amount of energy reserve becomes large enough, the system cost can be reduced by adding a hydrogen-fueled battery-charging subsystem that reduces the required battery-bank size and the total system cost. This subsystem consists primarily of a small electrolyzer, a hydrogen-storage tank, and a small motor generator set.

The electrolyzer, which operates the majority of the time on the excess energy available, produces hydrogen for storage. During times of extended low-wind periods, the batteries can be kept charged by means of the engine generator powered by hydrogen. Since a WECS of this design operates mostly in its battery-based mode, the higher efficiency of a battery-based WECS with direct energy use is maintained.

Of the loads surveyed in this study, only the home/dairy combination required enough energy storage (1,268 kWh = 3-day supply) to use a hydrogen-fueled battery-charging subsystem. In this application, the total installed system cost was reduced from $172,500 to $167,000 (about 3%).

Homes in Three Regions – In order to get some insight into the possible effect of different geographic regions on WECS designs and costs, WECS for homes in the Southeast (Atlanta), the Southwest (San Diego), and the Midwest (Chicago) were evaluated. For this comparison, the WECS design shown in Figure 6.17 was used. Other WECS designs were rejected from a cost standpoint. While the wind data obtained for all three regions showed an annual mean wind speed of less than 4 m/s, the lowest wind regime only provides one with a less economic system. Since computations for the higher wind regimes were performed first, it was elected not to use the lower wind regime. Table 6.17 contains the results of these evaluations.

Table 6.17: WECS Costs (Installed Cost, \$/kWh-yr Required) for Atlanta, San Diego, and Chicago Area Homes in 6 and 8 m/s Wind Regimes

	Battery-Based System			
	Without Direct Energy Use (Figure 6.8)		With Direct Energy Use (Figure 6.17)	
Home Area	$\bar{V}$ = 6 m/sec	$\bar{V}$ = 8 m/sec	$\bar{V}$ = 6 m/sec	$\bar{V}$ = 8 m/sec
	\$/kWh-yr			
Southeast (Atlanta)	1.97	1.84	1.87	1.77
Southwest (San Diego)	1.91	1.77	1.83	1.73
Midwest (Chicago)	.97	1.85	1.78	1.69

Source: PB 259 318

For any combination of wind regime and WECS design for these regions, little difference in system cost on a \$/kWh-yr basis results. Each region's wind-and energy-requirement profiles have about the same amount of similarity; this coupled with the different amount of energy storage required in each region (still using a 3-day minimum), produces these results. One hesitates, however, to generalize these results to all areas, because a wind profile similar to the Argonne data in an area where heating loads are small and air-conditioning loads are high would produce different results.

Future Technology WECS — Using the equipment specifications previously discussed for application in future technology WECS designs, all preliminary WECS designs were reevaluated. With the reduced wind-turbine cost used, long-term energy storage was rejected at all wind speeds (3.4, 6, and 8 m/s). For comparison purposes, a design which utilized the fuel cell as the basic output component for converting hydrogen to electricity was included (see Figure 6.18). The efficiency of the WECS designs shown in Figures 6.18, 6.8 and 6.9), plus these systems with the hot-water subsystem added, are as follows:

	Efficiency	
	Wind to ac	Wind to Thermal
	percent	
Assuming no direct energy use		
Figure 6.8, battery-based	45	49
Figure 6.18, hydrogen-based	22	37
Figure 6.9, hydrogen-based	18	37
Assuming 30% direct energy for ac load and 75% direct energy for thermal load		
Figure 6.17, battery-based	48	64
Figure 6.18 + Figure 6.12 (similar to Figure 6.15), hydrogen-based	32	62
Assuming 75% direct energy for thermal		
Figure 6.15, hydrogen-based	22	62

Figure 6.18: WECS Design with Fuel-Cell Storage

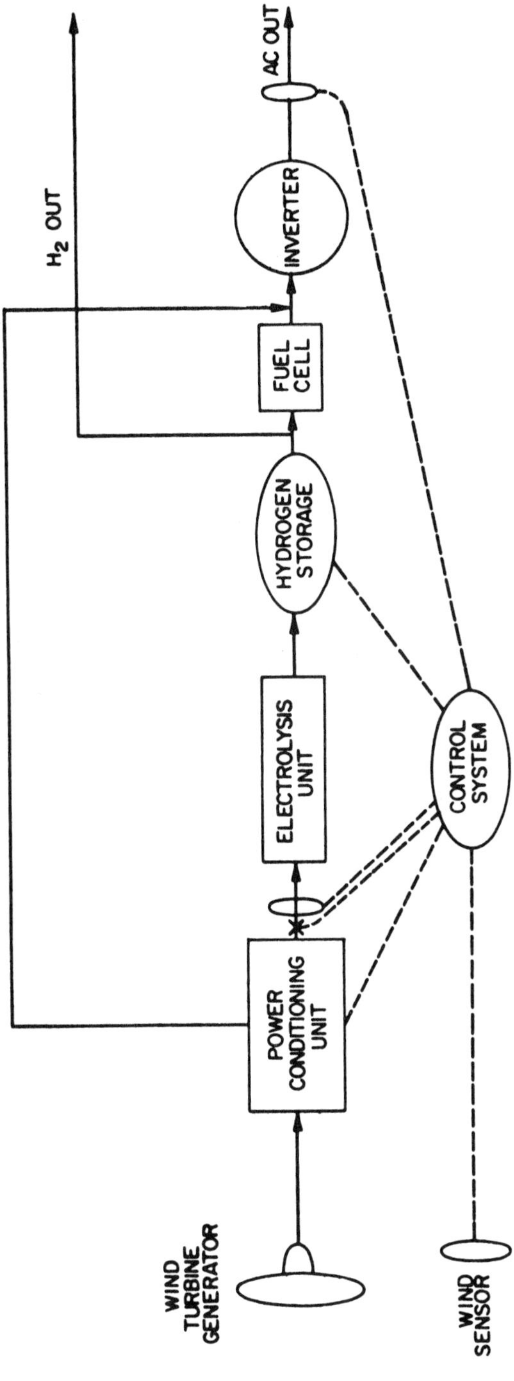

Source: PB 259 318

Table 6.18: Total Installed System Costs for Various Combinations of Loads and Wind Regimes

V, m/sec (mph) =	3.4 (7.63)	6 (13.4)				8 (18)					
V_R, m/sec (mph) =	8.9 (20)	8.9 (20)	11.2 (25)	13.4 (30)	15.6 (35)	8.9 (20)	11.2 (25)	13.4 (30)	15.6 (35)	17.9 (40)	20.1 (45)
	$ x 10^3										
Load	Current Technology, Hydrogen Based System (Figure 6.9)										
Dairy operation	482.9	236.3	–	–	163.4	147.3	–	–	123.9	–	–
Swine operation	283.3	140.7	–	–	109.3	115.1	–	–	80.9	–	–
Poultry operation	423.2	200.8	–	–	155.4	148.4	–	–	110.4	–	–
Home and dairy	713.7	343.7	289.8	267.0	259.0	236.9	210.5	164.3	195.6	202.2	231.5
Home and swine	537.7	259.6	–	–	184.4	179.5	–	–	137.8	–	–
Home and poultry	630.8	314.9	–	–	222.1	210.2	–	–	157.7	–	–
Home (Chicago)	279.0	163.0	–	–	143.0	136.8	–	–	109.7	–	–
	Current Technology, Battery Based System (Figure 6.8)										
Dairy operation	236.6	119.8	–	93.5	88.4	100.5	–	81.7	78.3	–	–
Swine operation	129.2	70.8	–	57.5	55.0	61.4	–	51.2	50.4	–	–
Poultry operation	180.5	88.5	–	67.8	63.9	73.1	–	58.5	55.8	–	–
Home and dairy	383.9	214.8	192.3	183.3	172.5	191.4	171.6	161.1	155.5	153.0	153.4
Home and swine	249.2	131.0	–	–	112.2	111.4	–	109.9	98.4	–	–
Home and poultry	298.0	146.9	–	–	106.6	121.7	–	97.7	93.4	–	–
Home (Chicago)	163.0	120.1	–	100.6	96.7	106.8	–	90.1	86.9	–	–
Home (Atlanta)	–	–	–	–	74.3	–	–	–	69.5	–	–
Home (San Diego)	–	–	–	–	56.8	–	–	–	52.7	–	–
	Current Technology, Battery Based System (Figure 6.17)										
Dairy operation	–	–	–	–	84.6	–	–	–	76.2	–	–
Swine operation	–	–	–	–	53.5	–	–	–	49.0	–	–
Poultry operation	–	–	–	–	61.5	–	–	–	54.5	–	–
Home and dairy	–	–	–	–	163.7	–	–	–	149.5	–	–
Home and swine	–	–	–	–	88.8	–	–	–	83.5	–	–
Home and poultry	–	–	–	–	88.9	–	–	–	83.9	–	–
Home (Chicago)	–	–	–	–	90.9	–	–		83.3	–	–
Home (Atlanta)	–	–	–	–	70.8	–	–	–	67.0	–	–
Home (San Diego)	–	–	–	–	54.5	–	–	–	51.3	–	–
	Future Technology, Hydrogen Based System (Figure 6.15)										
Home and dairy	–	–	–	–	113.7	–	–	–	91.5	–	–
	Future Technology, Hydrogen Based System (Figures 6.18 and 6.12)										
Home and dairy	–	–	–	–	128.6	–	–	–	101.4	–	–
	Future Technology, Battery Based System (Figure 6.17)										
Home and dairy	–	–	–	–	145.0	–	–	–	137.6 (88.4)*	–	–

*With 1 day storage.

Source: PB 259 318

Table 6.19: Installed System Costs on kWh-yr Required Basis for Various Combinations of Loads and Wind Regimes

V, m/sec (mph) =	3.4 (7.63)	6 (13.4)				8 (18)					
V_R, m/sec (mph) =	8.9 (20)	8.9 (20)	11.2 (25)	13.4 (30)	15.6 (35)	8.9 (20)	11.2 (25)	13.4 (30)	15.6 (35)	17.9 (40)	20.1 (45)
	$/kWh-yr										
Load	Current Technology, Hydrogen Based System (Figure 6.9)										
Dairy operation	9.03	4.42	–	–	3.06	2.75	–	–	2.32	–	–
Swine operation	12.86	6.39	–	–	4.96	5.22	–	–	3.67	–	–
Poultry operation	14.80	7.02	–	–	5.44	5.23	–	–	3.86	–	–
Home and dairy	6.96	3.35	2.83	2.60	2.52	2.31	2.05	1.60	1.91	1.97	2.26
Home and swine	7.56	1.65	–	–	2.54	2.52	–	–	1.94	–	–
Home and poultry	8.12	4.05	–	–	2.86	2.70	–	–	2.03	–	–
Home (Chicago)	5.68	3.32	–	–	2.91	2.78	–	–	2.23	–	–
	Current Technology, Battery Based System (Figure 6.8)										
Dairy operation	4.42	2.24	–	1.75	1.65	1.88	–	1.53	1.46	–	–
Swine operation	5.87	3.21	–	2.61	2.50	2.79	–	2.32	2.28	–	–
Poultry operation	6.31	3.10	–	2.37	2.23	2.55	–	2.04	1.95	–	–
Home and dairy	3.74	2.09	1.87	1.78	1.68	1.87	1.67	1.57	1.52	1.49	1.50
Home and swine	3.50	1.84	–	–	1.58	1.56	–	1.54	1.38	–	–
Home and poultry	3.83	1.89	–	–	1.37	1.56	–	1.26	1.20	–	–
Home (Chicago)	3.32	2.44	–	2.05	1.97	2.17	–	1.83	1.78	–	–
Home (Atlanta)	–	–	–	–	1.97	–	–	–	1.84	–	–
Home (San Diego)	–	–	–	–	1.91	–	–	–	1.77	–	–
	Current Technology, Battery Based System (Figure 6.17)										
Dairy operation	–	–	–	–	1.58	–	–	–	1.42	–	–
Swine operation	–	–	–	–	2.42	–	–	–	2.22	–	–
Poultry operation	–	–	–	–	2.15	–	–	–	1.90	–	–
Home and dairy	–	–	–	–	1.59	–	–	–	1.46	–	–
Home and swine	–	–	–	–	1.25	–	–	–	1.17	–	–
Home and poultry	–	–	–	–	1.14	–	–	–	1.08	–	–
Home (Chicago)	–	–	–	–	1.85	–	–	–	1.69	–	–
Home (Atlanta)	–	–	–	–	1.83	–	–	–	1.77	–	–
Home (San Diego)	–	–	–	–	1.83	–	–	–	1.73	–	–
	Future Technology, Hydrogen Based System (Figure 6.15)										
Home and dairy	–	–	–	–	1.11	–	–	–	0.89	–	–
	Future Technology, Hydrogen Based System (Figures 6.18 and 6.12)										
Home and dairy	–	–	–	–	1.25	–	–	–	0.98	–	–
	Future Technology, Battery Based System (Figure 6.17)										
Home and dairy	–	–	–	–	1.34	–	–	–	–	–	–

Source: PB 259 318

Table 6.20: Unit Energy Costs in ¢/kWh for Various Combinations of Loads and Wind Regimes

V, m/sec (mph) =	3.4 (7.63)	6 (13.4)				8 (18)					
V_R, m/sec (mph) =	8.9 (20)	8.9 (20)	11.2 (25)	13.4 (30)	15.6 (35)	8.9 (20)	11.2 (25)	13.4 (30)	15.6 (35)	17.9 (40)	20.1 (45)
	¢/kWh										
Load	**Current Technology, Hydrogen Based System (Figure 6.9)**										
Dairy operation	104	–	–	–	38	–	–	–	29	–	–
Swine operation	149	–	–	–	60	–	–	–	46	–	–
Poultry operation	169	–	–	–	65	–	–	–	47	–	–
Home and dairy	79	–	–	–	30	–	–	20	23	–	–
Home and swine	86	–	–	–	31	–	–	–	24	–	–
Home and poultry	93	–	–	–	34	–	–	–	24	–	–
Home (Chicago)	66	–	–	–	35	–	–	–	27	–	–
	Current Technology, Battery Based System (Figure 6.8)										
Dairy operation	50	–	–	–	19	–	–	–	17	–	–
Swine operation	67	–	–	–	29	–	–	–	23	–	–
Poultry operation	72	–	–	–	26	–	–	–	23	–	–
Home and dairy	43	25	22	21	20	22	20	19	18	18	18
Home and swine	40	–	–	–	19	–	–	–	16	–	–
Home and poultry	44	–	–	–	16	–	–	–	14	–	–
Home (Chicago)	38	–	–	–	23	–	–	–	21	–	–
Home (Atlanta)	–	–	–	–	23	–	–	–	22	–	–
Home (San Diego)	–	–	–	–	22	–	–	–	21	–	–
	Current Technology, Battery Based System (Figure 6.17)										
Dairy operation	–	–	–	–	19	–	–	–	17	–	–
Swine operation	–	–	–	–	28	–	–	–	26	–	–
Poultry operation	–	–	–	–	25	–	–	–	22	–	–
Home and dairy	–	–	–	–	19	–	–	–	17	–	–
Home and swine	–	–	–	–	15	–	–	–	14	–	–
Home and poultry	–	–	–	–	14	–	–	–	13	–	–
Home (Chicago)	–	–	–	–	22	–	–	–	20	–	–
Home (Atlanta)	–	–	–	–	22	–	–	–	21	–	–
Home (San Diego)	–	–	–	–	22	–	–	–	20	–	–
	Future Technology, Hydrogen Based System (Figure 6.15)										
Home and dairy	–	–	–	–	14	–	–	–	12	–	–
	Future Technology, Hydrogen Based System (Figure 6.18 and 6.12)										
Home and dairy	–	–	–	–	16	–	–	–	13	–	–
	Future Technology, Battery Based System (Figure 6.17)										
Home and dairy	–	–	–	–	17	–	–	–	16 (10)*	–	–

*With 1 day storage.

Source: PB 259 318

Table 6.21: Fixed Component Costs

Load	Equipment Type						
	Output		Storage				
	Engine Generator	Inverter	Hydrogen Tank	Batteries	Control	Water Treatment	Building
	$						
Home (Chicago)							
Hydrogen based	10,000	–	15,000	–	4,000	480	2,700
Battery based	–	14,600	–	43,200	N/A	N/A	6,500
Swine operation							
Hydrogen based	10,000	–	8,900	–	4,000	400	2,700
Battery based	–	18,000	–	17,500	N/A	N/A	3,600
Poultry operation							
Hydrogen based	10,000	–	13,400	–	4,000	400	2,700
Battery based	–	16,300	–	20,900	N/A	N/A	3,600
Dairy operation							
Hydrogen based	13,000	–	15,700	–	4,000	480	2,700
Battery based	–	18,400	–	35,000	N/A	N/A	5,100
Home and swine							
Hydrogen based	13,000	–	18,100	–	4,000	520	2,700
Battery based	–	18,400	–	44,150	N/A	N/A	5,800
Home and poultry							
Hydrogen based	13,000	–	18,100	–	4,000	550	2,700
Battery based	–	18,400	–	44,150	N/A	N/A	5,800
Home and dairy							
Hydrogen based	15,000	–	25,000	–	4,000	600	2,700
Battery based	–	27,500	–	77,200	N/A	N/A	9,100
Home (San Diego)							
Battery based	–	14,600	–	22,000	N/A	N/A	3,600
Home (Atlanta)							
Battery based	–	14,600	–	33,100	N/A	N/A	5,000

Source: PB 259 318

Again, both from an efficiency and from a cost standpoint, the battery-based WECS design (Figure 6.17) was optimum.

WECS Designs Cost Summary – All WECS design, load, and wind-regime combinations that were examined to the point of final costing are presented in Tables 6.18 and 6.19. Values are given in 1976 dollars. Table 6.18 is the estimated total cost of system x $1,000. Table 6.19 is the installed cost on a $/annual kWh required basis. Table 6.20 is the estimated energy cost in ¢/kWh. (This value was not calculated for all systems.)

The entries in Table 6.18 were arrived at by combining the cost of all systems components + 10% for installation and the cost of the required building (building cost included installation). Table 6.21 shows the fixed component (storage and output) estimated cost for each load when long-term energy storage was not required. The cost variable for each load is, therefore, the sum of the wind-turbine and electrolyzer (if required) costs, which are dependent on the component size required for different WECS designs and wind regimes.

Table 6.20 entries were calculated by dividing the estimated annual cost of the system by the annual kWh requirement, where the annual cost was determined using the following data:

Interest rate–8%
System life–10 years for engine generators (two units, dual operation assumed), 15 years for batteries, 20 years for all other components
Maintenance cost–7% of installed cost/year for engine generators, 1% of installed cost/year for all other components

In the 6-m/s wind regime, energy costs ranging from 17 to 25¢/kWh are shown (Table 6.20) for the battery-based system (Figure 6.17). Operation in the 8-m/s wind regime is shown to result in about a 1¢/kWh cost reduction. This small reduction occurs due to the large portion of the system's cost attributable to the system's storage and output components that are unaffected by wind speed. The 10¢/kWh shown for the home-plus-dairy operation under future technology (Table 6.20) represents a 6¢/kWh cost reduction achieved through a reduced storage-capacity requirement (or improved batteries). Additionally, the future-technology, hydrogen-based designs are potentially more economical than the battery-based designs, if improved batteries are not developed and storage requirements are large.

Conclusions

Technical, Economic, and Future Feasibility: Wind-energy conversion systems (WECS) operating independent of other energy sources for supplying the needs of homes and farms are technically feasible. Most components necessary within the systems require no modifications for their implementation. Others, such as the wind-turbine electrolyzer/battery-bank interface, will require some work to ensure maximum system efficiency. However, the hydrogen-based systems, which require the most modifications (wind-turbine electrolyzer, electrolyzer, and engine generator), are also the least efficient and economical in this application.

Wind-energy conversion systems that provide the energy needed for grain drying or mobile fuel for the tractor are economically unattractive. Grain drying consumes a large quantity of energy over a relatively short period of time. WECS dedicated to this application require considerable energy storage (approximately one-half of the annual requirement). The large capital investment for this storage, coupled with a low energy throughput (about twice the storage capacity), yields very high energy costs.

Mobile fuel production (hydrogen) is technically feasible using wind energy, but here again the major problem is energy storage: first, the depot requirements to meet varying use rates throughout the year, and second, the onboard fuel storage. While the depot requirement may impose cost restrictions for reasons similar to those experienced with grain drying, the major obstacle is providing the tractor with an adequate fuel supply. The full load run time of 1 to 2 hours, which results from current-technology hydrogen storage on-board the tractor, would probably be unacceptable to most farmers. WECS are better suited for applications supplying the energy requirements of homes and confinement-type farming operations. These loads have an energy-use pattern and a total annual requirement that can be met without a large, crippling storage system.

For the WECS and farm loads investigated in this study, implementation in the near term as a hedge against inflation may be practical. At 14¢/kWh, the home and poultry operation being supplied by a WECS in a 6-m/s wind regime would be cost effective if the cost of commercial electricity (1974 average equal to 3.05¢/kWh) were to increase 7.2% per annum for the life of the system (20 years). With the advanced systems projected, the picture is somewhat more favorable, with economic viability occurring with only a 5.4% per annum increase in commercial electricity costs.

The primary reason for this somewhat bleak forecast of WECS use is the 3-day storage requirement imposed on the WECS designs. If improved information on the wind's speed and its availability could reduce this requirement to 1 day or less, system costs (for battery-based systems) could be reduced by about 50%. Earlier implementation of WECS would then be possible on an economic basis.

Earlier implementation of WECS could also result if loads other than those surveyed here could be identified that have an energy requirement closely matching the wind energy available. The energy available but not used in the farm systems which were examined generally exceeded the load requirements by more than 100% on an annual basis. The use of this energy for supplementing a grain-drying operation, soil warming, or for sale to others could greatly reduce the unit cost of electricity from these systems.

The feasibility of WECS from a wind-regime standpoint was not clearly defined during this study. It is true that for the loads and WECS designs examined herein, an annual mean wind speed of 6 m/s was required for the systems to approach the realm of economic feasibility. It was not felt that this is a binding criteria for all WECS, however, because a load that closely tracked the wind energy available or through some means was able to use all the available energy could result in economical operation of WECS in lower wind regimes.

Observations: The major observations made during the course of this study are as follows.

> WECS designs using short-term battery storage are preferable to those using hydrogen storage.

Component development needed for battery-based WECS designs is minor; most components are off-the-shelf.

Development of a wind turbine producing 60-Hz ac power is not warranted for use in total-energy WECS designs for farms. The energy system's output components are sized to provide the required 60-Hz ac power; interfacing a wind turbine's output with the system's output would increase the WECS' complexity and cost.

Thermal loads are more efficiently satisfied by WECS than ac loads.

The cost of energy storage is the most expensive element in the WECS.

THE UNIVERSITY OF MASSACHUSETTS WIND FURNACE PROJECT

The wind furnace is a system intended to use wind energy or a combination of wind plus photothermal energy to provide heating at the residential level in climates characterized by about 6,200 degree days of heating requirement (based on 65°F). The concept stems from at least three considerations: (a) there are extensive regions in which a plot of wind energy convertible by a practical momentum exchanger follows the same shape against time as does the degree-day requirement; (b) in those same regions, photothermal energy convertible by a practical collector tends to be a very flat curve against time; and (c) at forty cents or more per gallon heating oil costs, mortgage loans between $4,800 and $8,000 might be offset by oil savings.

It also appears that there are enough installation opportunities in the U.S. to permit aggregation of as many as 600,000 barrels of petroleum per day savings. Finally, the required hardware could be produced in huge quantity, very near term, using a very broad base within existing U.S. industrial capability.

Five different models of the first generation wind furnace, the Mark I, were tested. Figure 6.19 shows four of those models. Model 5, not shown, is simply model 3 with the heat pump left out. Many other models will become apparent as the project progresses.

The work done in 1975 included: (a) completion of design and construction of the system required to demonstrate models 1, 2, 3, and 5; (b) the acquisition and fitting-out of a suitable test and demonstration facility; and (c) the completion of analytical and experimental work that will permit recommendation of all parts of a second generation system.

By cooperating with a Department of Agricultural Engineering program, supported by USDA, a test and demonstration facility was acquired, the Johnson Building. This building had been conceived as a very well insulated semiportable residence, constructed primarily from wood. The calculated heat loss from the building is a small fraction of that of the ASHRAE standard well-insulated house. This building will be placed atop a concrete basement laboratory; combined building and basement have a standard house heat loss. Redundant thermal storages (water) will be cast into the concrete foundation to facilitate tests, trials, and demonstration.

Figure 6.19: Four Models of the Mark I Wind Furnace

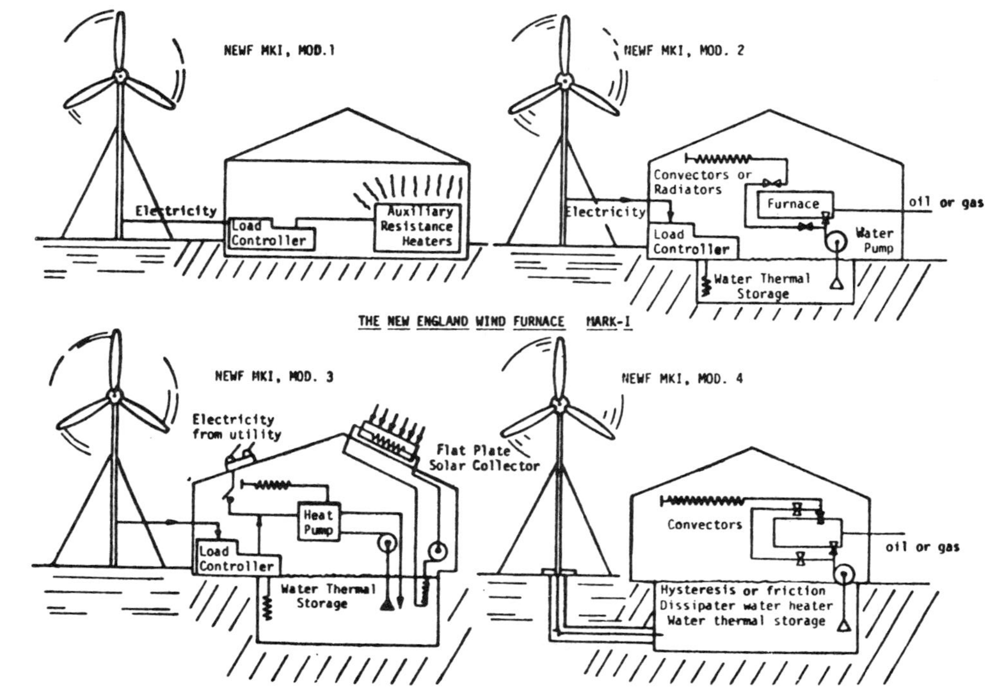

Source: NSF-RA-N-75-050

All subsystems and components have been sized using a computer simulation for parametric studies. Again, earlier work supported by others is being used to advantage. The first generation configuration derived from those studies can be summarized:

(1) Wind regime extrapolated from Bradley Field data; solar insolation data for Blue Hills used without modification.
(2) Wind wheel 10 meters diameter, three blades, 50 shaft horsepower at 157 rpm in a 26.1 mph wind.
(3) Wind shaft speed up to 1,800 rpm to drive a 3 wire wye connected synchronous alternator, rated voltage 480 volts, rated output 31.1 kVA, 24 kW.
(4) Resistive load connected in balanced delta with ten-step load matching controller between generator and load.
(5) Pitch control or separate excitation control to maximize momentum exchange.
(6) Machine axis height 60 ft, probably on a stayed pole mast support.
(7) 200 sq ft of photothermal collector, vertically oriented, liquid coiled.
(8) 2,000 gal of water thermal storage.

Predicted results, Figure 6.20 for model 5 of this system suggest that 30% of the heating required at Amherst will have to be supplied by an auxiliary heater, gas fired or wood fired. In a hill or coastal or plain siting, Q auxiliary would be much lower than 30%. Model 1 has shown much less attractive results in the computer simulations, Figure 6.21, due primarily to significant mismatch between wind availability and heating demand. The thermal storage is an excellent low-cost addition to the system.

Figure 6.20: Predicted Results for Model 5

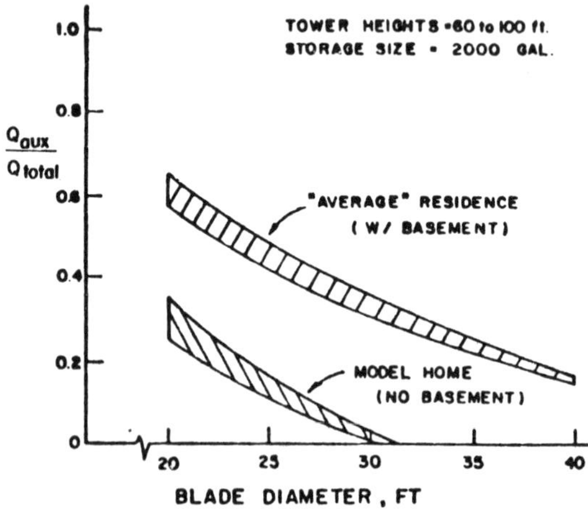

Source: NSF-RA-N-75-050

Figure 6.21: Predicted Results for Model 1

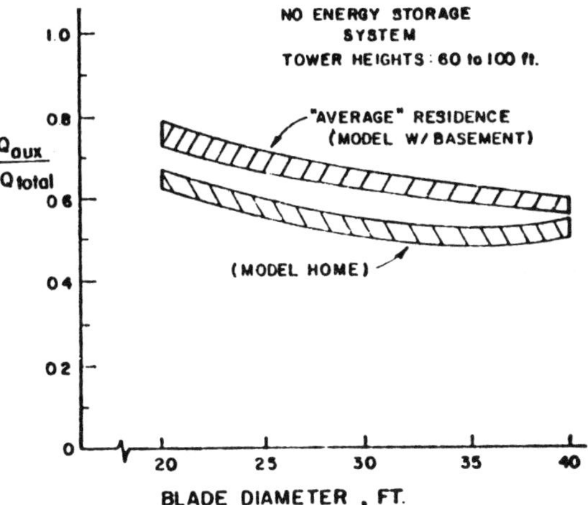

Source: NSF-RA-N-75-050

WIND POWERED AERATION FOR REMOTE LOCATIONS

A project on wind-powered aeration was sponsored by NSF/RANN at Colorado State University with cooperative work at the Colorado Division of Wildlife. This application of aeration is an example of methods of direct mechanical utilization of wind power. The simplicity of a direct mechanical linkage between the wind turbine and the air compressor may reduce costs.

The compressed air was injected into the bottom of a sewage lagoon to improve biological waste treatment efficiency and into the bottom of a high mountain lake to try to avoid the problem of fish winterkill. Fish winterkill may occur when a lake is covered with ice during the winter. During the period of coverage, no oxygen can transfer into the lake. If the oxygen demand is sufficiently high, the oxygen inventory of the lake will be consumed, and the fish will winterkill.

Background

The Colorado Division of Wildlife has several years of prior experience with mountain lake aeration. It was determined that trout could survive at 4°C with oxygen concentrations as low as 1.6 mg/l. Experiments with electrically-powered air compressors supplying two Helixors and a perforated pipe on the bottom of Road Canyon Reservoir, a 160 surface acre lake in southwest Colorado, proved not only that winterkill could be prevented, but that the lake oxygen levels were raised more than the calculated oxygen input from the compressors. The explanation for this is that the rising bubbles brought the warmer 4°C water from the bottom of the reservoir and maintained large openings in the ice in all weather except when the temperature reached -50°F. These openings were the source of increased oxygen transfer. The success of the electrically driven compressors prompted the trial of a wind power compressor at Regen Lake, many miles from an electric power line. The device used was a 6 ft diameter Aeromotor driving a single piston compressor 6 inches in diameter. A hole was maintained in the ice, but it was small because the perforated pipe broke, releasing all the air in one location.

Since the oxygen improvement was insufficient for that size lake, six compressor cylinders were connected to the wind turbine. This put too much strain on a $\frac{3}{16}$ inch pin in the drive, and the pin sheared. Later trials with four compressors also loaded the pin too much and it sheared again. In the winter of 1974-75, two compressors were used. The equipment did not fail, but low wind velocities contributed to the failure of the lake to avoid winterkill. The wind velocity was lower than other years. During one 640-hour period, the wind exceeded the cut in speed of the system only in 64 separate hours and this does not necessarily indicate 64 full hours of operation.

Objective

The objective of this project is to apply wind-powered aeration to the solution of biological problems. The matching of the power characteristics of a wind turbine to the load characteristics of an air compressor is an important consideration. Energy storage may not be a problem since the time constant of oxygen depletion of a lake may be a month or more and the lake can store the effects of a large amount of compressed air from a windy period as dissolved oxygen. The past work of the Colorado Division of Wildlife indicates that a larger capacity wind-powered compressor may be required or a smaller lake chosen to permit sufficient oxygen per liter to be absorbed by the lake. A lower cut-in design wind speed for the wind turbine may increase aeration in less windy periods. Site selection at the location needs to be studied to maximize the power generated.

Site Considerations

Three locations were selected for further application of wind-powered aeration: Cobb Lake Annex near Fort Collins, Lost Lake near Red Feather Lakes, and the sewage treatment lagoons of Keenesburg. All sites are in Colorado. The Cobb Lake and lagoon sites have little topographical interference. The terrain is flat to slightly rolling. At both sites, a shallow depression leads to the northwest. This feature is expected to assist in obtaining good wind power from prevailing winds in winter.

The wind turbine (near Cobb Lake Annex) has been sited part way up the side of the depression. At the lagoon, the wind turbine is to be located at the mouth of the depression. The lagoon site was chosen because two identical lagoons are present. Thus, one lagoon will be aerated and one not. The flow of sewage will be divided equally between the lagoons, and BOD and suspended solids analyses should indicate whether wind aeration produces measurable effects.

The siting of the wind turbine at Lost Lake is more complex since a bluff about 400 ft higher than the lake obstructs the northwest exposure. Closer examination of the site revealed a wind-swept area with flag trees, which are trees with branches only on one side due to wind action. This area indicates strong winds leading to a draw going up the bluff in a northwesterly direction. Although the site seems promising, wind tunnel tests with a terrain model of the Lost Lake region were planned. The model was scaled to 1 inch equals 80 ft and since it was 70 inches in diameter, it represented terrain over a mile in diameter.

The terrain model was tested in a laboratory wind tunnel. The wind velocity at various points around the lake is measured while the model is rotated to simulate different prevailing wind directions, and the wind tunnel air velocity is changed to simulate different wind intensity. Based on the results of the laboratory wind tunnel tests, the local site for the Lost Lake area was chosen.

Equipment Considerations

If matching the load characteristics of the wind turbine and the compressor were the prime requisites, then a centrifugal blower would be considered. The centrifugal blower develops pressure proportional to the rotational speed squared and requires power proportional to the speed cubed just as the power potential of the wind speed. However, the pressure required to inject the air at the bottom of the lake or lagoon equals the value of the liquid head to be overcome plus increases due to pressure losses in the supply line with the establishment of air flow. Thus a positive displacement type of compressor is the only possibility which can overcome the liquid head and start air flow at low wind velocities and slower shaft rpm. Two types of positive displacement air compression devices were considered. One is the piston reciprocating type already employed by the Colorado Division of Wildlife, and the other is a rotary lobe blower.

One of the disadvantages of a positive displacement air compressor working against a liquid head is high starting torque. To gain the high starting torque a high solidity turbine and high blade activity factor would be desirable. This suggests the use of an American multiblade-type turbine already employed by the Colorado Division of Wildlife. A power coefficient of 0.3 and tip speed ratio of one is often assumed for such wind turbines.

An alternative turbine with a solidity of nearly 40% and activity factor of one is an 8 ft turbine, sometimes called the Chalk bicycle wheel turbine, manufactured by the American Wind Turbine Company of Lyman, Nebraska. Higher power coefficients and tip speed ratios are claimed for this turbine, which may be due to the use of airfoil blades. Both of these multiblade turbines were considered for this application.

Alternative governing and weathercocking devices were also being considered. The use of a multiblade wind turbine with low to moderate speed airfoils may provide some self-governing characteristics. If the turbine and compressor can take it, it may be best to collect severe wind power, since the oxygen storage capacity of the water is high.

It is hoped that the study will be useful for comparison of aeration generated by wind power with alternative mechanical systems, sites, and applications.

REFERENCES

(1) Renolds, R.E., Deere and Co., private communication (July 1975).
(2) *Doanes Agricultural Report* 37, 1-5 (January 4, 1974).

(3) *Doanes Agricultural Report* 37, 12-6 (March 22, 1974).
(4) Pimental, D. et al, "Food Production and the Energy Crisis," *Science* 182, 443-49 (November 2, 1973).
(5) Escher, W.J.D., "Survey and Assessment of Contemporary U.S. Hydrogen-Fueled Internal Combustion Engine Projects," *Proceedings of the 10th IECEC*, Newark, Del. (August 1975).
(6) Flood, C.A., Jr., School of Agriculture, Auburn University, private communication, no date.
(7) Biederman, N.P., Darrow, K.G., and Konopka, A.J., *Utilization of Off-Peak Power to Produce Industrial Hydrogen,* Final report for IGT Project 8793 (EPRI 320-1), Chicago: Institute of Gas Technology (August 1975).
(8) Hittman Associates, Inc., *Residential Energy Consumption in Single-Family Housing,* Report No. HUD-HAI-2, Columbia, Md. (March 1973).
(9) Moyers, J.C., *The Value of Thermal Insulation in Residential Construction: Economics and the Conservation of Energy*, ORNL-NSF-EP-9, Oak Ridge National Laboratory, Oak Ridge, Tenn. (December 1971).
(10) Altman, L.B. and Charity, L.F., "Demand Estimation for Sizing Distribution Transformers and Secondary Services in Rural Areas," *IEEE Transactions on Industry and General Applications,* Vol. IGA-3, No. 3 (May-June 1967).
(11) Uhl, B., Commonwealth Edison Co., private communication (September 1975).
(12) Carpenter, T.G., Charity, L.F., and Altman, L.B., "Farmstead Wiring for Large and Scattered Loads," Paper presented at the winter meeting of the American Society of Agricultural Engineers, Chicago, Ill. (December 1963).
(13) *Total Energy Handbook*, Caterpillar Tractor Co. (July 1967).
(14) Moses, H. and Bogaer, M.A., "15 Year Climatological Report, January 1, 1950 to December 31, 1964," Argonne National Laboratory, Argonne, Ill.
(15) *Storage Battery*, Exide Power Systems Division, Philadelphia, Pa., no date.
(16) Edelson, R.D., Autodynamics, Inc., private communication (February 5, 1976).
(17) James, D., Exide Power Systems, private communication (October 1975).
(18) Johnson, J., Gould, Inc., private communication (September 1975).
(19) "Exide Stationary Lead Acid Battery Systems for Industrial and Utility Operations," ESB, Inc. (Copyright 1972).
(20) Escher, W.J.D., "The Hydrogen-Fueled Internal Combustion Engine, A Technical Survey of Contemporary U.S. Projects," Escher Technology Associates Report PR-51, St. Johns, Michigan (September 1975).
(21) Kramer, L., Winco-Division of Dyna Technology, Inc., private communication (January 1976).
(22) La Rocco, J.C., White Engines, Inc., private communication (January 1976).
(23) Lobree, C.D., Electro-Motion Pacific, Inc., private communication (January 1976).
(24) Mijanovich, R.T., Soleq Corporation, private communication (August 1975).
(25) Hart, H.A., Control Products Corporation, private communication (September 1975).
(26) Iliff, J.E., Modular Buildings, Inc., private communication (March 1, 1976).
(27) McKee, J.C., Morgan Portable Building Co., private communication (March 1, 1976).
(28) Vorgias, S.K., Stan's Construction, private communication (March 2, 1976).
(29) Harris, F.W., "An Economic Analysis of Proposed Hydrogen Fuel Systems for Farm Applications," Paper presented at Frontiers of Power Technology Conference, Oklahoma State University, Stillwater, Oklahoma (October 1-2, 1975).
(30) Heronemus, W.E., in *Proceedings of the Second Workshop on Wind Energy Conversion Systems*, Eldridge, F.R., Ed., 406, The Mitre Corp., McLean, Va. (1975).
(31) Titterington, W.A., General Electric Co., private communication (February 1975).
(32) Corley, W.G., Director of Engineering Development Department, Portland Cement Association, private communication (August 1975).
(33) Schwartz, H.J., "Batteries for Storage of Wind-Generated Energy," National Aeronautics and Space Administration, Lewis Research Center, Cleveland, Ohio.
(34) Sudworth, J.L., "Sodium/Sulfur Batteries for Rail Traction," Paper presented at the 10th IECEC, University of Delaware, Newark, Del. (August 18-22, 1975).

LEGAL, SOCIAL AND ENVIRONMENTAL ISSUES

Material for this chapter has been based upon papers presented at the Second Workshop on Wind Energy Conversion Systems, Washington, DC, June 9-11, 1975 (NSF-RA-N-75-050) and a report by Battelle Columbus Laboratories (ERDA/NSF/07378-75/1).

LEGAL-INSTITUTIONAL IMPLICATIONS OF WIND ENERGY SYSTEMS

At the Second Workshop on Wind Energy Conversion Systems, Dr. L.H. Mayo, Director of the Program of Policy Studies in Science and Technology at George Washington University proposed a project to examine the existing legal and institutional constraints on, and the corresponding changes necessary to optimize development of a variety of possible wind power configurations in certain defined social and geographical environments. This broad task will naturally necessitate consideration of a host of nonlegal concerns.

Research Plan

The research plan incorporates the following components: (1) specification of provisional hypotheses concerning the relationship between the existing legal-institutional structure and the development and utilization of wind power; (2) identification of the most promising data sources (both legal and empirical) relevant to these hypotheses; (3) conduct of legal-institutional research regarding these assumptions and hypotheses, such research being guided by a systematic matrix representation of the principal elements of a wind power legal-institutional context; and (4) testing of the resulting (or revised) hypotheses by means of workshops attended by representatives of affected institutions or communities. These components will be briefly elaborated upon below.

At this point, the Program of Policy Studies is able to posit the following, necessarily general, preliminary hypotheses: (1) the United States lacks a well-organized system of judicial doctrine or legal-institutional arrangements directed specifically to the development and utilization of wind power resources, and the existing judicial-institutional regime in fact seems to pose serious impediments to such development and utilization; (2) however, there does exist much judicial precedent arising in different contexts which could be applied to the wind power situation; (3) there also exist numerous federal and state regulatory schemes which might be relevant to the wind power situation; (4) but there exist no authoritative statutory directives establishing a federal-state or state-state distribution of regulatory control over wind power, and problems of federal preemption which could arise if and when the national government asserts final control over some aspect of this area. Because of the deficiencies in the existing legal-institutional regime, there is presently no basis for either a social benefit-cost judgment on the implementation and operation of particular wind power configurations, or for an estimate of the ultimate distribution of such benefits and costs; nor is there an appropriate institutional methodology or structure for the continuing monitoring, evaluation and modification of wind power operations.

These several hypotheses can assume some substance through examination of a few legal examples based on research already performed by the Program of Policy Studies. First, the relative unreliability of wind power systems conflicts with the current public utility regulatory requirement of reliable power, especially in peak demand periods. Second, a system of small, individual wind power systems is even more poorly adapted to current regulatory schemes since the requirement of reliability in peak demand situations would remain, and such a system might infringe upon the legal monopoly now granted public power companies. Thirdly, a variety of other governmental regulatory requirements (e.g., the FAA requirement of warning lights on towers) may be relevant to various wind power configurations.

Fourth, the physical possibility of competing demands upon the amount of wind energy available within a given space probably cannot be adequately balanced through existing water and wind doctrines such as "prior appropriation," easement, eminent domain, and zoning. Fifth, some wind power configurations may present hazards or difficulties (e.g., interference with television reception) which the existing doctrines of nuisance are inadequate to handle, and thus may necessitate new regulatory schemes. Sixth, wind power operations in coastal or offshore areas could pose jurisdictional problems as among the states, the Federal government and foreign countries; moreover, such installations may raise a host of as yet unresolved environmental questions.

Finally, there are a number of less important, miscellaneous legal questions whose resolution must receive some attention, for example: local zoning restrictions based on aesthetic criteria; the standard of tort liability (strict or negligence) governing wind power operations; and the availability of federal financing for small, individual wind power systems by the FHA, VA, and SBA.

The basic data source for the project is, of course, the vast array of standard legal research materials. In addition to the more or less expected sources, these should include legal precedents on the early use of wind power in the United States; precedents on the allocation of water resources; and the legal standards applied by foreign nations with systems of wind power. Also relevant to the project are public and private energy project reports, as well as certain newsletters, digests, abstracts, and annotations on energy questions generally.

Aims and Benefits

The aim of the project is to produce a final report on the "Legal-Institutional Implications of Wind Energy Conversion Resources" which will meet the tests of adequacy, namely: (1) presentation in a manner and format facilitating a ready understanding by the interested wind power communities; (2) warrantability of the comprehensiveness, relevance and reliability of the research and follow-on analysis; and (3) usefulness of the research products to the various interested wind power communities.

More concretely, the report can serve these communities in at least the following ways: (1) advising potential owners and operators of wind power installations how best to perfect, protect and develop ownership rights and to maintain such installations with minimum risk and maximum productivity; (2) alerting governmental agencies at all levels to the legal and regulatory considerations relevant to any deliberate approach toward wind power policy; and (3) assisting wind energy conversion researchers to understand the legal-regulatory constraints on the development and use of wind power resources. Even more tangibly, the report could enable wind power cost analysts to consider legal expenses in their analyses, wind power planners to make better design and siting decisions, and wind power engineers to consider only those technical problems which fall within the bound of likely present and future reality. Substantial progress in meeting these objectives can constitute an essential contribution to the developments of wind power resources and thus to the national initiative toward energy independence.

PILOT STUDY ON PUBLIC REACTIONS TO WIND ENERGY DEVICES

At the Second Workshop on Wind Energy Conversion Systems, R. Ferber of the University of Illinois described a study intended to explore, in a preliminary way, reactions of the general public in various parts of the country toward the construction of different types of windmills for generating electric energy. It should throw light on the extent to which people may object to the placement of windmills in different types of landscapes, as well as the extent to which different types of people are likely to raise such objections.

The study would be carried out at locations of high wind power in different parts of the country. Insofar as possible, different design aspects of windmills relating to possible public acceptance of them would be tested. In addition, the study would incorporate, with the cooperation of the National Park Service, the construction of a small windmill at one location in a national park and make use of this facility for obtaining people's reactions while viewing the windmill, and obtain, as a by-product, data on energy generated and other operating experiences with this windmill.

The specific objectives of the study were twofold. One objective was to provide a preliminary body of substantive information on public acceptance of different types of wind energy devices in alternative locations. The second objective was to furnish a methodological base for more intensive and more definitive studies of public acceptance of such devices at a later time. In other words, this study should provide information on the extent to which different types of windmills are likely to be accepted by different groups of the population and the extent to which particular types of windmills are likely to meet public opposition, while at the same time exploring different approaches to obtaining this information and evaluating the validity and efficiency of these different approaches for more intensive work at a later time.

Research Plan

Since initial construction of windmills for generating electric energy is likely to take place in rural areas and in relatively isolated smaller urban areas of high wind power, it seemed logical to focus the main part of the study in these areas. At the same time, some preliminary coverage should be provided to major urban areas of the country, since if the efficiency of windmill power is established, its use, at least for supplementary purposes, in the major urban areas of the country could be only a matter of time. Therefore, it seemed desirable to begin to collect information on the reaction to different types of windmills of people in these areas as well.

Taking these considerations into account, it was planned to carry out this study in six parts of the country that are very different both geographically and in terms of the socioeconomic characteristics of the population. These areas are highly diverse, including the flat plains of the Midwest, a national park area on the Atlantic Coast, a hilly densely settled area in the Pacific Northwest, a flat heavily settled area on the East Coast, and one of the major metropolitan areas of the country (Chicago).

With the exception of southwestern Florida, all of these are areas of relatively high wind power. Southwestern Florida is the site of the Everglades National Park, where electric cars are in use by the National Park Service and a windmill would be constructed to help power them and get public reactions.

In all of these areas but the national park, a statistical probability sample of 400 households would be selected by area probability methods so that every household in the area has an equal chance of falling into the sample. These households would then serve as the basis for the data collection, the objective being to obtain at least 300 interviews in each area. In addition, for pretest purposes an extra 60 households would be selected in the Chicago area and in the Kansas-Oklahoma-Texas area.

In the national park, the sample selection scheme would be different to allow for a before-and-after design in conjunction with the construction of the windmill. Sample members would be selected in that area by using time probability sampling, so that over a particular period of time sample members would be chosen from among those visiting the national park in such a way that every visitor during that period would have an equal chance of being selected. The same procedure would be carried out in a second stage of data collected at that site, with that second stage taking place after the windmill is in operation.

A number of approaches would be used in these interviews to ascertain public awareness and reaction to windmills. Underlying these approaches would be the use of three-dimensional (3D) viewing devices so that respondents can be shown landscapes and pictures of different types of windmills in a highly realistic manner to enable them to make better judgments. Different means of using these devices would be explored in the pretest.

The data collection would be carried out in two stages. The first stage would be interviews with the sample members in all six areas, obtaining their reactions to various uses and designs of windmills. The second stage would apply only to the area where a windmill is constructed, and would involve using essentially the same questionnaire, with a few modifications, to ascertain people's reactions when they actually see a windmill. Since the windmill would be in a national park, a different sample would have to be interviewed. Nevertheless, the difference in the replies of the two samples at the same location could serve as a basis for calibrating and making more meaningful the reactions obtained to windmills at the other five locations.

The pilot windmill to be constructed in a national park as part of this project would be of a standard design. Its purposes would be to provide a unique source of validation for the results obtained in the first stage of the study as well as to yield engineering data on performance characteristics. It would be of the usual horizontal-axis form, for no attempt would be made to develop any aerodynamic features, since this is not an objective of the study and might interfere with the collection of the behavioral data. It was also planned to record data on the operating experience of the windmill, with particular emphasis on the amount of energy generated over a period of time, the costs of this energy, and noise levels and types.

Relation to Overall WECS Program

The growing public interest in all phases of the environment suggests that, apart from questions of operating efficiency, before wind energy devices can be implemented on a wide scale, attention needs to be given to public reaction to such devices and the extent to which the public is willing to accept this type of technology. Numerous instances already exist of technological developments that are feasible and economical but which for reasons of public dissatisfaction had to be either modified or discarded altogether.

This question would seem to be particularly relevant in the case of wind energy devices, which for maximum efficiency would have to be located in large numbers at various locations in the country and which would occupy highly prominent positions in those locations. The willingness of the public, and its elected representatives, to accept such a modification of the landscape is likely to be a major determinant of whether the plans for wind energy generation can be implemented in their present form, and of the extent to which such plans may have to be modified to gain the necessary public acceptance.

EVALUATION OF THE POTENTIAL ENVIRONMENTAL EFFECTS OF WIND ENERGY SYSTEM DEVELOPMENT

The research program described here represents a first approach review for the environmental component of the NSF/ERDA coordinated Federal Wind Energy Program under the Legal/Social/Environmental Issues area. The general approach of this project has been to identify potential environmental effects of wind energy development in the United States and, where possible, to evaluate the significance of the effect. This research program was limited to conventional, horizontal axis, tower-mounted, terrestrial wind energy systems. A large number of independent turbines within a large multimegawatt system was not evaluated in detail because it was assumed that spacing would be adequate to provide a normalized airflow for each rotor. Advanced air stream augmentation systems either through blade design or terrain alteration were not evaluated. The potential environmental effects of energy transmission or storage were excluded from this study because storage and transmission are inherent components of all energy conversion methods, not just wind energy conversion and many of these effects have already been evaluated.

This initial research effort addressed the following objectives: (1) literature review of the environmental effects of winds and the alteration of winds on the other microclimate variables and on the biota; (2) conduct of a baseline field survey at a future wind energy system site; and (3) design of a longer term field program to monitor the operational effects of a wind energy system at the site in (2) above. The site of the ERDA/NASA 100 kW Experimental Wind Turbine at NASA Lewis Research Center's Plum Brook Station near Sandusky, Ohio, was chosen as the wind turbine site and system around which this program was designed. Follow-on field studies to evaluate effects through operational experience will be conducted at that location. Additionally, the design for this system and a theoretical analysis of its influences on the physical environment were used as the model system around which the literature review syntheses were made.

Aerodynamic Flow Analysis

In order to evaluate the potential environmental effects of wind energy development, an aerodynamic flow analysis of a representative system was required to serve as a model of the zone of influence of an operating wind energy system (WES). The ERDA/NASA 100 kW Experimental Wind Turbine, which is serving as the focus for field studies, met the general requirements of being a large, conventional, horizontal axis, tower-mounted, terrestrial WES, and was adequate to serve as the generalized model for the literature review. The results of the aerodynamic flow analysis and descriptive material about the 100 kW WES follow. This generalized description was used as the basis for the literature review. Specific details of the analysis formed the basis for design of future field programs to evaluate operational effects.

The rotor of the NASA Lewis Research Center's wind turbine generator consists of two blades 125 feet in diameter operating at 40 rpm and generating approximately 100 kW of useful power at wind velocities in excess of 18 mph (8 m/sec). Transmission train losses require that the rotor extract 133 kW from the available wind energy to ensure 100 kW output.

Between 8 and 18 mph, the power generated by this rotor is a function of wind velocity. From 18 to 60 mph, the variable pitch blades are rotated such that a constant 100 kW output is maintained. Below 8 mph and above 60 mph, the turbine blades are placed in the feathered position, extracting no useful power. The power extracted from the wind as a function of wind speed is shown in Figure 7.1. The power coefficient (ratio of actual power extracted to total available wind energy in the disk area) is plotted in Figure 7.2 as a function of λ (ratio of rotor tip speed to wind speed) (1).

Figure 7.1: Rotor Power as Function of Wind Velocity

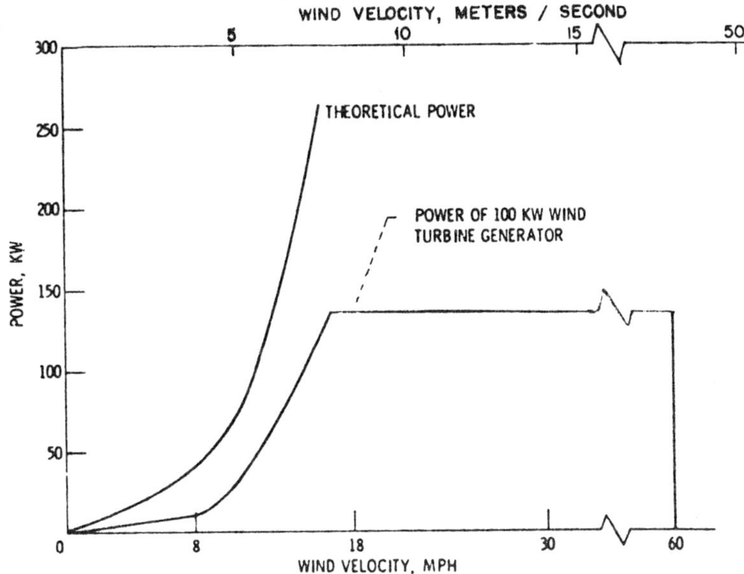

Figure 7.2: Power Coefficient as Function of λ

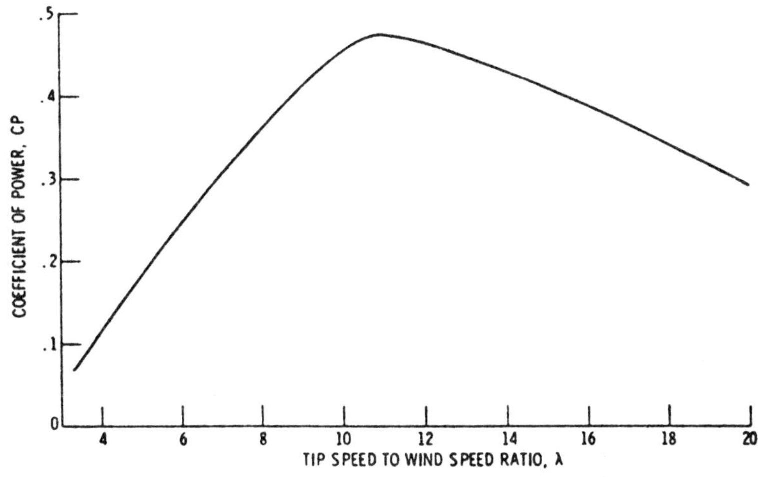

Source: ERDA/NSF/07378-75/1

Considering a cylindrical control surface of radius R around the rotor, as shown in Figure 7.3, simple one-dimensional flow theory can be applied to determine the wake characteristics far downstream of the rotor. Applying the principles of continuity, momentum, and energy to the flow across the rotor disk, both the thrust and power extracted can be easily determined if the flow is assumed to be entirely axial (2).

Figure 7.3: One-Dimensional Control Surface

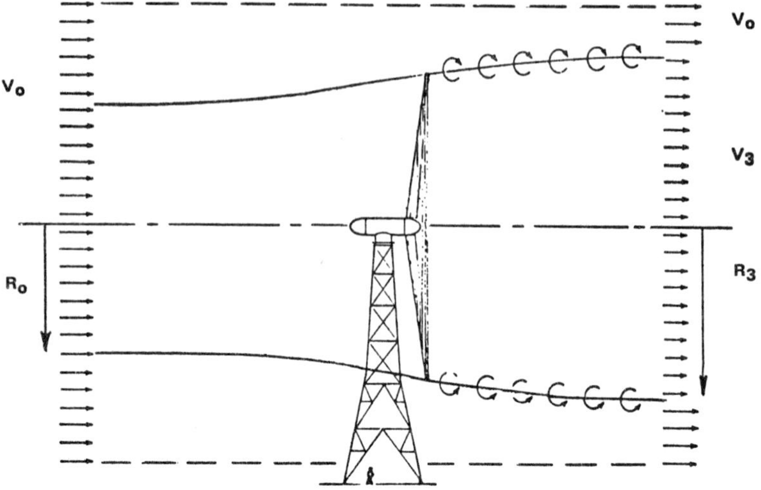

Source: ERDA/NSF/07378-75/1

An important result of the momentum theory of propellers states that the velocity through a propeller is equal to the average of the velocities infinitely far in front of and behind the propeller. Furthermore, the maximum power output, P_{max}, is obtained when the velocity far downstream of the rotor is one-third of the free stream velocity. Power coefficient is defined as the ratio of power extracted from the air to the ideal energy input through the area equivalent to an area swept by the blades. It can be shown that the maximum power coefficient achievable is

$$(1) \qquad C_{P_{max}} = \frac{P_{max}}{\frac{1}{2}\rho A V_o^3} = 0.593$$

where ρ is the air density, V_o is the free stream wind velocity, and A is the area swept by the rotor blades (11,910 ft^2 or 1,179 m^2). Note from Figure 7.2 that the maximum C_P actually achieved is only 0.47 (NASA 100 kW system will operate at a maximum 80% efficiency or 0.47/0.59).

In order to determine the extent of low-speed wake expansion downstream of the rotor, two distinct situations must be investigated. First is the rotor operation in winds between 8 and 18 mph (3.8 and 8 m/sec). For these cases the power coefficient is a function of the tip speed to wind speed ratio λ, which is itself a function only of wind speed, the rotational speed being constant.

The functional relationship for λ and wind speed between 8 and 18 mph is shown in Figure 7.4. Using Figure 7.2, the power extracted from the airstream is determined from

$$P = C_P \frac{\rho A V_o^3}{2} \quad (2)$$

Figure 7.4: Tip Speed to Wind Speed Ratio Versus Wind Speed

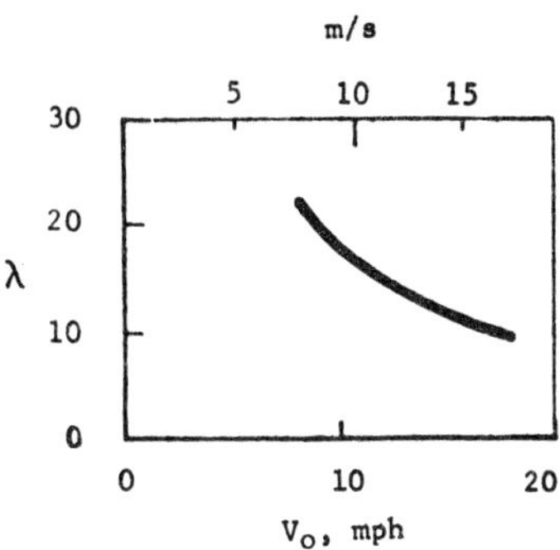

Source: ERDA/NSF/07378-75/1

It can also be shown that

$$\frac{4P}{\rho A} = (V_o + V_3)^2 (V_o - V_3) \quad (3)$$

from which the velocity far downstream V_3 and, by continuity, the downstream wake area can be determined.

The second case involves wind speeds above 18 mph (8 m/sec) but less than 60 mph (27 m/sec). Within this regime the turbine blades are rotated and blade pitch controlled such that a constant 133 kW are extracted from the airstream. Thus, equation (3) can be used directly by setting P equal to a constant 133 kW.

Results of calculations over the entire spectrum of operating wind speeds are displayed graphically in Figures 7.5 and 7.6. The maximum wake diameter occurs at a wind speed just less than 18 mph and increases the radial zone of influence by only 7.5 feet. Since the rotor hub with this design is located 100 feet above the ground, little or no aerodynamic effect of the wind rotor at ground level is anticipated due to wake expansion. Also, the decrease in wake velocity is minimal, especially at the higher wind speeds.

In Figure 7.6, it has been assumed that no rotational energy is imparted to the flow. However, since the free stream is irrotational, some rotational kinetic energy will be imparted to the wake by the rotor. This means that in order to extract a net 133 kW from the flow, more than 133 kW must be extracted from the translational kinetic energy. Thus, in reality, the V_3 values should be slightly less than indicated in Figure 7.5 and the R_3 values slightly larger. The induced flow which would be caused by both the bound and free vortices has also been considered a second order effect.

Figure 7.5: Wake Velocity

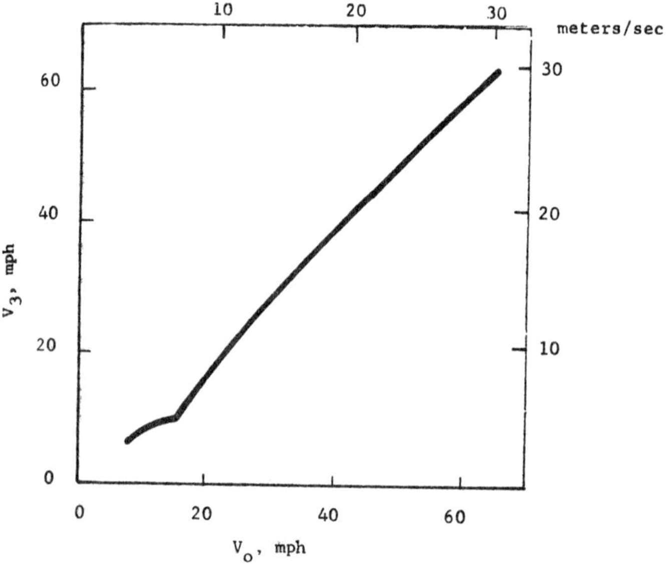

Figure 7.6: Wake Radial Dimension

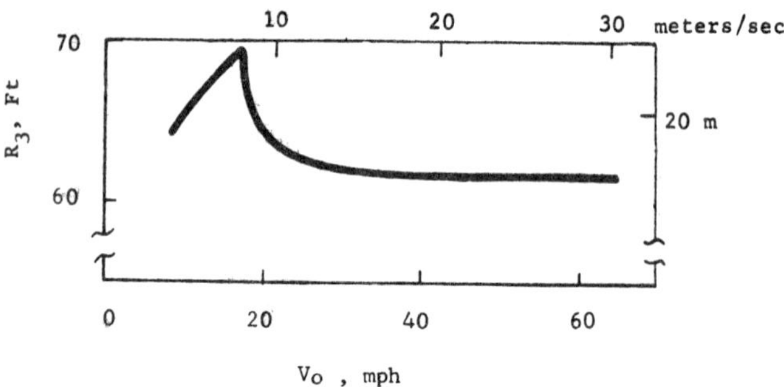

Source: ERDA/NSF/07378-75/1

No consideration has been given to the axial development of the wake downstream of the rotor. If, indeed, it had been determined that the rotor's wake would intersect the ground, it would be necessary to calculate the point of contact.

Application of the energy equation indicates a negligible decrease in total temperature across the disk of 0.03°F (0.05°C) and a pressure drop of 0.07 in (0.18 cm) H_2O at the 18 mph (8 m/sec) design point, together with a 6 mph (2.7 m/sec) reduction in wind velocity.

Potential Effects Literature Review

A review of the literature dealing with specific topics relevant to the environmental effects of wind energy generation was made. Subjects reviewed include microhabitat alterations and/or adaptations to wind, effects of surface winds on natural and agricultural systems, and ecological and meteorological information relative to regions which have been identified as potential wind energy conversion areas. Operational meteorological data for previously operating systems, where available, were also noted.

Very little, if any, information was available concerning the environmental effects of wind energy conversion. In many cases implications had to be drawn from facilities similarly affecting winds, such as the wind barriers of agriculture and TV and radio towers. The aerodynamic flow analysis of the ERDA/NASA 100 kW WES was used as a conceptual model for environmental considerations.

Environment of ERDA/NASA 100 kW Experimental Wind Turbine

An environmental baseline description of the vicinity of the ERDA/NASA 100 kW Experimental Wind Turbine at Sandusky, Ohio, has been prepared in support of field studies carried out within this research program and future environmental studies to be focused at this facility. Environment, as defined for this study, includes the physical environment—climatology, micrometeorology, soils, and biological environment—plants, birds, mammals, and insects.

The aerodynamic flow analysis represents the basis for defining the zone of influence of the WES. Field studies were designed to extend beyond this zone in order that adequate controls would be available in the event of a gap between the theoretically determined zone and the actual, operational zone of influence.

Area Climatology: Climatological data for the Plum Brook Station are available from several sources. Currently, the nearest first-order National Weather Service stations are at the airports in Toledo and Cleveland. These stations make hourly weather observations. Between 1936 and 1962 there was a weather station at the Post Office Building in Sandusky (Erie County) and hourly observations were made from 7:00 AM to 7:00 PM. This station is now operated as a National Weather Service cooperative-observer station making daily measurements of precipitation, maximum temperature, and minimum temperature. Other nearby cooperative-observer stations are at Put-in-Bay on South Bass Island (Erie County) and at Norwalk in northern Huron County, just south of Erie County.

Measurements of rainfall, temperature, pressure and wind have been made for various periods of time and for various purposes at several sites on Plum Brook Station. The Plum Brook data include the tower measurements of wind speed and direction that are being made in support of the wind energy program.

The prevailing (most frequently occurring) wind direction at Toledo is southwest while at Cleveland it is south. Records (3) from the Toledo Express Airport show that winds from the southwest predominate during all months of the year except January and April, when the prevailing direction is west-southwest, and May, when winds from the east-northeast predominate. March is the only month at Cleveland's Hopkins Airport when south is not the prevailing wind direction; winds are out of the west-northwest during this month. At the Sandusky Post Office during the hours of record, southwest winds prevailed for all months of the year.

Taking the number of days between the last freezing temperature of spring and the first freezing temperature in fall as the crop-growing season, this season averages 198 days in Sandusky (4). However, there is a sharp gradient in the length of the growing season across Erie County as a consequence of the tempering influence of Lake Erie on the areas close to the lake shore. At Put-In-Bay the average growing season is 208 days while at Norwalk it is only 155 days.

Precipitation in Erie County is well distributed throughout the year with fall being the driest season. Thunderstorms occur on about 35 days each year, being most frequent from April through August. The heaviest rain on record in Ohio through 1966 fell between 2:00 AM and 10:00 PM on July 12, 1966, at Sandusky. Total rainfall during this period was 26.7 cm. Three years later on July 4-5, 1969, 23.4 cm of rain was measured in the Plum Brook Station rain gage maintained by the Department of Interior (5).

During the period of record of the Department of Interior rain gage (Nov. 14, 1968, to Nov. 25, 1974), rainfalls of over 2.5 cm occurred several times each year between April and September. The frequency ranged from two in 1970 to eight in 1969.

Mean rainfall for Sandusky and Norwalk is between 85 and 90 cm/yr. Snowfall at these two cities averages 74 cm/yr. In the winter the predominant amount of precipitation throughout Ohio falls as rain, not snow.

Micrometeorological Environment: Two rain-gage networks have been operated in the vicinity of the Plum Brook 100 kW wind generator during part of this study. The first network consisting of nine farm-type clear plastic gages was operated between September 25 and December 10, 1975. These gages were placed around the 100 kW WES at various distances. The network was discontinued with the onset of freezing temperatures to avoid damage to the gages. The second network was set up on November 6, 1975, and consisted of seven "Clearvu" type butyrate plastic gages having a 4 inch diameter opening funneling to an inner tube. These gages were located at various distances to the northeast of the 100 kW generator. Each gage for both networks was attached to a wooden or metal fencepost so that the gage opening was 13 to 15 cm (5-6 in) above the top of the post and 1.8 to 2.0 meters (5.5 to 6.0 ft) above the ground.

The first network (M-network) was operated with the intention of gathering information about rainfall patterns in the area of the WES which might be caused by the presence of trees, buildings, or the turbine and tower itself. Data show a general uniformity except for gage M-1 which was located 30 meters southwest of the wind turbine tower. The total rainfall at this gage for the period was 1 cm below any other gage.

On six of the 13 rain days during this period, this site recorded as low or lower rainfall than any other site. However, during the rain of October 17, its catch was 0.5 cm less than the catch of any other rain gage. An observer present during this rain reported that it was accompanied by a strong northeasterly wind so that the raindrops had a large horizontal movement. This could account for the lower catch in a gage located near the southwest corner of the WES tower.

Observations made during the first month of operation of the second rain-gage network (P-network) have shown that the site within 30 meters of the WES tower (gage P-1) consistently records slightly less rain than the other sites further away.

A comparison check was made on the rain gages of the M-network and it was verified that gage M-1 was identical in its measuring scale to the other gages. Thus the low values measured at this site are real. A similar check is planned for the P-network. A review of wind records from the meteorological tower will be made to determine whether the low rainfalls in the rain gages nearest to the WES tower on the southwest and northeast sides are correlated with wind-driven rains from these two directions.

Vegetation: A vegetation (habitat) map (Figure 7.7) was prepared with the assistance of the Branch of Population Management, U.S. Fish and Wildlife Service (6). The cover or habitat types utilized were modified from a technique of Cannon, et al.

The environment in the immediate vicinity of the WES can be described as an old field dominated by grasses and forbs. The area has been out of cultivation for several decades. Mowing of the old fields every several years prevents the encroachment of woody species. Species composition across the field is very irregular, apparently reflecting the variable soil conditions. Areas of sandy, well-drained soils are occupied by goldenrod, dock, Queen Anne's lace, thistles, ragweeds (*Ambrosia* spp.), various berry vines (*Rubus* spp.), and several grasses (Poaceae). Low lying and poorly drained soils are vegetated by the above in reduced importance with localized colonies of sweet fern (*Comptonia* sp.), horsetails (*Equisitum* sp.), and increased abundance of sedges (Cyperaceae).

The nearest forested area to the WES is approximately 730 meters in a southerly direction. To the east and west forested lands are not encountered within the first kilometer. Within the NASA-Plum Brook Station, forest communities are of two basic types. Both are composed solely of hardwood species. One type occupies the less well drained areas and/or more recently forested types. These are most frequently dominated by cottonwoods, willows, dogwoods, and maples. The other types, occupying land uncut for longer periods, are dominated by elms, hickories, ashes, and oaks. The groundcover and understory of all wooded areas at Plum Brook are very poorly developed because of the excess deer population. Wildflowers and deer-palatable shrubs are virtually absent. There is a noticeable browse line.

Soils: Soils in the vicinity of the 100 kW WES are depicted in Figure 7.8. The loamy fine sands are descendants of Lake Erie beach soils. They are described as "deep, nearly level to moderately sloping, well drained to moderately well drained soils that have a subsoil of loamy fine sand and fine sand on hills and ridges" (7). These soils are productive, good for vegetable production, but are not dependable for pasture through late summer due to excessive drainage. Under-cultivation wind erosion can be a hazard in these soils.

The silty-clay-loam soils were formed along drainage features developed from the ancient Lake Erie bottom. These soils can be described as "deep, nearly

Figure 7.7: Plant Communities in the Vicinity of ERDA/NASA 100 kW Experimental Wind Turbine

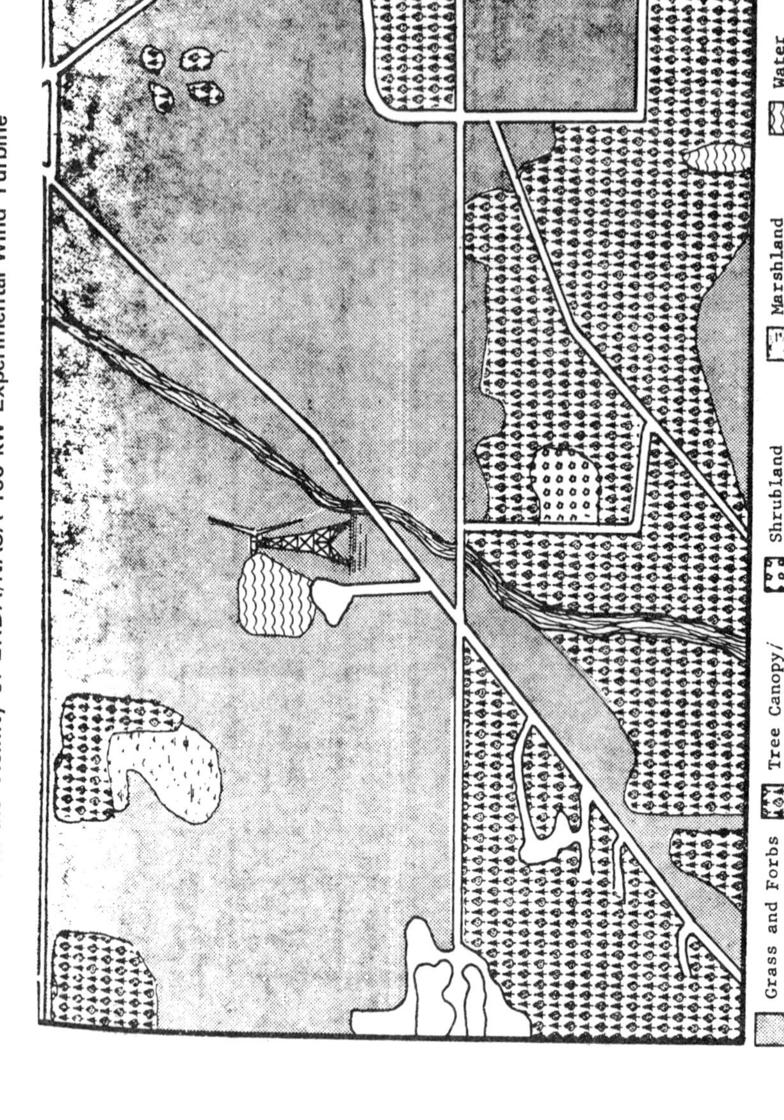

Source: ERDA/NSF/07378-75/1

Figure 7.8: Soils in the Vicinity of the ERDA/NASA 100 kW Experimental Wind Turbine

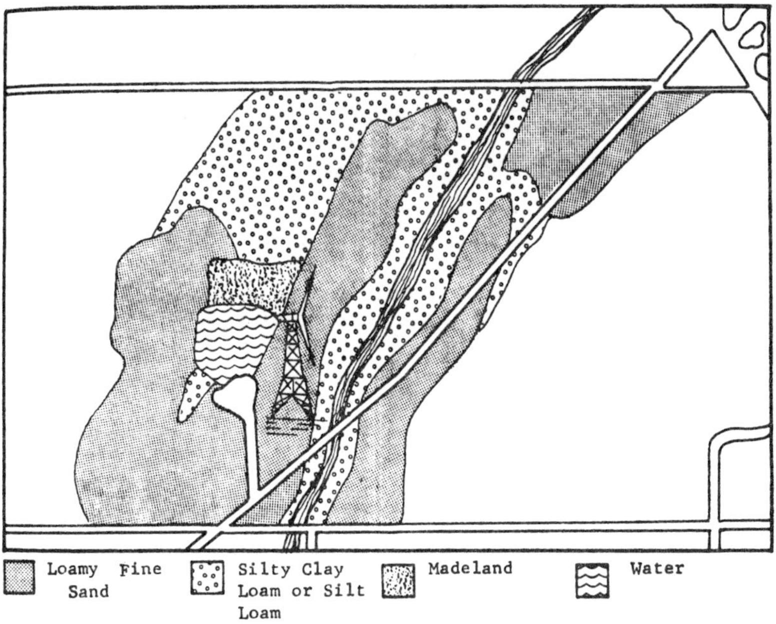

Loamy Fine Sand
Silty Clay Loam or Silt Loam
Madeland
Water

Source: ERDA/NSF/07378-75/1

level, somewhat poorly drained to very poorly drained soils that have a subsoil of silty clay to silty clay loam" (7). These soils are productive with poor drainage being the main limitation. They are not susceptible to wind erosion.

Mammals: Due to the low probability of any negative impacts of the wind turbine at NASA Plum Brook on mammals, no formal field surveys have been conducted by Battelle. However, communication with personnel of the U.S. Fish and Wildlife Service and a review of pertinent literature have resulted in a general picture of NASA-Plum Brook's mammals.

Of the 42 mammals likely to occur at NASA-Plum Brook, eight can be considered very common. The white-tailed deer is by far the most visible mammal, with recent population estimates ranging from 2,100 in 1974 to 2,500 in 1975. The lack of natural population control in this large deer herd has resulted in a controlled hunt to bring the number of deer into an equilibrium with their food supply.

Birds: Several excellent references are available which describe the 286 species of birds usually occurring in Ohio during one or more seasons (8). Two of these references contain recent information particularly pertinent to the southwestern Lake Erie area (9)(10). Additional baseline information on avifauna specific to NASA-Plum Brook was obtained by the two brief surveys described below; detailed field surveys of nocturnal migrants will be conducted in a second phase.

Breeding Bird Surveys — Breeding bird surveys on June 26 and 27, 1975, involved a roadside survey similar to that used by the U.S. Fish and Wildlife Service (11) and walking transects around the wind turbine and in a nearby woods. The roadside survey consisted of 50 stops spaced about 0.8 km (0.5 mi) apart along a winding course covering most of NASA-Plum Brook and lightly traveled secondary roads in the vicinity. Three minutes were spent at each stop, where all birds seen or heard within a 0.4 km radius were recorded. Location of stops were recorded on a topographic map for future use as an index of population changes.

Transect surveys were designed to record bird species at varying distances close to the wind turbine. The 2.75 hours of actual survey time prevented any analysis of relative abundance.

The five most abundant of the 43 bird species recorded during the roadside survey were the house sparrow, red-winged blackbird, American robin, indigo bunting, and mourning dove. Of special interest among those recorded are six species whose populations are presently declining in Ohio (12). Three species seen on this survey are also on the "Blue List", which is intended as an early warning of potentially dangerous, apparently noncyclical population declines in the bird species listed.

Seventeen species of birds were recorded along four transects in the fields and around the pond near the wind turbine. Thirteen species of birds were seen in a mature woods to the northwest of the wind turbine. Presence of the grasshopper sparrow near the wind turbine is interesting due to its inclusion on the "Blue List" (13). The rough-winged swallows seen at the pond are presently declining in Ohio (12).

Although not recorded during Battelle's brief surveys, six additional bird species have been seen at NASA-Plum Brook during the breeding season that are either rare or declining in Ohio and/or on the "Blue List".

Migratory Bird Surveys — Preliminary surveys of migratory birds were made at NASA-Plum Brook and other areas in the vicinity which are heavily used by migrants. Two observers conducted surveys during September 15 through 18, 1975, which is considered to be during the peak period for migration of small, night-migrating birds (9). The following three survey techniques were tested to see if they would be useful during the second phase of this project in identification and quantification of nocturnal migrants in proximity to the wind turbine: (1) ceilometer (narrow-beam spotlight) survey of low-level nocturnal migrants, (2) daytime species identification of grounded nocturnal migrants, and (3) study of night migration using the ARSR-2 radar at the en route air traffic control facility in Oberlin, Ohio. In addition, NASA-Plum Brook personnel made daily searches around the meteorological tower to look for any migrants which collided with the tower during the previous night.

Ceilometer Surveys: Surveys of low-level nocturnal migrants were made on September 15-17, 1975, between 1 and 2.5 hours after sunset (2030 to 2200 hours), using a portable ceilometer and the techniques described by Gauthreaux (14) and Able and Gauthreaux (15).

Rain ended observations on two of the evenings and was falling during at least part of the night on all three nights. Rain plus warm fronts over, or north of, northern Ohio were undoubtedly responsible for the extremely low migration traffic rates observed on all three nights. Migration traffic rates expressed in birds per kilometer of front per hour were 1,150 for September 15, and 94 for both September 16 and 17. Since Lowery and Newman (16) found an average of 4,250 birds per mile of front per hour during the same time of the night in their continent-wide study of bird migration using the moon-watching technique, it is obvious that the values obtained at NASA-Plum Brook indicate very light to negligible migration. The surveys at NASA-Plum Brook support the observations of other investigators, who found that rain can either delay migration or even stop it in progress (17)(18). Future ceilometer surveys will be for a sufficient number of nights to insure the observation of several nights of heavy migration.

Radar Surveys: Displays from the radars based at Detroit and Brecksville (near Cleveland) were viewed at the Oberlin Airways Facility Sector on the nights of September 15 through 17 between 2300 and 0100 hours using the techniques described by Gauthreaux (19). These two radars have overlapping ranges permitting a view of migration over all of southwestern Lake Erie, and thus give a broad view of migratory activity in the vicinity of NASA-Plum Brook.

Although some isolated flocks of waterfowl or shorebirds, represented by large, bright echoes on the radar screen, were seen, almost none of the small, mist-like echoes associated with small landbirds flying individually were noted.

Review of synoptic weather maps by the National Weather Service indicated that rain and warm fronts passed over the study area and moved northeast bringing even more rain over all of Ohio and Lake Erie during the survey period. These weather conditions are just the opposite of those described by Able (20) as the type of weather in which heavy migratory flights occur during autumn. He found that the largest flights occurred in the southward movement of dry polar air which immediately follows the passage of a cold front. Radar observations made of the southwestern Lake Erie area under these optimum conditions, would undoubtedly have resulted in a much better picture of nocturnal migration over NASA-Plum Brook. Detailed surveys are planned for a sufficient number of nights to insure that these optimum conditions will occur during radar observations.

Daytime Surveys of Grounded Migrants: Areas heavily used by migrants were surveyed at NASA-Plum Brook and Magee Marsh Wildlife Area (northwest of Port Clinton, Ohio) to document species of grounded nocturnal migrants. Magee Marsh was chosen as a survey area for shorebirds, since NASA-Plum Brook does not have suitable habitat for these migrants, and marshes in the immediate vicinity that are connected to Lake Erie were flooded from the high lake level. Surveys at both sites were hampered by rainy and misty weather.

Sixty species of birds were observed at NASA-Plum Brook during almost 6 hr of intensive searching on September 16 and 17, 1975. Among those species recorded were four species of thrushes, one species of vireo, and seven species of warblers, which are considered to be low-level night migrants (19)(21).

Seventy species of birds were observed at Magee Marsh Wildlife Area during 2 hours of intensive searching on September 18, 1975. Included in this list were two species of geese, four species of ducks, and ten species of shorebirds, which are normally high-level nocturnal migrants (19)(22). Many species considered low-level nocturnal migrants were also seen, including three species of thrushes, three species of vireos, and fifteen species of warblers. These three families of low-level migrants (Parulidae, Vireonidae, and Turdidae) accounted for the greatest percentage of all birds killed in the widespread series of mass mortalities reported from twenty-five sites in the eastern United States (23).

Bird-Tower Collisions: Searches for tower killed birds indicate that the meteorological tower had a minimal impact on nocturnal migration during the period when searches were made. One bird, a female blackburnian warbler (*Dendroica fusca*), was collected near the base of the meteorological tower during searches in late May by NASA-Plum Brook personnel. This species is considered a common transient, but not a breeding species in north-central Ohio (9)(8). The lack of data during peak migration periods, particularly since the wind turbine was not operational at night, prevents any additional generalizations about impacts.

Future searches are planned to coincide with nights when the wind turbine will be operational. The search period will include at least one test of the effect of scavengers on tower-killed birds. However, the negligible number of birds found to date indicates that scavengers are not likely to patrol the area with any regularity.

Flying Insects: Insects close to the wind turbine were sampled by sweep net in mid-August. Six samples were taken, three in the old field, one on the bank of Plum Brook, one along the reservoir, and one in mowed grass adjacent to the 4.1 kW wind turbine. Samples 1 to 4 and 6 consisted of 100 sweeps and sample 5 of 50 sweeps. The adult insects from each sample were sorted to taxonomic groups except for wingless forms, which were excluded from the analyses. Aerial arthropods other than insects, such as spiders, were also excluded.

The most common groups of insects collected were dipterans (flies), homopterans (hoppers), and orthopterans (grasshoppers). Dipterans, particularly midges, were the most common insect. Midges, the larvae of which are usually aquatic, are generally associated with environments having wet areas such as ditches, streams, ponds, and lakes. The other major categories of insects present were homopterans (leafhoppers and froghoppers) and orthopterans (grasshoppers and crickets). Those insects present in these two orders are strictly herbivorous and were common to all samples; they were most numerous in the drier areas and are to be expected in herbaceous vegetation.

Some froghoppers and leafhoppers feed on clover, alfalfa, strawberries, apples, potatoes, and sugarbeets, and can be economic pests (24)(25). Grasshoppers and crickets are found in nearly all herbaceous communities; tree crickets may be encountered in trees, shrubs, and weeds. Tree crickets lay their eggs in bark and stems which can result in damage to the trees (24); adults may become pests by eating holes in ripe fruit (25).

Numbers of species and individuals recorded at Plum Brook varied considerably in the field, which supported both high and low densities and diversity. This reflects differences in the environmental conditions, food, and cover within the field. The grassy area provided good habitat for midges and leafhoppers. The reservoir edge provided the best habitat for beetles; midges were also numerous along the water's edge.

Conclusions and Implications

The natural environment is inherently extremely variable even at a given location. This variability is particularly obvious when one deals with harnessing wind in a "windy" location. The biological components occupying these windy environments are generally adapted to fluctuations in wind and other wind-influenced microclimatic parameters such as temperature and humidity. Wind turbines sited within such environments generally are not expected to alter the local environment beyond the normal fluctuation extremes. Furthermore, identification of wind-turbine-induced environmental effects will be difficult to key out and quantify within this variable environment. The environmental studies proposed to evaluate the operating influences of the ERDA/NASA 100 kW WES should provide experience over the time frames adequate to record natural environmental fluctuations, while focusing on the identification of any turbine-influenced environmental effects.

Wind energy systems appear to represent a potential power supply which will result in minimal alteration or intrusion into the environment. Some land will be required for the tower and control building. At present the area required to provide an adequate "windview" for the rotor is not well defined, but under most conditions other land uses with certain height restrictions will be compatible in this zone. The direct generation of electricity utilizing wind will result in no chemical effluents and only minimal heat evolution (that resulting from any energy state change—in this case from kinetic to electrical through mechanical means).

While no major effluents, land consumption, mining, or fuels processing and transportation are associated with wind-electrical conversion, these wind energy facilities will influence the physical environment into which they are placed and may also affect the biological environment occupying the same area. The results of this preliminary study indicate that under many, and, in fact, most environmental configurations the effects will be minimal and often not measurable. In two cases, further studies are being made to more clearly define these potential effects through observation of an operating wind turbine. Generally, these effects are (1) ground-level microclimatic alterations induced by the wind turbine and tower and (2) potential collisions of migratory organisms, particularly birds, with the rotating blades.

The microclimate in an area adjacent to the wind turbine and tower will be altered by the presence of this structure, the size and shape of this area varying with the changing wind. Minor changes will be observed in wind speed, temperature, available moisture and various other wind-influenced atmospheric parameters. The degree of change, size, and shape of the zone of influence are largely dependent on the design solidity of the rotor and penetrability and size of the tower. These anticipated microclimatic changes are quite small and are not likely to induce measurable secondary effects to flora and fauna, including agricultural species occupying the area.

The theoretically determined zone of influence for the 100 kW WES design indicates a one-third reduction in wind velocity within the downstream wake at the 18 mph design speed for this system. This reduced wind velocity is accompanied by minor reductions in air temperature and pressure. The wake expands slightly in the downstream direction with a maximum expansion of 7.5 ft beyond the disk diameter. Thus, under normal operating conditions the altered microclimate within the wake would be no closer than approximately 34 ft to a level ground surface. Assuming that any trees have been cleared or are shorter than this height within the downwind proximity, little or no effect to the surface biota and microclimate is anticipated.

The potential ground level effects of slightly altered solar insolation (due to shadow of tower) and precipitation (due to interception and altered wind currents) together with any other alterations in microclimate parameters within the zone of influence are anticipated to be quite small for WES designs of similar height and permeability to the 100 kW unit. In fact, these effects may not be measurable due to the natural variability of atmospheric parameters in locations of uneven terrain and/or frequent windy conditions suitable for WES siting.

Should wind energy systems be designed and sited such that the reduced-velocity wake passes over the ground surface, several additional factors must be considered. Geiger (26) notes in his discussion of windbreaks that the same effect may be favorable in one microclimate and unfavorable in another. More specifically reduced wind velocity passing over crops which exhibit reduced productivity when exposed to high winds is a positive effect. Reduced wind and, therefore, reduced evaporation is a plus in arid or semiarid environments. This same increased moisture under humid conditions may provide the necessary vehicle for plant disease as a negative factor.

These examples demonstrate the types of effects which can result from wind-sheltering of a location. Because wind turbines require a largely unobstructed windview and because state-of-the-art airfoil design can be only partially successful in converting wind to another energy form, it is reasonable to assume that high degrees of ground level sheltering by wind turbines will not occur in the near future. Therefore, effects either positive or negative of the degree mentioned above are not likely to occur.

In dealing with potential siting or wind energy systems in such diverse environments as the Aleutians or Hawaiian Islands, the North Atlantic or Pacific Coasts, or the Great Lakes Shores or Great Plains a vast number of environmental variables will be encountered. In heavily forested, mountainous terrain where clearcutting of forests may be required to provide an adequate "windview" for the turbine care will be required in designing the clearcut area so as to minimize windthrow of adjacent forests. Experience is available within the forestry industry to assist in determining the windthrow potential and its minimization for a particular wooded site.

Siting of a large wind energy system within an orchard, vineyard, or other areas where crops are grown which have a particularly environmentally sensitive period, e.g., the spring flowering/frost consideration for fruit trees, should be accompanied by a consideration of the microclimate of the proposed operating system to determine if any potential conflict would occur at that site. The anticipated localized degree of wind diminution by an operating wind turbine is not expected to significantly reduce either wind erosion of soils or evaporation of irrigation waters in areas where these conditions are a concern. There is no evidence that even vast numbers of large conventionally designed wind turbines for electrical power production would significantly influence these conditions.

The significance of birds, bats, and insects colliding with the rotating blades of a wind energy conversion system varies with the location, time of day, season of the year, and prevailing climatic conditions. Migratory songbirds are more likely to be affected on dark foggy nights at the peaks of migration in spring and fall. Daytime migratory birds can be expected to avoid the blades as would most nighttime migrants in fair weather. Wind turbine towers greater than 500 ft in height increase the number of birds exposed to potential collision with the blades. Similarly, wind turbines sited along the crests of mountain ridges or on isolated peaks increase the effective height of the blades above the ground surface with respect to the birds.

Generally, birds avoid obstacles which they can see; however, behavioral information concerning birds confronted with the blades is only presently being collected as a part of the second year of study. Particular attention should be taken in siting WES facilities in the breeding or wintering territory or along migratory concentrations of threatened or endangered bird or bat species.

In general, the environmental effects of a wind turbine facility will be analogous to those of other towers of similar design and height. The rotating blades represent few unique aspects. The potential for collision with airborne organisms can largely be controlled by design and siting. Because of the variability of winds at many potential wind turbine sites, the existing biological environment is adapted to fluctuating conditions. This adaptation will serve to minimize the effects of minor changes in microclimate. Turbine-induced changes will therefore tend to be "lost" in the normal background variability.

REFERENCES

(1) Putholff, R.L. and Sirocky, P.J., "Preliminary Design of a 100 kW Wind Turbine Generator," NASA TM X-71585. NASA Technical Memorandum. Paper presented at International Solar Energy Society Conference, held in Fort Collins, CO, (Aug. 21-23, 1974).

(2) Wilson, R.E., and Lissaman, P.B.S., *Applied Aerodynamics of Wind Power Machines,* NSF/RA/N-74-113, National Science Foundation, Research Applied to National Needs, Washington, DC, PB-238 595 (1974).

(3) U.S. Department of Commerce, Weather Bureau, "Climates of the States—Ohio," *Climatography of the United States* No. 60-33 (1959).

(4) U.S. Department of Commerce, Environmental Science Services Administration, "Climatological Summary—Sandusky, Ohio," *Climatography of the United States* No. 20-33-76 (1969).

(5) Dolbeer. R., personal communication, U.S. Department of the Interior (December 10, 1975).

(6) Andrews, D.A., "Habitat Cover Map of NASA Plum Brook Station, Sandusky, Ohio," U.S. Department of the Interior, Fish and Wildlife Service, Branch of Population Management, Sandusky, OH (1975).

(7) Redmond, C.E., Hole, T.J.F., Innis, C.H. and Wachtman, M., "Soil Survey of Erie County, Ohio," U.S. Department of Agriculture, Soil Conservation Service (1973).

(8) Trautman, M.B. and Trautman, M.A., "Annotated List of the Birds of Ohio," *Ohio Journal of Science* 68(5):257-332 (1968).
(9) Campbell, L., *Birds of the Toledo Area,* Blade, Toledo, OH (1968).
(10) U.S. Department of the Interior, Fish and Wildlife Service, "Birds of the Ottawa, Cedar Point, and West Sister Island National Wildlife Refuges," U.S. Department of the Interior, Fish and Wildlife Service, Refuge Leaflet 249 (1970).
(11) Robbins, C.S. and VanVelzen, W.T., "The Breeding Bird Survey 1967 and 1968," U.S. Department of the Interior, Fish and Wildlife Service, Special Scientific Report, Wildlife No. 124, Washington, DC (1969).
(12) Smith, H.G., Burnard, R.K., Good, E.E. and Keener, J.M., "Rare and Endangered Vertebrates of Ohio," *Ohio Journal of Science* 73(5):257-271 (1973).
(13) Arbib, R.B., "The Blue List for 1975," *American Birds* 28(6):971-974 (1974).
(14) Gauthreaux, S.A., Jr., "A Portable Ceilometer Technique for Studying Low-Level Nocturnal Migration," *Bird-Banding* 40(4):309-320 (1969).
(15) Able, K.P. and Gauthreaux, S.A., Jr., "Quantification of Nocturnal Passerine Migration with a Portable Ceilometer," *Condor* 77(1):92-96 (1975).
(16) Lowery, G.H., Jr. and Newman, R.J., "A Continent-Wide View of Bird Migration on Four Nights in October" *Auk* 83(4):547-586 (1966).
(17) Gauthreaux, S.A., Jr., "A Radar and Direct Visual Study of Passerine Spring Migration in Southern Louisiana," *Auk* 88(2):343-365 (1971).
(18) Bellrose, F.C., "The Distribution of Nocturnal Migrants in the Air Space," *Auk* 88(2): 397-424 (1971).
(19) Gauthreaux, S.A., Jr., "Radar Ornithology: Bird Echoes on Weather and Airport Surveillance Radars," Clemson University, Department of Zoology, Clemson, SC (1975).
(20) Able, K.P., "The Role of Weather Variables and Flight Direction in Determining the Magnitude of Nocturnal Bird Migration," *Ecology* 54(5):1031-1041 (1973).
(21) Berger, A.J., *Bird Study,* John Wiley and Sons, Inc., New York (1961).
(22) Pettingill, O.S., Jr., *Ornithology in Laboratory and Field,* Fourth Edition, Burgess Publishing Co., Minneapolis (1970).
(23) Johnston, D.W. and Haines, T.P., "Analysis of Mass Bird Mortality in October, 1954," *Auk* 74:447-458 (1957).
(24) Borror, D.J., and DeLong, D.M., *An Introduction to the Study of Insects,* 3rd edition, Holt, Rinehart, and Winston, New York (1971).
(25) Metcalf, C.L., Flint, W.P. and Metcalf, R.F., *Destructive and Useful Insects—Their Habits and Control,* 4th edition, McGraw-Hill Book Co., New York (1962).
(26) Geiger, R., *The Climate Near the Ground,* 4th edition, translated by Scripta Technica, Inc., Harvard University Press, Cambridge, MA (1965).

APPENDIX

This appendix is a list of names and addresses of manufacturers, researchers and distributors currently involved in the development of wind energy conversion systems. The list is based upon an index prepared by the American Wind Energy Association for Rockwell International Corporation (DOE/RF/3533-78/1)

Aeolian Energy Co.
RD #4, Ligonier, PA 15658

Aerolectric Co.
13517 Winter Lane, Cresaptown, MD 21502

Aerovironment Inc.
145 Vista Ave., Pasadena, CA 91107

Alternative Energy Systems
Box 497 A, Kennebunkport, ME 04046

American Energy Alternatives, Inc. (Amerenalt Corp.)
5420 Arapahoe, Boulder, CO 80302

American Institute of Architects
1735 New York Ave., NW, Washington, DC 20006

American Wind Energy Association, National Office
54468 CR 31, Bristol, IN 46507

Applied Solar Technology, Inc.
5523 Mystic Ct., Columbus, MD 21044

Eldon J. Arms
Box 7, Woodman, WI 53827

Athabasca Research Corp.
11210 143rd St., Edmonton, Alberta, Canada, T5M IV5

Baker Mfg. Co.
133 Enterprise St., Evansville, WI 53536

Battelle Columbus Laboratories
505 King Ave., Columbus, OH 43201

Battelle, Pacific Northwest Lab.; Atmos. Sci. Dept.
P.O. Box 999, Richland, WA 99352

Karl J. Bea Associates
128 Dewitt St., Syracuse, NY 13203

Brookhaven National Laboratory, Associated Universities, Inc.
Upton, NY 11973

Bureau of Reclamation, Div. of Atmos. Water Res. Mgt., Eng. and Res. Center,
Denver, CO 80225

California State University
Northridge, CA 91330

Center for Natural Resources, Energy & Transportation
United Nations, Rm. DC 828, New York, NY 10017

Central NE Public Power & Irrigation District
P.O. Box 356, Holdrege, NE 68949

Colorado State University, Fluid Dynamics and Wind, Dept. of Ag. Eng. and Dept. of Atmos. Sci.
Fort Collins, CO 80523

Coulson Wind Electric
RFD #1, Box 225, Polk City, IA 50226

Crews, MacInnes & Hoffman
4111 Minnesota Dr., Anchorage, AK 99503

Disco, Inc.
1010 Greenwood Lake Tpk., Ringwood, NJ 07456

Mark Drabick
322 North 7th St., #11, Allentown, PA 18102

David Drozd
558 Churchill St., Southington, CT 06489

Edmund Scientific Co.
101 East Gloucester Pk., Barrington, NJ 08007

Energy Alternatives, Inc.
69 Amherst Rd., Leverett, MA 01054

Energy Development Co.
179 E. Rd. #2, Hamburg, PA 19526

Enertech Corp.
Box 420, Norwich, VT 05055

Environmental Energies, Inc.
Box 73, Front St., Copemish, MI 49625

Eugene Water & Electric Board
Box 10148, Eugene, OR 77401

Foundation for Rural Technology
903 Pine St., Boulder, CO 80302

Franklin Inst. Research Labs.
20th St. & the Parkway, Philadelphia, PA 19103

General Electric, Space Div.
Box 8555, Philadelphia, PA 19101

General Electric Co., Valley Forge Space Center, King of Prussia Park
P.O. Box 8661, Philadelphia, PA 19101

Gentle Energy Systems, Inc.
Ireland Rd., Starksboro, VT 05487

The Georator Corp.
9016 Prince William St., Manassas, VA 22110

George Washington University, Program of Policy Studies in Sci. and Tech.
Washington, DC 20052

Georgia Institute of Technology, Eng. Exp. Station
Atlanta, GA 30332

Dr. Robert Gordon
620 West Huntington Dr., Arcadia, CA 91006

Martin L. Greenwald
RD #1, Thompson Ridge, NY 10985

Grumman Aerospace Corp., Fluid Dynamics Research Dept.
Bethpage, NY 11714

John B. Haldiman d.b.a. Haldi Ranch, Gila Co., AZ
P.O. Box 791, Phoenix, AZ 85001

Heller-Aller Co.
Box 29, Napoleon, OH 43545

Holyoke Gas & Electric Dept.
70 Suffolk St., Holyoke, MA 01040

Home Spun Power Co.
1085 Columbus, Fredericksburg, TX 78624

Honeywell Inc., System & Research Center MN17-2688
2600 Ridgway Pky., Minneapolis, MN 55413

Hydro-Catylator Corp.
3579 East 10th Ct., Hialeah, FL 33013

Independent Energy Systems
6043 Sterrettania Rd., Fairview, PA 16415

Independent Power Developers, Inc.
Box 1467, Noxon, MT 59853

Institute for Environmental Studies
1225 West Dayton St., Madison, WI 53706

Institute of Gas Technology, IIT Center
Chicago, IL 60616

Iowa State University, Elec. Eng. Dept.
Ames, IA 50011

I.R.E.Q.
P.O. Box 1000, Varennes, Quebec, Canada, JOL 2PO

JBF Scientific Corp.
1701 K St., NW #905, Washington, DC 20006

Robert M. Jeffreys
1204 Goodview Ave., Fayetteville, NC 28305

Kaman Aerospace Corp.
Old Windsor Rd., Bloomfield, CT 06002

Kaman Sciences Corp.; Energy Systems Devel. Group
P.O. Box 7463, Colorado Springs, CO 80933

Kedco, Inc.
9016 Aviation Blvd., Inglewood, CA 90301

Kelsey Wind Systems
P.O. Box 465, Zuni, NM 87327

Lawrence Livermore Lab., Univ. of California
P.O. Box 808, Livermore, CA 94550

Lehigh University, Dept. of Mech. Eng. & Mechanics
Bethlehem, PA 18105

Bob McBroom
217 Kansas, Holton, KS 66436

McDonnell-Douglas Aircraft Co.
Box 516, St. Louis, MO 63166

John L. Marple
633 Pleasant Ave., Saugatuck, MI 49453

Martin Marietta Corp.
1450 South Rolling Rd., Baltimore, MD 21227

Merritt Windmill, Inc.
Box 1374, Merritt, BC, Canada, VOK 2BO

Merton Eng. Co. Inc.
119 Stuart Rd., Racine, WI 53406

Mitre Corp., Westgate Research Park
McLean, VA 22101

N.A.S.A. - LeRC
21000 Brookpark Rd., Cleveland, OH 44135

Natural Power Corporation
6031 St. Clair Ave., Cleveland, OH 44103

Natural Power, Inc.
Francestown Twp, New Boston, NH 03070

Natural Power Systems, Inc.
3316 Augusta Ave., Omaha, NE 68144

Naval Construction Battalion Center, Civil Eng. Lab.
Port Hueneme, CA 93043

New Mexico State University, Physical Science Lab.
Las Cruces, NM 88003

Nielsen Engineering & Research
510 Clyde Ave., Mountain View, CA 95030

North Wind Power Co., Inc.
Box 315, Warren, VT 05674

Northwestern University, Civil Eng. Dept.
Evanston, IL 60201

O'Brock Windmill Distributors
9435 12th St., North Benton, OH 44449

Oceanographic Services, Inc.
135 East Ortega St., Santa Barbara, CA 93105

Oconto Electric Cooperative
Oconto Falls, WI 54154

S. Ohba & Associates
5969 North Elston Ave., Chicago, IL 60646

Oklahoma State University, School of Elec. Eng.
Stillwater, OK 74074

Owens-Corning Fiberglas
Fiberglas Tower, T7, Toledo, OH 43659

Pacific Gas & Electric, Dept. of Eng. Research
3400 Crow Canyon Blvd., San Ramon, CA 94583

Ted Pheiff
1103 Hillside Dr., Bettendorf, IA 52722

Pinson Energy Corp.
Box 7, Marstons Mills, MA 02648

Polytechnic Institute of New York
Rt. 110, Farmingdale, NY 11735

James A. Potter
12 Green House Blvd., West Hartford, CT 06110

Prairie Sun and Wind Co.
4408 62nd St., Lubbock, TX 79414

Princeton University, Dept. of Aerospace and Mech. Sci.
Princeton, NJ 08540

J. P. Rameau Inc.
Box 117, Bailey Harbor, WI 54202

Real Gas & Electric Co., Inc.
Box 193, Shingletown, CA 96088

Rockwell International; Atomics International Division, Wind Systems Prog.
Box 464, Golden, CO 80401

Tony Sadar
4346 Spring Garden Rd., Pittsburgh, PA 15212

Sandia Laboratories
Albuquerque, NY 87115

Ralph Schupback
321 13th St., Alva, OK 73717

Sheehan Consultants
5320 Ken Caryl Rd., Littleton, CO 80123

Sigma Engineering Co., Inc.
Box 5285, Lubbock, TX 79417

Solaero Research
Box 11613, Tucson, AZ 85734

Solar/Wind Center
RFD #1, Exeter, NH 03833

Soleq Corp.
5969 N. Elston Ave., Chicago, IL 60646

Structural Composites Industries, Inc.
6344 North Irwindale Ave., Azusa, CA 91702

Teledyne Aer-Cal
528 East Mission Rd., San Marcos, CA 92069

Chuck Tellas
3075 Judd Rd., Milan, MI 48160

Toward Tomorrow Fair
105 Hills North, University of Massachusetts, Amherst, MA 01003

T.R.C. of New England
125 Silas Deane Hwy., Wethersfield, CT 06109

United Technologies Research Center; Aerodynamics Center
Silver Lane, East Hartford, CT 06108

University of Dayton, Research Institute
Dayton, OH 45469

University of Illinois
Urbana, IL 61801

University of Massachusetts, Dept. of Eng.
Amherst, MA 01003

University of Rhode Island; Dept. of Mech. Eng.
Kingston, RI 02881

University of Wisconsin-Milwaukee
Milwaukee, WI 53201

Wadler Manufacturing Co., Inc.
Rt. #2, Box 76, Galena, KS 66739

West Texas State University
Box 248, Canyon, TX 79016

West Virginia University, Dept. of Aerospace Eng.
Morgantown, WV 26506

Westinghouse Electric Corp.
700 Bradock Ave., Bldg, 7L27, East Pittsburgh, PA 15112

Westinghouse Electric Corp., Advanced Energy Systems
P.O. Box 10864, Pittsburgh, PA 15236

Wind Power Digest
54468 CR 31, Bristol, IN 46507

Wind Power Systems, Inc.
P.O. Box 17323, San Diego, CA 92117

Wind-De-Go, Inc.
1940 West Cheryl, Phoenix, AZ 85021

Windependence Electric Co.
Box M1188, Ann Arbor, MI 48106

Windlite-Alaska
3701 Mt. View Dr., Anchorage, AK 99510

Windworks
Box 329, Rt. 3, Mukwonago, WI 53149

Windy-Ten Ltd.
150 Orchard View, Box 111, Shelby, MI 49455

Winpower Corp.
Newton, IA 50208

Alfred O. Wurdelman
Huron City Rd., Port Hope, MI 48468

SOURCES UTILIZED

The following reports were used in the preparation of this book:

BNL 50736

The Role of Wind Power in Electric Utilities by H. Davitian of Brookhaven National Laboratory, Associated Universities, Inc., for the Wind Energy Branch, Solar Electric Division, U.S. Department of Energy, September 1977.

BNWL-WIND-5

Synthesis of National Wind Energy Assessments by D.L. Elliot of Battelle Pacific Northwest Laboratories for the Energy Research and Development Administration, July 1977.

COO/2613-2

Segmented and Self-Adjusting Wind Turbine Rotors by P.F. Jordan and R.L. Goldman of Martin Marietta Laboratories prepared for ERDA, Division of Solar Energy, April 1976.

COO/2614-76/2

Self-Regulating Composite Bearingless Wind Turbine by M.C. Cheney and P.A.M. Spierings of United Technologies Research Center for ERDA, Division of Solar Energy, September 1976.

COO/2616-2(Pt. 1)

Investigation of Diffuser-Augmented Wind Turbines by R.A. Oman, K.M. Foreman and B.L. Gilbert of the Fluid Dynamics Research Department, Grumman Aerospace Corporation prepared for ERDA, January 1977.

COO/4130-77/1

Electrofluid Dynamic (EFD) Wind Driven Generator by J.E. Minardi, M.O. Lawson and G. Williams of University of Dayton Research Institute prepared for the National Science Foundation, October 1976.

DOE-NASA-1010-77

Large Experimental Wind Turbines—Where We Are Now by R.L. Thomas of Lewis Research Center, technical paper presented at Third Energy Technology/ Exposition sponsored by Government Institutes, Inc., Washington, DC, March 29-31, 1976.

DOE/NASA/9404-76/1

Design Study of Wind Turbines 50 kW to 3000 kW for Electric Utility Applications, Executive Summary, by Kaman Aerospace Corporation prepared for NASA, Lewis Research Center, for the U.S. Department of Energy, July 1977.

DOE/RF/3533-78/1

An Index of Manufacturers, Researchers, and Distributors Currently Involved in the Development of Wind Energy Conversion Systems, by the American Wind Energy Association prepared for Rockwell International, Atomics International Division, February 1978.

ERDA/NASA/1004-77/2

Dynamic Blade Loading in the ERDA/NASA 100 kW and 200 kW Wind Turbines by D.A. Spera, D.C. Janetzke and T.R. Richards of NASA, Lewis Research Center, prepared for ERDA, Division of Solar Energy, August 1977.

ERDA/NASA/9403-76/1

Design Study of Wind Turbines 50 kW to 3000 kW for Electric Utility Applications, Volume 1—Summary Report by General Electric Company, Valley Forge Space Center, prepared for NASA, Lewis Research Center, as part of ERDA, Division of Solar Energy, Federal Wind Energy Program, September 1976.

ERDA/NSF/07378-75/1

Evaluation of the Potential Environmental Effects of Wind Energy System Development by S.E. Rogers, M.A. Duffy, J.G. Jefferis, P.R. Sticksel and D.A. Tolle of Battelle Columbus Laboratories prepared for ERDA, Division of Solar Energy, August 1976.

N77-31614

Investigation of Excitation Control for Wind-Turbine Generator Stability by V.D. Gebben of NASA, Lewis Research Center prepared for ERDA, Division of Solar Energy, August 1977.

NSF-RA-N-75-050

Proceedings of the Second Workshop on Wind Energy Conversion Systems, Washington, DC, June 9-11, 1975 prepared by the Mitre Corporation for the National Science Foundation with assistance of ERDA and the American Institute of Aeronautics and Astronautics; edited by F.R. Eldridge of the Mitre Corporation. Papers by R.E. Donham of the Lockheed-California Company, Division of Lockheed Aircraft Corporation; M.C. Cheney, Jr. of United Technologies Research Center; R.E. Walters of West Virginia University; R. Ramakumar and W.L. Hughes, of the School of Electrical Engineering, Oklahoma State University; T.S. Jayadev, of the College of Engineering and Applied Science, University of Wisconsin-Milwaukee; W.E. Heronemus, Department of Civil Engineering, University of Massachusetts; W.H. Babcock, Colorado Division of Wildlife, A.L. Frey, Department of Agricultural Engineering and W.L. Somervell, Department of Atmospheric Science, Colorado State University; L.H. Mayo, Program of Policy Studies in Science and Technology, George Washington University; and R. Ferber, University of Illinois.

PB 249 936
Wind Machines by F.R. Eldridge of Mitre Corporation prepared for Research Applied to National Needs (RANN), National Science Foundation, October 1975.

PB 259 318
Wind-Powered Hydrogen Electric Systems for Farm and Rural Use by R.R. Tison, N.P. Biederman, T. Donakowski, R.H. Elkins, J.B. Pangborn, M.A. Turnquist, R. Young and B. Yudow of the Institute of Gas Technology, prepared for the National Science Foundation, April 1976.

PB 259 898
Optimization and Characteristics of a Sailwing Windmill Rotor by M.D. Maughmer of Princeton University, Department of Aerospace and Mechanical Science prepared for Research Applied to National Needs (RANN), National Science Foundation, March 1976.

PB 263 749
Vortex Augmentor Concepts for Wind Energy Conversion by P.M. Sforza of Polytechnic Institute of New York, Department of Aerospace Engineering and Applied Mechanics prepared for RANN, National Science Foundation, January 1976.

PB 273 582
Cost-Effective Electric Power Generation from the Wind: A System Linking Windpower with Hydroelectric Storage and Long-Distance Transmission by C.J. Todd, R.L. Eddy, R.C. James and W.E. Howell of the Division of Atmospheric Water Resources Management, Engineering and Research Center, Bureau of Reclamation, Denver CO, August 1977.

SAND 76-0650
Engineering Development Status of the Darrieus Wind Turbine by B.F. Blackwell, W.N. Sullivan, R.C. Reuter and J.F. Banas of Sandia Laboratories for ERDA, March 1977.

SAND 76-5535
Low-Cost Blade Design Considerations by L.V. Feltz of Sandia Laboratories for ERDA, 1976.

SAND 76-5683
Status of the ERDA/SANDIA 17-Metre Darrieus Turbine Design by B.F. Blackwell of Sandia Laboratories for ERDA, 1976.

ENERGY FROM SOLID WASTE 1979
RECENT DEVELOPMENTS

Edited by Francis A. Domino

Energy Technology Review No. 42
Pollution Technology Review No. 56

The emphasis in the field of solid waste has shifted from disposal to utilization. Vast stores of energy are waiting to be tapped through competent technology.

When we consider that every day each resident of the United States generates about four pounds of waste from products of great diversity, it is apparent that solid waste may someday serve as a sizeable source of energy. But first technical and environmental aspects, as well as social, legal and economic factors, will have to be examined in depth. This book reviews the many phases of waste disposal and energy recovery, discussing recently tested technology. It describes plants presently producing power from refuse while also salvaging material. Technological difficulties are also discussed. It also presents a proposed waste utilization system for the city of New York, as an example that could be applied to other cities.

The final chapter offers greatly detailed recommendations for municipal officials on waste processing and energy recovery plants.

Examples of some important subtitles are found along with chapter headings in the partial, condensed table of contents given below:

ISBN 0-8155-0750-X **321 pages**

INSULATION GUIDE FOR BUILDINGS AND INDUSTRIAL PROCESSES 1979

Edited by L.Y. Hess

Energy Technology Review No. 43

The question today is not whether to insulate but "How?" and "How much?" This definitive guide provides the answers. It assesses thermal insulation materials and systems for residential and industrial building applications, appraising building insulation as used for different purposes in Part A.

Recent rapid rises in fuel prices have confirmed the economic value of insulating as a fuel conserving practice. At the same time that maximum insulation of buildings effects considerable savings in costs, it confers the added benefit of increasing the availability of fuel, thus providing industry with more fuel to be used for productive purposes.

Part B of this book explores the use by industrial plants and utilities of thermal insulation to conserve energy, protect personnel and maintain process temperatures. It considers the effective temperature range and optimum thickness of insulating materials, their applicability in industrial processes, and the resultant economic rewards, noting the high rate of return on investment from insulating. It also evaluates sources of information on heat transmission.

The following condensed table of contents with chapter headings lists examples of **some important subtitles.**

ISBN 0-8155-0752-6 **200 pages**

ELECTRIC AND HYBRID VEHICLES 1979

Edited by M.J. Collie

Energy Technology Review No. 44

This year several hundred electric vehicles, operated by five private companies, will take to the road under an Energy Research and Development Administration program. These will be evaluated for city and suburban driving. Vans similar to those used in England since World War II to carry milk noiselessly "from the dairy to the doorstep" will appear here more often. With such stimulus from government and industry abetted by the petroleum scarcity, this industry should mature in the decade ahead. The number of all-electric vehicles and their hybrid counterparts for passenger and commercial uses should increase a great deal.

This up-to-date assessment of electric vehicles summarizes data on characteristics, costs, maintenance and energy consumption gleaned from actual tests and from the literature and trade. It includes a survey of electric bus operations in 19 foreign and domestic locations and a survey of 11 different types of cars and delivery vans in the United States.

Energy storage devices and power systems were analyzed as alternatives to current automobile propulsion systems, with the following batteries covered: Pb/acid, Ni/Zn, Ni/Fe, Li/Fe sulfide, Na/S, Zn/Cl, Ni/H, Zn/Br, Fe/air, Li/metal sulfide, and others.

The partial table of contents below lists chapter headings and **examples of some** subtitles. Bibliographies and a glossary complement the text.

ISBN 0-8155-0754-2 **633 pages**

PASSIVE SOLAR ENERGY DESIGN AND MATERIALS 1979

Edited by J.K. Paul

Energy Technology Review No. 41

Passive solar energy is a dynamic system whereby a building collects and stores energy to heat and cool itself. Unlike active solar systems employing complicated technology, passive systems collect, store and distribute thermal energy by natural radiation, conduction and convection. Sophisticated design and wise selection of building materials eliminate the need for fans, pumps or heat exchangers.

This volume describes over 100 buildings using specially designed components for the 3 passive systems, direct gain, indirect gain and isolated gain. Movable insulation for nocturnal shielding of glass areas and polymer ceiling tiles containing material to store and release heat are innovations discussed. Hybrid arrangements, utilizing simple mechanical aids to operate an otherwise passive system, are included.

A select group of case studies gives the reader an in-depth view of passive solar design. This book will interest producers of architectural and building materials, chemicals, and components, as well as architects and designers.

A partial and condensed table of contents with chapter headings and **examples of some** subtitles follows. Over 200 illustrations.

ISBN 0-8155-0746-1 **386 pages**

ENERGY SAVING BY INCREASING BOILER EFFICIENCY 1979

Edited by Lee Yaverbaum

Energy Technology Review No. 40

The boiler plant manager or operator will find this definitive text furnishes the precise assistance he needs to implement efficiency improvement techniques. With the sharp boiler maintenance described he can realize fuel savings of from 10 to 30%.

To keep efficiency high each device in the system must be adjusted for optimum performance. This book presents the necessary information regarding adjustments, modifications, equipment and maintenance that should be considered, from the simplest adjustment of fuel/air ratio (which can effect remarkable fuel savings of several percentage points) to the more complex auxiliary equipment. It contains material on installing combustion controls, minimizing heat losses, adjusting combustion conditions, and removing impurities.

Not to be disregarded are the data on alternate fuels. The testing so essential to better boiler functioning is also discussed in detail, and in the last chapter the economic effects of the suggested boiler changes are evaluated.

Following is a condensed table of contents with **examples of some** important subtitles.

ISBN 0-8155-0745-3 **226 pages**

HEAT PUMP TECHNOLOGY FOR SAVING ENERGY 1979

Edited by M.J. Collie

Energy Technology Review No. 39

With energy costs escalating and fossil fuel supplies diminishing, the heat pump is becoming attractive for residential space heating and cooling because of its high efficiency. This book compares its overall efficiency to that of electrical resistance and fossil fuel heating, covering air-source and water-source pumps, and single-package and split-system units.

It reports on computer-simulated studies of residential heating in 9 cities in which a heat pump replaced a gas or oil furnace, and on tests of heat pumps in buildings in Washington, D.C., Hanover, New Hampshire, and Albuquerque, New Mexico. Data on the Annual Cycle Energy House, in which a heat pump, thermal storage and solar assistance provide space heating, cooling and water heating, attest to energy saved, as do the solar-assisted heat pumps in other areas.

Because early heat pumps had design deficiencies, improvements were mandated. The latter chapters examine modifications such as capacity control to alter refrigerant flow.

Following is a condensed table of contents with **examples of some** important subtitles.

ISBN 0-8155-0744-5 **315 pages**

COGENERATION OF STEAM AND ELECTRIC POWER 1978

Edited by Robert Noyes

Energy Technology Review No. 29

The cogeneration concept, as presented in this book, refers to the generation of both process steam and electricity on site by heating water with a primary fuel, such as coal, oil or gas.

When fuel was cheap, industries were not interested in capital outlays necessary for cogeneration equipment. Now, however, the picture has changed. Refineries, chemical plants, paper mills and other big energy users can get maximum value from their fuel by a "topping cycle," viz. producing steam and running it through a turbine generator before using the steam for processing operations. Also, cogeneration can be accomplished through a "bottoming cycle," utilizing process steam prior to electrical power generation. Both methods are discussed in detail.

In the past few years, the U.S. Government has been placing much emphasis on the need for conserving energy. One of the ways of saving considerable amounts of energy is the development of cogeneration plants by those industries which use significant amounts of electricity and thermal energy for process heat and/or space heating. Since electric power generation produces heat as a by-product, any thermal energy which can be used produces a fuel savings of from 10 to 30%, depending on the applications and methods involved. The studies indicate that for the greatest potential development of cogeneration, government support may be required.

Various technological studies, such as material prepared for the National Science Foundation and the Federal Energy Administration, were the basis for this review. A partial, condensed table of contents follows here.

ISBN 0-8155-0706-2 **258 pages**

INDUSTRIAL AND INSTITUTIONAL WASTE HEAT RECOVERY 1979

Edited by P.G. Stecher

Energy Technology Review No. 37

Here is an excellent guidebook to productive management of discarded or waste heat and a guide to technology for institutional integrated energy/utility systems.

The first part of the book discusses industrial waste heat recovery. Simplified examples are explored, e.g., the use of medium- to high-temperature exhaust gases to preheat combustion air for gas turbines using regenerators. Effective waste heat management including its measurement technology and economics are discussed, with emphasis on energy conservation to increase profits.

What are the benefits and costs? How does a manager evaluate equipment? Answers to these questions—and many more—are furnished in this book. Commercial firms of all kinds can benefit from the financial guidelines given to determine the feasibility of installing a waste heat recovery system.

Of great value are the descriptions of 14 successful installations in U.S. industries such as glassmaking. Design specifications for recovery devices, cost studies, and resultant energy savings combine to form a complete case study.

There is a chapter devoted to equipment to aid the reader in assessing the merits of various heat transfer devices. Instrumentation techniques for measuring the essential parameters of a waste heat recovery system are elaborated. Engineering data for analyzing waste heat problems and a list of referral organizations are also provided.

Architect-engineers, as well as industrial and power plant owners and suppliers, will find the integrated utility system (IUS) described in the second part a promising way to conserve energy. By combining the utilities of electrical power, water, sewage treatment and waste disposal previously purchased from outside sources into one within an institution, the wastes from each individual system are used to furnish another needed utility. Power generation yields energy for heating and cooling; solid waste is burned for heat content; sewage effluent is reused for equipment make-up water and irrigation.

A partial and condensed table of contents with chapter headings and subtitles follows:

ISBN 0-8155-0738-0 **362 pages**

EUROPEAN TECHNOLOGY FOR OBTAINING ENERGY FROM SOLID WASTE 1978

Edited by D.J. De Renzo

Energy Technology Review No. 34

Pollution Technology Review No. 54

The recovery of energy through the combustion of municipal solid waste is by now a well recognized and established technique for conserving energy. Combustion units can produce electricity, hot water for domestic use and steam for district heating and for industrial processes or for the drying of sewage sludge.

Western Europe is definitely the leader in this field. Energy recovery from waste began, in a small way, in France, Denmark, and Italy before the second world war. During the 1950s it was introduced on a large scale in West Germany, France and Holland. The largest single furnace is located in Paris-Ivry and has a capacity of 50 metric tons/hr, while the largest European plant near Rotterdam has 6 units of 20 metric tons/hr.

Much can be learned from these European waste-to-energy systems. For each country a national overview is given, followed by a description of particularly significant developments and case histories: Household sorting and collection methods, combustors, furnaces, incinerators, air pollution control, latest plant designs, operation and economics.

A summary of key findings is also included. Serious study of this book should make it possible to reveal and identify solid waste/energy processes (both existing and under development) that appear to offer potential advantages over processes currently employed in the U.S. Maps and a tabulation of the European systems are included. This book is based on several studies produced under contract for EPA and ERDA (now DOE). A summarized table of contents follows here:

ISBN 0-8155-0730-5 **281 pages**

GEOTHERMAL ENERGY

Recent Developments

1978

Edited by M.J. Collie

Energy Technology Review No. 32

The heat underneath the Earth's crust seems a highly desirable energy source alternative to oil and gas. The Earth does not give up her energy readily, still geothermal energy activities are progressing on a broad front worldwide wherever possible.

Operational fields are increasing in numbers, and new and improved technologies, especially new turbine designs and deep-well pumps, coupled with heat exchangers at the surface, are leading to much greater operating efficiency.

Yet, as a rule, geothermal power stations, with a few exceptions, are inefficient in converting thermal energy to electric energy, because of the overall low temperatures. Also the highly mineral-laden waters in the geothermal reservoirs increase operating expenses through scale formation, thus complicating the process of extracting the heat.

Geothermal energy is not expected to play a large overall role in the U.S. But it can have regional significance, especially in the West. For developing countries, however, geothermal energy, especially in volcanic regions, may be the only economic form. Electricity systems there are too small to justify nuclear power stations, but the comparatively small size of geothermal power stations fits the scale of electricity supply systems in these countries.

This energy Technology Review is based on studies conducted under the auspices of various government agencies and the last chapter constitutes a survey of the recent patent literature.

ISBN 0-8155-0727-5 **445 pages**